'*A Species In Denial* is a superb book...it brings out the truth of a new and wider frontier for humankind.'

John Morton, Emeritus Professor of Zoology, Auckland University, New Zealand; Lay Canon Emeritus of Holy Trinity Cathedral, Auckland; Fellow St. John's Theological College, Auckland

'A breakthrough in understanding the human condition.'

Dr John H. Champness, Australian psychologist and educator

'Nothing less than the manual now for humankind.'

Tim Macartney-Snape AM OAM,
distinguished Australian mountaineer and biologist

'A must read for every human.'

Brian Carlton, Executive Producer, Radio 2GB, Sydney, Australia

'A most enlightening treatment of the human situation.'

Dr Arthur Jones, theologian and
former Anglican Bishop of Gippsland, Australia

'A rewarding journey.' *The Australian*, July 2003

'Jeremy Griffith is the latest and one of the most challenging thinkers...[his book] goes against so much of the resigned don't-rock-the-boat attitude of the Western mind...There are not many books offering as much. It offers so many insights into our divided selves.'

Ronald Conway OAM, distinguished Australian psychologist,
Australian Quarterly Journal of Contemporary Analysis, Jan-Feb 2004

'A landmark in understanding.'

Geelong Advertiser, Australia, August 2003

'There is no doubt that this book is an important one and breath-taking in its breadth...The psychological and biological stages of life are examined with great insight...'

Reading Time, Journal of Children's Book Council of Australia, May 2004

'This book is a must read for all...most rewarding.'

Southland Times, New Zealand, November 2003

'A book that confronts the way we think about life...People like [Griffith] used to be drummed out of town by the vicar...Griffith gives the serious reader plenty to ponder...There is never any doubt of the courage of [Griffith's] stance in writing this book because of his commitment to his fellow man and the future of the planet.'

Wairarapa News, Wellington, New Zealand, December 2003

'Jeremy Griffith is an Australian biologist but his range of interests and his store of knowledge seem almost infinite...The chapter called *Resignation* is brilliant in its insight into human nature and what we call the idealism of the young...It's worth reading the book for this essay alone but, of course, there's so much more. Those who need brain food will find it here. It can't be said of many books that the world looks different after you've read them. It can be said of this book.' *Toowoomba Chronicle,* Australia, February 2004

'A must-read for anyone vaguely interested in the subject of who we are and what we are doing on this earth...a heroic work.'

Townsville Bulletin, Australia, October 2003

A SPECIES IN DENIAL

Jeremy Griffith

with a Foreword by
Charles Birch

Foundation for Humanity's Adulthood
The crux problem on Earth has always been the human condition, our capacity for good and evil. The discovery of its cause is the key to our freedom. Humanity can now end its insecure, upset adolescence and enter its peaceful adulthood.

Published by FHA Publishing and Communications Pty Ltd
First published in Australia 2003
Revised UK edition published 2004

All inquiries to
Foundation for Humanity's Adulthood
GPO Box 5095, Sydney NSW 2001, Australia
Phone: + 61 2 9486 3308
Fax: + 61 2 9486 3409
Email: info@humancondition.info
Website: www.humancondition.info

ISBN 1-74129-001-5
CIP — Psychobiology

Cover: *Cringing in Terror* (c.1794-96) and *Albion Arose* (c.1794-96) by William Blake; coloured impression of *Albion Arose* by Carol Marando. Design by Judi Rowe, Guy Jeffery and the FHA.

Edited by Fiona Cullen-Ward and Sally Kaufmann,
FHA Publishing and Communications Pty Ltd

Typeset in New Baskerville by Annie Williams,
FHA Publishing and Communications Pty Ltd

Printed in Australia by
Griffin Press Pty Ltd, Netley, South Australia

On behalf of the future, this book is dedicated to
the Members of the Foundation for Humanity's Adulthood.

Notes to the Reader

- The reader will notice a few sections with dotted lines in this book. These are sections that have been temporarily withdrawn because they touch on issues concerning defamation actions taken by this author and others against the Australian Broadcasting Corporation (ABC) and the Fairfax publishing group regarding a 1995 *Four Corners* documentary and also a *Sydney Morning Herald* newspaper article, both by a Minister of the Uniting Church. The human condition is a very contentious subject as it is the realm of inquiry where science and religion finally overlap. Those affected vigorously protested the portrayal in the ABC broadcast to the Australian Broadcasting Authority. After a two-year investigation, in the strongest ruling it had ever made, the Authority found the ABC had breached its broadcasting codes of practice. The Authority subsequently took the unprecedented step of advising the ABC that it was appropriate to apologise. The ABC did not exercise its right to challenge the ruling but despite this, has not apologised, and so the defamation actions ensued.

  In the few sections—mainly between pages 159 and 169—where the content of this book touched on issues before the court the options were to write around those sections so the reader did not realise the sections had been withheld, or insert dotted lines where the text is removed. Of the two options, the dotted line option was considered the most respectful of the reader.

  In the apology note, included where the sections have been withheld, it is explained that when the court case restrictions end—in six months to two years time from the launch of this book in mid 2003—reprints of the affected pages, with the text fully restored, will be made available on the FHA's website <www.humancondition.info>, or by contacting the FHA.

- The underlinings within quotes throughout this book are the author's emphasis unless otherwise stated. The underlinings in the author's writing is for emphasis or, in most cases, to indicate a particular subject is being addressed.

- There is no bibliography at the conclusion of this book because the source is given in small text at the end of each quote, where the reader can see immediately when and where the quote was made. Also, rather than give the particular edition and/or publisher of the book that the quote comes from, the page number where the quote appears and the total number of pages of the particular edition used for the source is given. This enables the reader to find the comparative place of the quote in any edition.

- There are references in this book to Jeremy Griffith's earlier book *Beyond The Human Condition* (1991). The full text of *Beyond The Human Condition* is presented on the Foundation for Humanity's Adulthood (FHA) website <www.humancondition.info>, or copies can be purchased by phoning the FHA on + 61 2 9486 3308.

Contents

Foreword

By Charles Birch 17

Introduction

Broaches the subject of the human condition—the issue of which the human species has had to live in psychological denial because historically it has been an extremely confronting and suicidally depressing subject.

The elephant in our living rooms 21

What is the human condition? 25

The agony of the human condition 27

Humans' historic denial of the issue of the human condition 33

Plato's 'cave' allegory for life in denial 35

The 'deaf effect': the difficulty of reading about the human condition 42

Unaware of the 'deaf effect' readers blame the presentation 52

Age and the 'deaf effect' 56

Intelligence and success also increase the 'deaf effect' 58

The non-falsifiable situation 63

Overcoming the 'deaf effect' 64

The need for patience and perseverance 67

Summary of the old and new highways of thought in response to analysis of the human condition 70

This move to a new human condition-ameliorated world has no relationship to the New Age Movement, or to a religious movement 72

Anticipations of the arrival of the human condition-ameliorated new world 74

The benefit of an indirect approach 78

The essays of 'A Species In Denial' 79

Deciphering Plato's Cave Allegory — And in the Process Explaining How The Human Condition is Resolved 83

Explains the allegory and in the process shows how biological understanding of the human condition liberates humanity from its human condition-afflicted cave-like existence.

The metaphor of the sun and fire, the meaning of existence and the demystification of God 85

The burning effect of the sun and fire 89

The illuminating effect of the sun and fire 94

The reason for the cave existence 94

Liberation from the cave 96

ASPECT A 97

A1: The difficulty of acknowledging the human soul and the spectrum of alienation, or denial of soul 99

A2: Plato's acknowledgment of the nature of our soul 101

A3: Recognition of the soul's world in mythology 104

A4: The biological explanation for how we acquired our soul, our instinctive orientation to integrative, cooperative meaning 106

A5: The contrived excuses for humans' divisive nature 113

A6: A brief description of how our soul became corrupted 117

A7: There were people who could live in the sun's light, but they were unbearably condemning for the cave prisoners 123

ASPECT B 130

B1: Science—the liberator 131

B2: Soul—the synthesiser 136

ASPECT C 143

C1: The difficulty of taking the truth back into the cave 144

C2: The cave allegory was the theory, this is what happened in practice 158

C3: Humanity's departure from the cave to life in the sun 176

Resignation 187

Examines the most important psychological event in human life. It is an event that has dictated the very nature of existence for adult humans and yet it is an event that has not previously been acknowledged. The essay explains that at about 12 years of age humans begin to try to understand the dilemma of the human condition. However, with humanity unable—until now—to explain this deepest of issues, adolescents eventually, at about the age of 15, learn they have no choice but to resign themselves to a life of denial of the dangerously depressing subject.

The elephant in the living room that only young people could see ... 188

The destructive effect of the silence from the world of resigned adults ... 194

Trying to confront the human condition within ... 200

The depression ... 206

Resignation poetry ... 208

'Life' leading up to resignation ... 211

Ships at sea ... 218

Olive Schreiner's extraordinarily honest recollection of the world of the pre-resigned mind ... 223

A history of analysis of resignation ... 229

The moment of resignation: how humans change from believing in a cooperative, selfless, loving world, to believing in a competitive, selfish, aggressive, egocentric, must-win, power-fame-fortune-and-glory-obsessed world ... 232

The extent of humans' fear of the issue of the human condition ... 243

The cost of resignation: 'fifty foot of solid concrete' between humans and their souls ... 256

The autistic state demonstrates how the mind can dissociate itself from unpleasant experiences ... 269

The necessary dishonesty of the resigned mind ... 277

The dysfunction of the resigned mind ... 282

The loneliness of humans' alienated state ... 287

The objective of the human journey was to solve the human condition ... 292

Renegotiating resignation ... 294

The end of resignation ... 309

Our species' fabulous future ... 311

Bringing Peace To The War Between The Sexes 317

and

The Denial-Free History Of The Human Race 349

Some of the deepest wounds in human life have been caused by the lack of understanding in the relationship between men and women. The heartache, bitterness, suffering, and the hurt to children has been immense. There are so many questions about the relationship between men and women that need answering. With understanding of the human condition it is now at last possible to answer all these questions and bring peace to the 'war' between the sexes.

With understanding of the human condition it is not only possible to talk honestly about the different roles of men and women, it is also possible to present an honest, denial-free history of the human race, and explain the roles different races have played in the human journey.

The Demystification Of Religion 375

Powerfully demonstrates how understanding the phenomenon of resignation demystifies previously impenetrable aspects of human life, in particular the world of religious metaphysics and dogma. In fact, it is a clear demonstration of how science can and should be a winnower of mystery and superstition.

Understanding the human condition makes it possible to acknowledge the immense differences in alienation between humans 376

'Homes' at different distances from the incinerating bonfire of truth 379

Prophets and the concept of the 'Virgin Mother' demystified 383

The story of Noah's Ark explained 393

Demystifying the role of religions 395

The Apocalypse and the Battle of Armageddon explained 398

The demystification of God 406

The difficulty of exposure that the demystification of God brings 407

Where is the spirituality in negative entropy? 416

If negative entropy ends with the 'heat death' or 'big crunch' end of the universe, where does that leave God? 421

Afterlife explained 423

The Trinity demystified 426

Jesus Christ demystified 428

Christ's miracles and resurrection demystified 433

Why the psychologically desperate do not suffer from the 'deaf effect' 438

Why prophets were 'without honour' in their 'own home' and 'own country' 439

Denial-free books 447

Humour, swearing and sex demystified 449

Recognising myself as a contemporary unresigned denial-free thinker or prophet 453

The authority of an unresigned prophet 454

A prophet's role is to say the truth 461

The role of contemporary prophets is to resolve the human condition, and by so doing make the need for religions obsolete 462

Denial-free thinkers are not 'brilliant' or 'clever' or 'geniuses' 468

Australia's role in the world ... 470

Expressions of anticipation of the arrival of understanding of the human condition ... 490

One cold, pure February night ... 493

Conclusion ... 496

The Foundation for Humanity's Adulthood (FHA) ... 499

A brief profile of the FHA written by FHA Vice-President Tim Macartney-Snape AM

Foreword

There are two main streams of thought concerning the nature of things including human beings. They can be called the objective stream and the subjective stream. The objective stream goes back at least as far as the Greek philosophers such as Thales and Democritus about 500 BCE. It finds its great burst forth in the scientific revolution in the 15th and 16th C. continuing to this day. The subjective stream goes at least as far back as Ikhnaton in Egypt about 1374-58 BCE. It has been the stream to which many poets, and some philosophers and theologians belong.

The objective stream of thought is typified by scientific understanding. It has had a revolutionary influence on our world, particularly the Western world. Its model is the machine. The Copernican universe is a huge machine. The atom is a smaller machine, though not the smallest. In biology the heart is interpreted as a machine, as is the DNA molecule. The theory of evolution of life is a mechanistic theory. An overriding principle of this stream is the interpretation of the higher in terms of the lower. Human life and for that matter all life is to be interpreted in terms of the atoms and molecules that make up the living organism.

The subjective stream of thought is not one but many streams that sometimes coalesce. Its emphasis is on the world of feeling. What is it like to be a human being with all its joys and sorrows, its values and disvalues, the good and the bad. If there is an overriding principle it is the interpretation of the lower in terms of the higher. Human feelings provide a clue to the nature of nature all the way down from the human being to the atoms and molecules that constitute the human being. The universe may have mechanical aspects, but it is also a feeling universe. The human is not the only pebble on the cosmic beach of feeling. This way of looking at things is a first person view of things as contrasted with the scientific mode which is a third person view of things. The discoverer of the electron just over one hundred years ago J.J.Thomson said to know what an electron really was you would have to be one! That has of course its difficulties. How can I have a first person view of an electron? We can best approach that sort of question by coming to it closer to home namely the life of a cat. In the Middle Ages the scholastic case was argued of a cat that stood to the left of a pillar. The cat may adjust itself to the pillar so that, for example, it gets shade from the very hot sun. Its quality of experience is definitely qualified by its relation to the pillar. We don't have the experience of the cat but by inference from our own experience we can reasonably infer some sort of feeling of the cat which changes as it changes its position in relation to the pillar. The cat we say is a subject. The pillar is an object, for nothing similar is obvious concerning the pillar's relation to the cat. The pillar, we say, has only external relations to the world (a bulldozer may destroy it). The cat has external relations but also internal relations, namely its feelings to the world. The question then arises—what are subjects and what are objects only. That is a topic beyond this preface, yet a very important one.

From time to time in history the two streams of understanding meet, sometimes in opposition, sometimes in coalition. At the height of the scientific revolution in England there was what has been called the romantic reaction against science. The relevant poets were Milton, Pope, Wordsworth and Tennyson. Wordsworth's characteristic thought was 'We murder to dissect' which is what he saw science as doing. He protested the scientist who 'botanised on his mother's grave.' The romantic reaction was a protest against mechanism on behalf of values and feelings.

It is my view that the greatest coming together of the two streams in modern times is that of A.N. Whitehead who was both scientist and philosopher. He owes a great deal to a subjective stream of thought from Plato onwards with the so-called process thinkers and their philosophy of organism. A famous statement of Whitehead is 'The philosophy of organism is mainly devoted to the task of making clear the notion of being present in another entity' (eg the pillar in the life of the cat or the cat in the life of a human being). It is a philosophy that brings science and the subjective together in a holistic view. It is an event way of thinking as opposed to a substance way of thinking. It is in the network of internal relations of persons (our feelings) that reality is seen to be most fully revealed.

In his book *A Species In Denial* Jeremy Griffith seeks to bring together the two streams of thought, the objective and the subjective. He is not, of course, the first person to try to do this. The questions the reader will ask are these. Is it true? What light does it throw on the human condition? I don't propose to answer these questions directly but to suggest to the reader to bear in mind that Jeremy Griffith's ideas have taken root in his Foundation for Humanity's Adulthood. Its many supporters testify to the transforming influence of the ideas they have discovered in this Foundation. That is the story that primarily interests me rather than to make a critical analysis of the ideas which every reader should be encouraged to do for themselves as they read this book.

Charles Birch

Dr Charles Birch is Emeritus Professor at the University of Sydney and was Challis Professor of Biology for 25 years. He was awarded the Templeton Prize for Progress in Religion in 1990. Dr Birch has written numerous books on science, religion and human existence.

Introduction

The elephant in our living rooms

Were someone to tell you that humans are unaware of something so significant in their lives that it is the equivalent of them not noticing an elephant in their living room, you would be likely to say they were being absurd. If you were also told that this 'elephant' was the underlying cause of all the problems and unhappiness in the world you would no doubt be even more disbelieving. And you would undoubtedly be totally dismissive if you were told humans can now rid themselves of this all-pervasive and troublesome 'elephant'.

Well, the existence of this 'elephant in our living room' is actually true. It is also true that it is the cause of all our fears and worries. In fact this 'elephant' stands squarely between where humanity is now and a future for humanity, if there is to be one. Astonishingly enough, it is also true that this mammoth impediment can now be eliminated from human life.

Incredulous as you may be, I and others who have been studying these ideas, know from experience with what is to be presented in this book that before you have finished reading the first few pages you will already have a strong intimation that this 'elephant' does indeed exist—and that by the time you have read 10 pages you will be deeply aware of its existence. In fact, after 30 pages you will be saying to yourself that, 'this is all too true for comfort'. Despite the discomfort, after 60 pages, you will be starting to see how we can at last expunge this monstrous problem from human life.

You are about to embark on a journey where humans have never been able to venture before, right into the very heart of the issue of what it is to be human. It is a truly astonishing journey, and it is exciting—but it is also deeply unsettling; so much so that it will demand all of your courage and capacity for perseverance. Indeed, someone who read the draft of this book said, **'You should call this book *The Truth—If You Want It*, because people are going to start reading the book and discover it *really is* the truth about humans, and they might well find that it is too much for them to cope with, that they prefer to stay living in denial of the truth.'** Retreating to denial will, without doubt, be an initial response for many people. However, it will be a temporary reaction. After some time walking around with the knowledge that, while humans have had no choice up until now other than to live in an extraordinarily dishonest, superficial and, ultimately, humanity-destroying state of denial, knowing that such an existence is no longer necessary, and, furthermore, knowing that living that way is now both pointless and irresponsible, they will return to the book and say, 'Let's get on with the task of facing the truth'. Or, if they are not able to confront the truth about humans themselves, they will realise that now that the truth has at last emerged all they have to do is support it over their inclination to live in denial of it, and in time humanity will be free from the denial, and from all the resultant suffering. At last somewhere on Earth the truth about humans exists, it is simply a matter of people defending it now and in time humanity will be free of its immense historic psychosis.

As mentioned, once the truth about humans has emerged denial becomes a pointless, futile occupation. No matter how much the practitioners of the denial try to maintain it, once the truth is out the denial has lost its power. In Hans Christian Andersen's 1837 fable, the *Emperor's New Clothes,* once the child broke the spell of the deception—pointed out that the emperor was indeed naked—the whole edifice of deception crumbled. A person cannot very well think or say or write anything from a basis of denial that has now lost its currency. Once obsolete, the strategy of denial inevitably becomes ineffective as a strategy for living. As the Jesuit priest and philosopher, Pierre Teilhard de Chardin, said, **'the Truth has to appear only once...for it to be impossible for anything ever to prevent it from spreading universally and setting everything ablaze'** *(Let Me Explain,* 1966; trs René Hague & others, 1970, p.159 of 189).

Further, the overall reality is that if humanity is to avoid self-destruction it has nowhere else to go but through the doorway of honesty.

The situation is that the truth about humans now exists on Earth, the spell is broken, the hold that humans' entrenched and once-necessary denial has had over humans is over. The 'elephant' is exposed. The truth is out. The 'game is up'.

1988 JEREMY GRIFFITH

The benefit of persevering with this journey of discovering what it really is to be human will be to arrive at a state of extraordinary freedom. It has to be immediately emphasised that this is not the sort of feel-good freedom that New Age and 'self-improvement' gurus promote through artificial forms of motivational reinforcement gained, for instance, from repeating mantras that 'you have every right to love yourself', and the like. Nor is it the reinforcement and comfort humans have long derived from religious faith and belief. This is a freedom achieved through being able to at last *understand* human nature, understand ourselves. It is not dependent on dogma or belief; in fact this ability to understand ourselves brings such a fundamental freedom to the human situation that it takes humans beyond the state where they need dogmatic forms of reinforcement and religious faith and belief. What is to be introduced is a scientific explanation, that is, first principle-based knowledge, but it is science of the most profound form; knowledge that so deeply penetrates the human situation, so completely ends human insecurity, that it actually ends the need for any dogma, mysticism, superstition or abstract metaphysics. It demystifies the concepts in religion and deciphers our myths. The whole human situation will at last be clarified. This is an astonishing claim, but nevertheless before you have read many pages into this book you will be discovering that it is true.

At the core of the issue of what it is to be human lies a subject that humans have traditionally found impossible to confront. It is this all-important yet unconfrontable subject that is the 'elephant in our living room'. While it is the core, all-important issue that must be addressed if the truth about humans is to be uncovered, it is nevertheless an issue that has been so deeply distressing and depressing that for almost 2 million years humans have had no choice other than to live in almost complete denial of it. In fact so dominating has this practice of denial been in human life that if there were any enlightened intelligence in outer space it would likely regard us as 'that species that is living in denial'. Indeed humans live in such denial and, as a result, are separated from their true situation and true selves to such an extent, that we could also be known throughout the universe as 'the estranged or alienated ones'.

The question then is, what is this overwhelmingly daunting subject, this 'elephant in our living room', that at a conscious level humans are almost totally unaware of? It is the subject of *the human condition.*

So daunting has this subject of the human condition been that we have rarely ever referred to it. For example, while 'human nature' appears in dictionaries, 'human condition' never does. Only in moments of extreme profundity did we even mention the topic, and even then it was only ever a glancing reference. For example, in the mission statement of a philanthropic organisation in America called the Fetzer Foundation there are lofty words about the foundation's dedication to research, education and service, and spliced in amongst them are these words: **'as we press toward unique frontiers at the edge of revolutionary breakthroughs in the human condition'**. Humans have lived in such deep denial of the issue of the human condition that when they encounter the term 'human condition' many think it refers to the state of poverty or disease that afflicts much of humanity. If you search 'human condition' on the Internet most references interpret it as being to do with humans' physical predicament rather than with humans' psychological predicament, which, as will become clear, is its real meaning.

What is the human condition?

The greatest of all paradoxes is the riddle of human nature. Humans are capable of immense love and sensitivity, but we have also been capable of greed, hatred, brutality, rape, murder and war. This raises the question, are humans essentially good and if so what is the cause of this evil, destructive, insensitive and cruel side? The eternal question has been why 'evil'? In metaphysical religious terms, what is 'the origin of sin'?

More generally, if the universally accepted ideals are to be *cooperative, loving and selfless*—they are the ideals accepted by modern civilisations as the basis for their constitutions and laws and by the founders of all the great religions as the basis of their teachings—*why* then are humans *competitive, aggressive and selfish? What* is the reason for humans' divisive nature? Does this inconsistency with the ideals mean that humans are essentially bad? Are we a flawed species, a mistake—or are we possibly divine beings?

The agony of being unable to answer this question of why humans are the way they are, divisively instead of cooperatively behaved, has been the particular burden of human life. It has been our species' particular affliction or condition, the *'human condition'.*

In fact the fundamental issue of human life, the issue of humans' divisive nature, has been so troubling and ultimately depressing that humans eventually learnt that the only practical way of coping was to stop thinking about it, block the whole issue from their minds. So depressing was the subject of the human condition that humans learnt to avoid even acknowledging its existence, despite the fact that it was the real issue before us as a species. Philosopher Ludwig Wittgenstein made the point in his now-famous line, **'About that which we cannot speak, we must remain silent'** *(Tractatus Logico-Philosophicus,* ch.7, 1921).

If we call the embodiment of the ideals that sustain our society 'God', then humans have been a 'God-fearing' species—people living in fear and insecurity, made to feel guilty as a result of their inconsistency with the cooperative, loving, selfless ideals. The human predicament, or condition, is that humans have had to live with a sense of guilt—although, as is explained in my earlier books *Free: The End Of The Human Condition* (1988) and *Beyond The Human Condition* (1991), a sense of guilt that was *undeserved.* (Note: these books will be referred to as *Free* and *Beyond* throughout this book. Also, the full text of *Beyond* is presented on the website <www.humancondition.info>.) Whenever humans tried to understand why there was such divisiveness and, in the extreme, 'evil' in the world, and indeed in themselves, they couldn't find an answer and were eventually forced to put the question out of their minds. Humans coped with their sense of guilt by blocking it out, sensibly avoiding the whole depressing issue. T.S. Eliot recognised our species' particular frailty, which was having to live psychologically in denial of the issue of the human condition, when he said that **'human kind cannot bear very much reality'** *(Four Quartets, Burnt Norton,* 1936).

It is a measure of how accomplished humans have become at overlooking the hypocrisy of human life and blocking out the question it raises of their guilt or otherwise that, although they are surrounded by that hypocrisy, they fail to recognise it or the question it raises. Revealingly, while adults now fail to recognise the paradox of human behaviour, children in their naivety still do. They ask, 'Mum why do you and Dad shout at each other?', 'why are we going to a lavish party when that family down the road is poor?'; 'why is everyone so lonely, unhappy and preoccupied?'; 'why are people so artificial and false?'; 'why do men kill one another?'; and 'why did those people fly that plane into that building?' The truth is that these are the real questions about human life, as this quote, attributed to George Wald, points out, **'The great questions are those an intelligent child asks and, getting no answers, stops asking'** (mentioned in Arthur Koestler's 1967 book *The Ghost in the Machine,* p.197 of 384). The reason children **'stopped asking'** the real questions—stopped trying to point out the 'elephant', the issue of the human condition—was because they eventually realised that adults couldn't answer their questions, and, in fact, were made distinctly uncomfortable by them.

The truth is, the hypocrisy of human behaviour is all around us. Two-thirds of the people in the world are starving while the rest bathe in material security and continually seek more wealth and luxury. Everywhere there is extreme inequality between individuals, sexes, races and even generations. When a woman pointed out on a radio talk-back program that, **'we can get a man on the moon, but a woman is still not safe walking down the street at night on her own'**, she was acknowledging the hypocrisy of human life.

Humans can be heartbroken when they lose a loved one but are also capable of shooting one of their own family. We will dive into raging torrents to help others without thought of self but are also capable of molesting children. We torture one another but are also so loving we will give our life for another. A community will pool its efforts to save a kitten stranded up a tree and yet humans will also **'eat elaborately prepared dishes featuring endangered animals'** (*Time* mag. 8 Apr. 1991). We have been sensitive enough to create the beauty of the Sistine Chapel, yet so insensitive as to pollute our planet to the point of threatening our own existence.

Good or bad, loving or hateful, angels or devils, constructive or destructive, sensitive or insensitive, what are we? Throughout our history, we have struggled to find meaning in the awesome contradictions of the human condition. Neither philosophy nor science has, until now, been able to give a clarifying explanation. For their part, religious assurances such as 'God loves you' may offer comfort but do not explain why we are lovable.

The real problem on Earth is humans' predicament or condition of being insecure, unable to confront, make sense of and deal with the dark side of human nature. The real struggle for humans has been a psychological one.

The agony of the human condition

So necessary has denial of the issue of the human condition been that it is only when moved to extreme profundity that humans have acknowledged this underlying insecurity about whether they are at base good or evil beings. There have been innumerable books written about humans' capacity for good and evil, but what will become clear as you read on is that very, very few people have been able to face down and grapple with the core issue of what it really is to be human; actually confront the dilemma of the human condition. The following few examples constitute almost the entire collection of such profound moments that I have found in the 27 years since 1975, when I first started to actively write about the human condition. (The handful of examples that are not included here, particularly examples from the writings of Olive Schreiner and Albert Camus, are mentioned elsewhere in this book.) The rarity of examples indicate just how difficult a subject the human condition has been for the human mind to approach. You will notice that it required the capabilities of some of the world's most gifted writers to manage even to allude to the subject.

In 1988 *Time* magazine invited Alan Paton, author of *Cry, the Beloved Country,* to write an essay about apartheid in South Africa. He instead provided a deeply reflective piece about some of his favourite pieces of literature. In the essay, which turned out to be the great writer's last written work, he said: **'I would like to have written one of the greatest poems in the English language—William Blake's "Tiger, Tiger Burning Bright", with that verse that asks in the simplest words the question which has troubled the mind of man—both believing and non believing man—for centuries: "When the stars threw down their spears/And watered heaven with their tears/Did he smile his work to see?/Did he who made the lamb make thee?"'** (*Time* mag. 25 Apr. 1988). Blake's poem poses the age-old riddle and fundamental question involved in being human: how could the mean, cruel, indifferent and aggressive 'dark side' of human nature—represented by the metaphor of the **'Tiger'**—be consistent and reconcilable with, and derivative of, the same force that created the lamb in all its innocence? It is the line, **'Did he who made the lamb make thee?'** that, if you allow yourself to dwell on it, is so disturbing. In this essay, the culmination of a lifetime of thoughtful expression, Paton finally brings his focus to bear on this line, a few words that take the reader into the realm where there resides the deep fear about what it really is to be human, about the core issue—that one day had to be addressed and solved—of are humans evil, worthless, meaningless beings, or aren't they? The opening lines of Blake's poem **'Tiger, Tiger, burning bright/In the forests of the night'**, refer to humans' denial of the issue of their divisive condition. It is a subject humans consciously repress but it is an issue that 'burns bright' in the 'forests of the night' of their deepest thoughts. As Paton pointed out, despite humans' denial of it, *the* great, fundamental, underlying question has always been, *are* humans part of God's **'work'**, part of his purpose and design, or *aren't* they?

In the following quote the poet Alexander Pope considers wisdom to be the ultimate **'system'** because it can make all things understandable or **'coherent'**. Like Paton and Blake, he believed that **'in the scale of the reasoning'** involved in becoming wise, the ultimate **'question'** to be answered (the one the human mind has **'wrangled'** or struggled with for **'so long'**) is this question of whether or not humans are a mistake. In his renowned 1733 work, *Essay on Man,* Pope wrote: **'Of systems possible, if 'tis confess'd/That Wisdom infinite must form the best/**

Where all must fall, or not coherent be/And all that rises, rise in due degree/Then in the scale of reas'ning life, 'tis plain/There must be, somewhere, such a rank as man:/And all the question (wrangle e'er so long)/Is only this, if God has placed him wrong?'

In his distillation of a lifetime of mentally grappling with what it is to be human, Australia's only literary Nobel laureate, Patrick White, also gave a rare, honest description of the core agony in human life of having to live with this unresolved question. The following is the key passage from White's 1981 autobiography, *Flaws in the Glass:* **'What do I believe? I'm accused of not making it explicit. How to be explicit about a grandeur too overwhelming to express, a daily wrestling match with an opponent whose limbs never become material, a struggle from which the sweat and blood are scattered on the pages of anything the serious writer writes? A belief contained less in what is said than in the silences. In patterns on water. A gust of wind. A flower opening. I hesitate to add a child, because a child can grow into a monster, a destroyer. Am I a destroyer? this face in the glass which has spent a lifetime searching for what it believes, but can never prove to be, the truth. A face consumed by wondering whether truth can be the worst destroyer of all'** (p.70 of 260).

What is so brave about what Patrick White has written is that he has managed to put down on paper the core fear he is living with. If you allow yourself to think deeply about what it is that White is daring to articulate you will see that it is a terrifying issue that he is facing. Gradually, as this book unfolds, the full horror of the issue of the human condition, and the enormous difficulty humans have had trying to plumb the depths where the issue resides, will become clear. The essay in this book titled *Resignation* will especially bring home how difficult, indeed impossible, it has been for humans to confront the issue of the meaningfulness or otherwise of the dark side of themselves and our species. The reader will be brought into full contact with the issue of the human condition and see that we have indeed been a 'God-fearing' species, a species in denial, with only a handful of people in recorded history able to confront the ideals or God, face to face. It will become clear that the denial and its evasion permeates every aspect of human life. Even science, which is supposed to be a rigorously objective discipline, free of personal bias, is full of subjective evasion, saturated with the denial.

Sharing the same sentiments and even using the same imagery as Patrick White, William Wordsworth wrote of the agony of the dilemma of the human condition in his celebrated 1807 poem, *Intimations of*

Immortality from Recollections of Early Childhood. With extraordinary honesty, Wordsworth recalled all the beauty in the world that humans were able to access before 'the fall'; before the human species departed from the fabled state of harmony and enthralment in which it lived prior to the emergence of the alienating state of the human condition. He ended the poem by alluding to the reason for humans' loss of innocence and sensitivity, adding, **'The Clouds that gather round the setting sun/Do take a sober colouring from an eye/That hath kept watch o'er man's mortality/...To me the meanest flower that blows can give/Thoughts that do often lie too deep for tears.'** The emergence of the human condition made humans red-eyed from being worried—or **'troubled'** as Paton said—about their life's value, meaning and worth. Wordsworth is saying that worrying about our mortality is ultimately due to being insecure about our life's value and worth—hence the reference in the poem's title to the **'intimations of immortality'** humans had during our species' pristine, uncorrupted **'early childhood'**. The thoughts that are now buried so deep that they are beyond the reach of humans' everyday emotional selves—they are **'too deep for tears'**—are the thoughts about humans' present corrupted state that the beauty of even the plainest flower has the ability to remind humans of, if they let it; if they have not practiced burying the issue deeply enough.

Morris West is another distinguished Australian writer. The author of 26 novels, including *The Shoes of the Fisherman,* he has been described as one of the 20th century's most popular novelists. Many times he was asked to write the story of his life and declined, until in 1996, at the age of 80, he reviewed the chronicle of his life and belief in *A View from the Ridge—the testimony of a pilgrim.* In this book he confided: **'Evil, you see, is not explainable. It is not even understandable. It is what the writers of the *Dutch Catechism* called "the great absurdity, the great irrelevancy"...brutalise a child and you create a casualty or a criminal. Bribe a servant of the state and you will soon hear the deathwatch beetles chewing away at the rooftrees of society. The disease of evil is pandemic; it spares no individual, no society, because all are predisposed to it. It is this predisposition which is the root of the mystery. I cannot blame a Satan, a Lucifer, a Mephistopheles, for the evils I have committed, the consequences of which have infected other people's lives. I know, as certainly as I know anything, that the roots are in myself, buried deeper than I care to delve, in caverns so dark that I fear to explore them. I know that, given the circumstances and the provocation, I could commit any crime in the**

calendar' (p.78 of 143). It is the **'caverns so dark'** that are going to be explored in this book, and, despite what Morris West has said, evil will be explained, it will be made understandable.

As Alan Paton, William Blake, Alexander Pope, Patrick White, William Wordsworth and Morris West bravely express, it took virtually all humans' courage merely to exist under the duress of the human condition. Having no answer to this core question in human life—of humans' meaningfulness or otherwise, of are humans part of God's **'work'** or aren't they—meant that trying to think about the problem led only to deep depression. As Australian comedian Rod Quantock once commented, **'Thinking can get you into terrible downwards spirals of doubt'** *(Sayings of the Week, Sydney Morning Herald,* 5 July 1986).

Another renowned Australian literary figure, Henry Lawson—whom Ernest Hemingway greatly admired, and whose work he referred to in his 1970 book *Islands in the Stream*—wrote extraordinarily forthrightly about the dangerous depression that awaits those who might try to confront the issue of the human condition. In his 1897 poem, *The Voice from Over Yonder,* Lawson wrote, **'"Say it! think it, if you dare!/Have you ever thought or wondered/Why the Man and God were sundered?/Do you think the Maker blundered?"/And the voice in mocking accents, answered only: "I've been there."'** Implicit in the final phrase 'I've been there' are the unsaid words 'and I'm not going *there* again'. The 'there' and the 'over yonder' of the title is a reference to the state of depression that resulted from trying to confront the issue of the human condition—trying to understand **'why the Man and God were sundered'**, why humans lost their innocence, departed from the cooperative, loving, ideal state, 'fell from grace', became corrupted, 'evil', 'sinful'. To avoid depression humans had no choice but to repress the issue of the human condition, block it out of their conscious awareness, and cease trying to decide whether **'the Maker blundered'**.

Alluding to the unconfrontable and horrifically depressing issue of the human condition, the Danish philosopher Søren Kierkegaard wrote about an underlying **'despair'** in human life, and succinctly described the depression that results from that despair with the title of his 1849 book *The Sickness Unto Death.* In it he wrote, **'there is not a single human being who does not despair at least a little, in whose innermost being there does not dwell an uneasiness, an unquiet, a discordance, an anxiety in the face of an unknown something, or a something he doesn't even dare strike up acquaintance with...he goes about with a sickness, goes**

about weighed down with a sickness of the spirit, which only now and then reveals its presence within, in glimpses, and with what is for him an inexplicable anxiety' (tr. A. Hannay, 1989, p.52 of 179). Kierkegaard described the depression that is like a living death, which the **'tormenting contradiction'** of the human condition has caused in humans, when he wrote that, **'the torment of despair is precisely the inability to die...that despair is the sickness unto death, this tormenting contradiction, this sickness in the self; eternally to die, to die and yet not to die'** (ibid. p.48).

In his 1931 book, *The Destiny of Man,* the Russian philosopher Nikolai Berdyaev referred to Kierkegaard's experience of deep depression when he described the **'terror'** and **'fear'** that trying to think about the tormenting contradiction of the human condition evokes in all but those with a **'clear conscience'**, those who are **'prophetic'**. Berdyaev says that while thinking about the human condition is terrifyingly depressing for most people—he refers to the **'deadly pain in the very distinction of good and evil'**—he emphasised that only by thinking about it can real knowledge be found. Clearly if you are living in denial of a problem you are not in a position to solve it. Berdyaev wrote: **'Knowledge requires great daring. It means victory over ancient, primeval terror. Fear makes the search for truth and the knowledge of it impossible. Knowledge implies fearlessness...Conquest of fear is a spiritual cognitive act. This does not imply, of course, that the experience of fear is not lived through; on the contrary, it may be deeply felt, as was the case with Kierkegaard, for instance...it must also be said of knowledge that it is bitter, and there is no escaping that bitterness ...Particularly bitter is moral knowledge, the knowledge of good and evil. But the bitterness is due to the fallen state of the world, and in no way undermines the value of knowledge...it must be said that the very distinction between good and evil is a bitter distinction, the bitterest thing in the world...Moral knowledge is the most bitter and** [requires] **the most fearless of all for in it sin and evil are revealed to us along with the meaning and value of life. There is a deadly pain in the very distinction of good and evil, of the valuable and the worthless. We cannot rest in the thought that that distinction is ultimate. The longing for God in the human heart springs from the fact that we cannot bear to be faced for ever with the distinction between good and evil and the bitterness of choice...Ethics must be both theoretical and practical, i.e. it must call for the moral reformation of life and a revaluation of values as well as for their acceptance. And this implies that ethics is bound to contain a prophetic element. It must be a revelation of a clear conscience, unclouded by social conventions; it must be a** ***critique***

of pure conscience' (tr. N. Duggington, 1955, pp.14–16 of 310).

This collection of quotes about the agony of the human condition shows how fearful humans have been of the issue. At the very end of his 1948 book *Cry, the Beloved Country,* Alan Paton alluded to humanity's dream of one day finding understanding of the human condition and by so doing free itself from this **'bondage of fear'**. He wrote: **'But when that dawn will come, of our emancipation, from the fear of bondage and the bondage of fear, why, that is a secret.'**

In truth, human self-esteem, which at base is the ability to defy the implication that they are not worthwhile beings, is so fragile that if a man loses his fortune in a stock market crash, or his reputation from some mistake he has made in the management of his life, he all-to-frequently will suicide or, if not suicide, then completely crumple as a person, lose the will to actively participate in life. The truth is that the limits within which humans feel secure and can operate are extremely narrow. Humans' insecurity from the dilemma of the human condition has been such that their comfort zone is only a tiny part of the vast, true world that humans are capable of living in. As the full extent of humans' insecurity under the duress of the human condition becomes clear it will become apparent that, in terms of the true potential of the universe, the world that humans have lived in has been merely a miserable, tiny, dark fortress of a corner. Humans' psychological circumstances—which, at base, arise from their struggle with the human condition—are extremely fragile. Although they have had to live in denial of this truth to cope with it, the fact is humans are an *immensely* insecure species.

Humans' historic denial of the issue of the human condition

Tragically, unable to explain and thus resolve this deepest of dilemmas of the human condition, humans had no choice other than to live in denial of the whole issue. While they lacked understanding of the human condition, denying it—extremely dishonest, false and, as we will see, limiting a response as that was—was their only sensible means of coping with it. The truly extraordinary aspect of humans, and a measure of their immense bravery, is that they have managed to keep a bright and optimistic countenance despite the awful realities of their circumstances. The courage to live in denial, despite the dishonesty of this behaviour and the extremely artificial and superficial existence it left humans with, has been the very essence of our species' immense bravery. Humans may be the most alienated species in the universe but they must also be among the bravest.

Humanity's historic denial of the issue of the human condition began when consciousness first emerged from the instinct-dominated state some 2 million years ago and has been reinforced ever since. (The emergence of the dilemma of the human condition with the emergence of consciousness in our human ancestors, and our resulting departure from the fabled Garden of Eden where humans lived instinctively in a state of cooperation in the so-called 'Golden Age', is summarised in the early part of the *Plato* essay, the first essay of this book, and is comprehensively explained in *Beyond*.) As a result of having practiced the denial of the issue of the human condition for so long, humans now live in a state of almost complete denial of the issue, to the point effectively of being unaware of it at a conscious level. The issue of the human condition is now deeply buried; a part of humans' subconscious awareness.

Common to all the human race at a subliminal, subconscious level is an immense insecurity, a deep sense of guilt about being divisively behaved.

On the face of it, to be told there is a crux, fundamental, all-important issue facing humans that they are currently not consciously aware of must seem absurd. It is not easy to accept that there is an 'elephant in the living room' that people have lived in such denial of that they are no longer aware of its existence. While this situation may sound unbelievable at first, the mental process involved is no different to that which takes place in the minds of, for example, incest victims who, after finding they cannot comprehend such violation, realise that their only means of coping is to block out any memory of it. 'Repressed memory', living in denial of an issue, is a common occurrence. In fact blocking thoughts from our mind has been one of humans' most powerful coping devices.

Plato's 'cave' allegory for life in denial

The great writers mentioned earlier were brave enough to refer to the agony of the dilemma of the human condition, but it was the Greek philosopher Plato who described more clearly than anyone else, albeit allegorically, the whole situation associated with the human condition. This astonishing description, which employed the allegory of a cave, appeared in Plato's great work, *The Republic,* which he wrote in about 360 BC.

Considering how penetrating Plato's description of the human condition is, a much more detailed analysis will be presented in the first essay of this book, *Deciphering Plato's Cave Allegory—And in the Process Explaining How The Human Condition is Resolved.* For the purposes of this *Introduction,* a brief interpretation is sufficient.

Since the full version of Plato's allegory goes for eight pages, the following summary from the 1996 *Encarta Encyclopedia* offers a succinct description: **'The myth of the cave describes individuals chained deep within the recesses of a cave. Bound so that vision is restricted, they cannot see one another. The only thing visible is the wall of the cave upon which appear shadows cast by models or statues of animals and objects that are passed before a brightly burning fire. Breaking free, one of the individuals escapes from the cave into the light of day. With the aid of the sun, that person sees for the first time the real world and returns to the cave with the message that the only things they have seen heretofore are shadows and appearances and that the real world awaits them if they are willing to struggle free of their bonds. The shadowy environment of the cave symbolizes for Plato the physical world of appearances. Escape into the sun-filled setting outside the cave symbolizes the transition to the real world, the world of full and perfect being, the world of Forms, which is the proper object of knowledge.'**

Plato's parable says that between the natural, radiant, all-visible, sunlit world and humans' **'cave'** existence stands a **'brightly burning fire'** that prevents them from leaving the cave. In the full text of the allegory in *The Republic,* Plato says that **'the light of the fire in the prison** [cave] **corresponds to the power of the sun'** (*Plato The Republic,* tr. H.D.P. Lee, 1955, p.282 of 405). What the **'sun'**, and its Earthly representation, the **'brightly burning fire'**, represent is the condemning cooperative ideals of life, the ideals that bring the depressing issue of the human condition into focus—the question of why are humans competitive, aggressive and selfish when the ideals are to be cooperative, loving and selfless. The **'sun/fire'** represents the confronting glare of the ideals and the burning heat of the issue of the human condition that those ideals bring into focus, the issue that humans have had to live in denial of, the issue that has forced humans to, metaphorically speaking, hide in a dark **'cave'**.

So intense was **'the power'** of the **'sun/fire'** to condemn and depress humans that they could not face it, le[illegible] approach it, and they were so held in bondage by the unconfrontable issue of the human condition, so **'chained'** up, as even to be estranged—alienated—from each other. As the *Encarta* summary translated it, they **'cannot see one another.'** (It is worth reca[illegible] Paton described humans as living in **'the bondage of fear'** of the issue of the human condition.)

Also, because the human condition is the crux issue before us as a species, living in denial of it meant humans were living an extremely fraudulent, artificial and superficial existence, they were living in a world that wasn't **'real'**, a world of delusion and illusion. Plato's **'shadows'** on the back wall of the cave symbolise this world of illusions that humans have been living in. In the full text of the allegory in the *Republic,* Plato says that, **'if he** [a prisoner in the cave] **were made to look directly at the light of the fire, it would hurt his eyes and he would turn back'** *(Plato The Republic,* tr. H.D.P. Lee, 1955, p.280 of 405). The prisoner had to face away from the fire and could only look at the shadows cast by the real world on the back wall of the cave.

I have always admired the 1976 Francis Bacon painting titled *Study for Self Portrait* (owned by the NSW Art Gallery in Australia) for its honest portrayal of the human condition; in fact I once unsuccessfully sought permission to use it on the cover of my first book, *Free.* Bacon depicted the human condition as honestly as anyone has ever managed to write about it. *Study for Self Portrait* shows one of Bacon's characteristic twisted, smudged, distorted—alienated—human faces (in this case his own, which makes the painting that much more honest), but it also shows the body's arms to be chained up behind the body, which is constrained in a box. The entire image is reminiscent of Plato's representation of the human predicament under the duress of the human condition. The two William Blake pictures that I have used on the cover of this book are also dramatic depictions of the story of humans' struggle with the human condition. The top picture marvellously represents Plato's sunlit, liberated state above ground, while the bottom picture dramatically depicts humans' tortured, cave-like existence below ground. (The top picture was a black and white etching by Blake that an artist friend colourised to complement the bottom picture, which Blake did in colour.)

The use of fire as a metaphor for the cooperative ideals of life that barred humans' escape from their **'restricted'**, alienated condition, is common in many mythologies. In Christian mythology, for example, in the story of the Garden of Eden, there is **'a flaming sword flashing back and forth to guard the way to the tree of life'** (Genesis 3:24). Later in the *Bible* it is recorded that the Israelites said, **'Let us not hear the voice of the Lord our God nor see this great fire any more, or we will die'** (Deut. 18:16). While fire is a metaphor for the cooperative, loving, selfless ideals of life that so condemned humans, the metaphysical, religious term humans have historically used for these ideals is 'God'.

This interpretation of God as the embodiment of the cooperative, selfless, loving ideals of life will be much more fully explained when Plato's allegory is more completely deciphered in the first essay of this book. It will be explained there that the cooperative ideals of life are a manifestation of the most profound—and confronting—of all truths, that of the negative entropy-driven, matter-integrating, cooperation-dependent, teleological, holistic purpose or design or meaning in existence.

The following quote offers an example of how the ancient Zoroastrian religion used fire to represent the Godly, upright, pure ideals of life: '[In the Zoroastrian religion] **Fire is** [considered] **the representative of God...His physical manifestation...Fire is bright, always points upward, is always pure'** (*Eastern Definitions*, Edward Rice, 1978, p.138 of 433). God, of whom fire is representative, is **'pure'**, the personification of the all-meaningful cooperative, loving, selfless ideals of life, however because these ideals are in such contrast to humans' apparently non-ideal and thus apparently non-meaningful competitive, aggressive and selfish nature, humans have naturally feared God. They have been a 'God-fearing' rather than a 'God-confronting' species. Humans have been fearful that they may be a horrible mistake: White wondered whether a human was **'a monster, a destroyer'**, Blake asked **'Did he** [God] **who made the lamb make thee?'**, Pope pondered **'if God has placed him** [man] **wrong?'**, and Lawson asked **'Do you think the Maker blundered?'**

In the biblical account, Job pleaded for relief from confrontation with the issue of the human condition that the Godly, meaningful, cooperative, loving, selfless ideals of life brought into issue when he lamented, **'Why then did you** [God] **bring me out of the womb?...Turn away from me so I can have a moment's joy before I go to the place of no return, to the land of gloom and deep shadow, to the land of deepest night** [depression]**'** (Job 10:18, 20-22).

Job's **'land of gloom and deep shadow...the land of deepest night'** perfectly equates with the analogy of life in a cave; that state of deepest and darkest depression caused by trying unsuccessfully to confront and make sense of the apparent extreme lack of ideality in the human make-up. Only by turning away from the **'fire'**, avoiding the condemning glare of the cooperative ideals of life, could humans find some relief from the criticism and resulting terrible—even suicidal—depression. Again, as the Israelite people said, **'Let us not hear the voice of the Lord our God nor see this great fire any more, or <u>we will die</u>.'** When reading the *Resignation* essay, it will become palpably clear how terrifyingly depressing the issue of the human condition has been for humans.

The reader may wonder how it is possible to confront and talk about the subject of the human condition so completely and with such ease if it is so dangerously depressing. The reason it is possible is that humanity has at last found the dignifying biological explanation for why humans have not been ideally behaved. It has found the explanation that liberates humans from the insecurity that the dilemma of the human condition has for so long caused them—and, thankfully, with the elimination of that historic insecurity all the products of that insecurity, in particular humans' angry, competitive, selfish, egocentric and alienated behaviour, subside and eventually disappear forever. In religious terms, that question of questions of what is the 'origin of sin' has been answered. The explanation is presented in *Beyond* and elaborated upon in this book.

As is made clear in *Beyond,* it is not me, but humanity as a whole that has found this explanation of the human condition, because it is only as a result of the discoveries of science, the peak expression of *all* human intellectual effort, that I have been able to synthesise the biological explanation of the human condition. In fact it is 'on the shoulders' of eons of human effort that our species' freedom from the human condition has finally been won.

The evidence that the human condition has been solved is that it is being talked about so openly and freely here. The human condition is such that you cannot talk about it until you have understood it, and, by so doing, broken free of it. The biological explanation for humans' 'corrupted', non-ideal condition has been found, and thus the criticism from the Godly ideals of life has been removed—humans' historic 'burden of guilt' has been lifted—the **'great fire'** has been doused.

The *Encarta* summary of Plato's cave allegory describes how **'Breaking free, one of the individuals escapes from the cave into the light of day. With the aid of the sun, that person sees for the first time the real world and returns to the cave with the message that the only things they have seen heretofore are shadows and appearances and that the real world awaits them if they are willing to struggle free of their bonds.'** As has been explained, the **'sun'** represents the condemning Godly, meaningful, cooperative ideals of life. The reference to needing **'the aid of the sun'** to decipher the human condition and thus **'see for the first time the real world'**, refers to the fact that you cannot solve the human condition if you are living in the **'cave'** of denial. As is stated in the *Encarta* entry, **'vision is restricted'** in the **'cave'**. To be able to synthesise the explanation of the human condition required that I confront and acknowledge, rather than live in denial of, the Godly, meaningful, cooperative ideals of life and the associated issue of the human condition. In Plato's terms, I needed to be living in the presence of the illuminating **'sun'**, rather than in the dark **'cave'** of denial. In *The Republic* Plato talked of **'objects** [being] **illuminated by daylight...** [just as they are] **by truth and reality'** (*Plato The Republic*, tr. H.D.P. Lee, 1955, p.273 of 405). If you are living in denial of the human condition you are in no position to assemble the truth about the human condition. You cannot assemble the truth from a position of lying, from a position within the **'cave'**. While science makes clarifying explanation of the human condition possible, the actual synthesis of the explanation requires that the truth of the Godly, meaningful, cooperative ideals of life and associated issue of the human condition be confronted. How it is that some people have been more able to confront and thus think about the human condition than others is examined in the *Plato* essay where human alienation and its many degrees are explained.

Considering how dangerously depressing the subject of the human condition has historically been for most humans, it needs to be emphasised that it is at last safe to confront the issue of the human condition; this assurance is supported by the fact that the subject is being so completely and freely confronted here in this book. Also, how could the many people who have been involved since the late

1980s in the Foundation for Humanity's Adulthood (FHA)—the organisation that has been established to develop and promote these understandings of the human condition—be actively supporting these understandings of the human condition, be so involved, if the understandings were not enabling them to safely confront the human condition. The evidence that 'the fire' has been 'extinguished' is that there are people who are not being 'burnt' by it.

In the words of the *Encarta* summary, Plato emphasised that **'the proper object of knowledge'** was to achieve **'the transition to the real world'**; that is, solve the human condition and end the alienated state of denial that humans have had to live in, get out of the **'cave'**. While humans have had to live in denial of the human condition in order to cope with it, that did not mean that our task was not ultimately to solve it. While most people, including almost all scientists, studiously avoided any confrontation with and thus any analysis of the human condition, confronting and analysing the human condition is in fact **'the proper object of knowledge'**. As poet Alexander Pope said in his *Essay on Man,* **'Know then thyself, presume not God to scan; The proper study of Mankind is Man.'**

Pope's admonition that we should not leave it to **'God to scan'** makes the point that faith was not going to be sufficient to sustain humanity. As has already been emphasised, while religious assurances such as 'God loves you' could comfort us we ultimately had to understand *why* we were lovable; answer the question, *are* humans part of God's **'work'**, part of his purpose and design, or *aren't* they? While religions played a crucial role in sustaining humans who were living under the duress of the human condition, they could not lift the burden of guilt from humanity, explain the human condition, produce the dignifying biological understanding of human nature. There had to be a biological explanation for humans' non-ideal, upset, corrupted, insecure, divisive competitive, aggressive, selfish, angry, egocentric and alienated behaviour and our responsibility as conscious animals has been to find that explanation. Historian Jacob Bronowski stressed this responsibility in his 1973 television series and book, *The Ascent of Man,* saying, **'We are nature's unique experiment to make the rational intelligence prove itself sounder than the reflex** [instinct]**. Knowledge is our destiny. Self-knowledge, at last bringing together the experience of the arts and the explanations of science, waits ahead of us'** (p.437 of 448).

The Australian biologist Charles Birch, whom I was fortunate to have as a teacher at Sydney University, has been described as **'Australia's leading thinker on science and God'** (*Sydney Morning Herald* article titled *God* by D. Smith, 27 Feb. 1998). In 1990 he was a joint winner of the prestigious Templeton Prize, which is awarded for **'increasing man's understanding of God'** (*The Templeton Prize,* Vol.3, 1988–1992, p.108 of 153). In his 1999 book, *Biology and the Riddle of Life,* Charles Birch emphasised that **'the onus is now on biologists to demonstrate the importance of self-organisation in biological evolution'** (p.110 of 158). As will be explained in the first essay of this book, 'self-organisation' is a reference to negative entropy's ordering or integration of matter, to the teleological, holistic purpose or design or meaning in existence that I mentioned earlier. Therefore what Birch is saying is that the responsibility of biologists in recent times has been to finally face the truth of integrative, cooperative meaning and, in so doing, confront the issue of the human condition. Solving the human condition *was* a problem for biologists because the human condition is about the behaviour of the human species.

It needs to be emphasised that finding understanding of humans' non-ideal, upset, corrupted, divisive behaviour does not condone such behaviour, it does not sanction 'evil'; rather, through bringing compassion to the situation, it allows the insecurity that produces such behaviour to subside, and the behaviour to disappear. As is explained in *Beyond,* humans' non-ideal, 'evil' behaviour is a result of a conflict and insecurity within themselves that arises from the dilemma of the human condition; resolve the dilemma and you end the conflict and insecurity. One of the greatest philosophers of our time, Sir Laurens van der Post, has written: **'Compassion leaves an indelible blueprint of the recognition that life so sorely needs between one individual and another; one nation and another; one culture and another. It is also valid for the road which our spirit should be building now for crossing the historical abyss that still separates us from a truly contemporary vision of life, and the increase of life and meaning that awaits us in the future'** (*Jung and The Story of Our Time,* 1976, p.29 of 275).

The task Sir Laurens wrote of as **'crossing the historical abyss'** was the task of confronting and solving the terrifyingly depressing subject of the human condition. Elsewhere in Sir Laurens' writings he acknowledges both the terrifying nature of that abyss where dark depression exists and the extent of humanity's denial of the human condition. He wrote of **'a world which does not recognise the reality of "these mountains of the mind and their cliffs of fall, frightful, sheer, no-**

man-fathomed" of which [the poet Gerard] **Manley Hopkins had spoken'** (ibid. p.156). At last this **'historical abyss'** of depression has been **'fathomed'**, the human condition has been explained.

Living in denial was a horribly dishonest, superficial and loathsome state to have to endure, but it allowed humans to live when their inability to understand their divisive condition would have otherwise led to epidemic suicidal depression—and the end of the human race. As Joseph Conrad wrote in his 1915 book *Victory*, **'For every age is fed on illusions, lest men should renounce life early and the human race come to an end'** (p.90 of 396).

Armed at last with the understanding of the human condition, the understanding of why humans are not fundamentally bad, all the denials, facades, delusions and illusions that humans have hidden behind can be exposed and dismantled. As the *Encarta* summary states, **'the message** [is brought back] **that the only things they** [the prisoners in the cave] **have seen heretofore are shadows and appearances and that the real world awaits them if they are willing to struggle free of their bonds.'** With understanding of the human condition everyone can leave the dark, immensely **'restricted'**, cave existence of living in denial—that is as long as they are prepared to **'struggle free of their bonds'** of denial of the human condition. The difficulty of struggling free of denial is the next issue to be examined.

The 'deaf effect': the difficulty of reading about the human condition

Given humans' justified extreme fear of the subject of the human condition, a very important question arises: How are humans going to be able to consider presentation and analysis of the human condition now? Won't their historic fear be triggered? Won't their habituated practice of denial come into play? Won't their mind start blocking out what is being said, effectively unable to absorb the information?

The answer to these questions is a resounding 'yes', humans are effectively 'deaf' to discussion of the issue of the human condition. Humans' historic denial prevents description and analysis of the human condition coming through into their conscious awareness. Such is the nature and purpose of a denial.

The following extract from a newspaper article is an illustration of the power of the human mind to block out or alienate itself from or deny what it does not want to hear. The article, headed **'Patients can't comprehend doctors' bad news'**, said that a study of cancer patients **'showed that many patients use denial as a way of coping...Professor Stewart Dunn, director of the medical psychology unit at the University of Sydney and Royal Prince Alfred Hospital, said the new findings suggested many patients did not want to hear their doctor because they were in "denial mode". "We found that denial is a way of dealing with severe distress...In a typical 28-minute consultation, a cancer patient received one new piece of information a minute—too much for traumatised people to absorb," he said...Ms Kay Roy, 50, who was diagnosed with breast cancer two years ago, recalls that she remembered nothing from what the doctor said in the initial consultation, other than "you can expect a normal lifespan". "So, I obviously picked up the words I wanted to," she said. "I think there is a general discrepancy between what the doctor says and the patient hears. I thought I was cool, calm and collected but I must have been in a state of shock. The words just seemed to flood over me." Ms Roy, of Wahroonga, said it took another two weeks before reality began sinking in'** (*Sydney Morning Herald,* 18 Aug. 1995).

It is true that as soon as discussion of the human condition begins, the human mind's highly practiced, virtually automatic denial asserts itself to protect it from the implicitly condemning criticism that it has come to expect will follow. In the full text of his cave allegory in *The Republic,* Plato described the situation perfectly when he wrote: **'if he** [a prisoner in the cave] **were made to look directly at the light of the fire** [as explained, the sun and its Earthly representation fire represent the cooperative ideals and the associated issue of the human condition]**, it would hurt his eyes and he would turn back and take refuge in the things which he could see, which he would think really far clearer than the things being shown him. And if he were forcibly dragged up the steep and rocky ascent** [out of the cave of denial] **and not let go till he had been dragged out into the sunlight, the process would be a painful one, to which he would much object, and when he emerged into the light his eyes would be so overwhelmed by the brightness of it that he wouldn't be able to see a single one of the things he was now told were real'** (*Plato The Republic,* tr. H.D.P. Lee, 1955, p.280 of 405).

Plato says that when the cave dweller **'emerged into the light his eyes would be so overwhelmed by the brightness of it that he wouldn't be able to see a single one of the things he was now told were real.'** Humans have had to live in such deep denial of the issue of the human condition that when they are confronted with the issue they, as Plato says, are not **'able to see a single one of the things'** they are being told. Not being able to take in **'a single thing'** when reading logical and simple to follow explanation is so against the normal person's experience that the reader is entitled to be extremely disbelieving at being told that this will occur with analysis of the human condition. However, despite the readers' expected initial extreme disbelief, Plato is right, the reader *will be* 'blind' or, depending on which metaphor you use, 'deaf' to discussion of the issue of the human condition. I hope the reader will not be offended by this assertion. The intention of this book is that the further someone reads into the book, the more aware they will become of just how frightening the issue of the human condition has been for humans, and thus how deeply humans have been living in denial of the subject. There truly *has* been an elephant in everyone's living room that virtually everyone has very deliberately not been able to see. The aim of this book is to bring the reader, step by step, into contact with the extent of the problem of humans' fear and thus denial of the issue of the human condition because it is only then that they will be in a position to appreciate the reconciling, dignifying, ameliorating and liberating explanation of the human condition. To be in a position to rid themselves of the 'elephant', humans had to first rediscover its existence.

Plato elaborated on the difficulty of trying to get people to confront the issue of the human condition, actually describing it as being like trying to **'put sight into blind eyes'** (ibid. p.283). As the *Plato* essay will explain in detail, people who have historically been referred to as prophets are individuals who had the rare good fortune to receive a sufficiently nurtured upbringing to remain uncorrupted enough in self to be able to confront the Godly, meaningful, cooperative, loving ideals of life and the associated issue of the human condition without experiencing condemnation and depression. They are individuals who have a clear conscience. As Berdyaev said in the aforementioned quote from his 1931 book *The Destiny of Man*, **'Moral knowledge is the most bitter and** [requires] **the most fearless of all for in it sin and evil are revealed to us...There is a deadly pain in the very distinction of good and evil, of the valuable and the worthless...** [for this reason the study of] **ethics is bound to contain a prophetic element. It must be a revelation of a clear conscience, unclouded by social conventions; it must be a *critique of pure conscience*.'** The biblical prophets Isaiah and Christ were two such fortunate people and both encountered the 'deaf effect' response to their unevasive, denial-free, human condition-confronting, truthful words. Using the same sensory metaphor as Plato, Isaiah said, **'"You will be ever hearing, but never understanding; you will be ever seeing, but never perceiving." This people's heart has become calloused** [alienated]**; they hardly hear with their ears, and they have closed their eyes'** (*Bible, New International Version,* 1978, Isaiah 6:9,10, footnote). Christ similarly said, **'why is my language not clear to you? Because you are unable to hear what I say...The reason you do not hear is that you do not belong to God** [you live in denial of the Godly, meaningful, cooperative ideals of life and the associated issue of the human condition]**'** (John 8:43–47).

T.S. Eliot once wrote of the **'terrifying honesty'** of the denial-free, unevasive, sun-confronting writings of William Blake (*William Blake Selected Poems,* ed. P.H. Butter, 1982, pxiii of 267). As we have seen, Blake dared to plumb the depths of the human condition.

The truth is that almost all humans have *had to* employ immense mental block-out from, and thus separation or split-off or alienation from, the truth of their corrupted state. The human condition is, to re-quote Kierkegaard, **'something'** a human **'doesn't even dare strike up acquaintance with'**. This deepest of issues of the human condition has been an anathema to most people. They have had the attitude of Albert the alligator in the old Pogo comic strip: **'The inner me? Naw, got no time fer him. Ah got trouble enough with the me whut's out cheer whar Ah kin get mah hands on 'im. Ez fer the inner me, he goes his way, Ah go mine'** (mentioned in Charlton Heston's autobiography, *In The Arena,* 1995).

Significantly, the longer the human journey had to continue without the reconciling understanding of the human condition, the more psychologically embattled and insecure, and thus the more in denial and alienated each generation of humans became. Once adults were corrupted and alienated, their children's pure, original, instinctive self or soul was going to be compromised, hurt, damaged and corrupted by the unsound, dishonest environment they were having to grow up in. 'Hurt' has unavoidably been causing hurt since the human condition emerged. As it is described metaphysically in the *Bible,* **'I...am a jealous God, punishing the children for the sin of the fathers to the third and fourth generation of those who hate me'** (Ten Commandments, Exod. 20:5, Deut. 5:9). Also, the level or amount of corruption has compounded as each generation contributed to the existing corruption with the corruption it incurred from its own conscious search for knowledge. (How the conscious search for knowledge created humans' corrupted angry, egocentric and alienated behaviour is the central explanation presented in *Beyond*.) While self-restraint could, to a degree, contain the corruption and its propagation, ultimately only reconciling, dignifying and ameliorating understanding of the reason for this corrupt behaviour could end the cumulative process.

Alienation has been exponentially accumulating. In fact a graph charting the extent of alienation over time would show its curve beginning to climb vertically in recent years, the rate of increase having become so rapid. The levels of alienation in the human race now are so extreme that we are fast approaching an end-point where the human body cannot endure any more suffering from depression and loneliness, and the human mind cannot become any more alienated without going completely mad.

Part of humans' strategy for avoiding the issue of the human condition has been to avoid the true extent of their devastation of the world around them and within them. Humans have *had to*, as is said, 'put on a brave face', 'stay positive', 'keep our chin up'—as the actor David Niven once commented, humans **'had a duty to be cheerful'**. The only way humans could do that in the face of their awful reality was to not allow themselves to see this reality. News bulletins report when someone is run over by a bus or is murdered but they never report the deaths that are occurring daily of the souls of humans, deaths that are just as real as physical deaths. Alienation is not a subject people want to know about. The Scottish psychiatrist R.D. Laing wrote that **'the only real death we recognize is biological death'** (*The Divided Self*, 1960, p.38 of 218), while Kierkegaard, in his book *The Sickness Unto Death,* noted that **'The biggest danger, that of losing oneself, can pass off in the world as quietly as if it were nothing; every other loss, an arm, a leg, five dollars, a wife, etc., is bound to be noticed'** (p.62 of 179). (Kierkegaard was certainly one of the rare people capable of, as the dustjacket of the 1989 Penguin edition of his book says, **'digging deep in the graveyard of denial, refusal and despair'** of the human situation.) As mentioned earlier, this practice of positive stoicism was the essence of our species' bravery, but the point here is that it has masked the true extent of alienation in humans from themselves.

The following quote from a review of Stanley Cohen's 2001 book, *States of Denial: Knowing about Atrocities and Suffering,* describes humans' inability to face the extent of the devastation of the world around them. It also provides another illustration of humans' capacity for denial of unconfrontable realities, acknowledging that **'denial has become the condition of our times'** and alienation is a **'wholesale pathology'** in our society.

The reviewer, Anthony Elliott, writes: **'Few topics can be as disquieting as the strategies we use to shield ourselves from administered atrocities—torture, political massacres, genocides. From the shadows cast by Auschwitz to recent terrors in Bosnia, Rwanda, Chechnya and Kosovo, the apparent indifference of Western publics to mass suffering is shocking, disturbing, haunting. With media images of violence relayed instantly across the globe, knowledge of organised cruelty seems increasingly threatening—such knowledge incapacitates us intellectually, drains us emotionally. Blocking out, shutting off, not wanting to know: denial has become the condition of our times...Cohen's *States of Denial* is out to show that the personal and political ways in which we avoid uncomfortable realities are deeply layered at the level of both personality and contemporary culture. When we deny painful knowledge, says Cohen, we use unconscious defence mechanisms to protect ourselves. Cohen finds the subterranean, ambivalent waters of the Freudian unconscious a useful explanatory tool for making sense of subtle variations in professed denials of fact or knowledge. In both the private realm and the public sphere, Cohen traces mechanisms of denial from "splitting" (where a person psychically detaches from painful knowledge) to "inner emigration" (a retreat inwards to shut out an unbearable outer reality). Intriguingly, Cohen nicely captures one of the most common modes of avoidance with the term "pseudo-stupidity". This is a way of both knowing and not-knowing, of suspecting but not seeking to check one's suspicions. While noting that some "switching off" is necessary in order to retain our sanity, Cohen argues that our rising inability to "face" or "live with" unpleasant truths is producing wholesale pathology in the form of alienated individuals and remote communities. In an exceptionally wide-ranging treatment of the topic, Cohen's timely book traces multiple forms of the denial of distant suffering. He analyses denial through the rich literature of its expression, including cognitive psychology and psychoanalysis, social and political sources, the reports of witnesses and bystanders, legal theory and literary texts'** (*The Australian,* 4 Apr. 2001).

The denial that humans are currently practicing towards the atrocities of our time is undoubtedly extreme, nevertheless it is but a tiny drop in the 2-million-years-deep ocean of denial and resulting alienation humans have developed to protect themselves from the trauma of the human condition. Friedrich Nietzsche recognised the phenomenal role lying has played in human life when he wrote, **'That**

lies should be necessary to life is part and parcel of the terrible and questionable character of existence' *(The Will To Power,* 1901; tr. A.M. Ludovici, 1909). In his major work *Faust* (1808), Goethe argued that lying is part of human nature while truth is not.

What is so significant about the 2-million-years-reinforced alienation humans now have from the issue of the human condition is that it means that the 'deaf effect', the amount of block-out or denial that humans are employing to protect themselves from confrontation with the issue, is now almost complete. In 1967 R.D. Laing described the situation when he wrote, **'There is a prophecy in Amos that there will be a time when there will be a famine in the land, "not a famine for bread, nor a thirst for water, but of *hearing* the words of the Lord." That time has now come to pass. It is the present age'** *(The Politics of Experience* and *The Bird of Paradise,* 1967, p.118 of 156). The **'famine'** that is now upon humans is their near total inability to access and acknowledge the issue of the human condition. As mentioned, **'the Lord'** or 'God' are the metaphysical terms we have used for the profoundly meaningful cooperative, loving, selfless ideals of life, a meaningfulness that humans have not been able to confront while they could not explain their competitive, aggressive and selfish nature. While humans have been unable to confront the truth of such meaningfulness, and have thus had to live in denial of it, there are many other important truths that bring into focus humans' apparent lack of compliance with the Godly ideals of life, and of which humans have therefore also had to live in denial. For example, humans have had to live in denial of the truth that our species lived in an instinctive innocent, all-sensitive, utterly cooperative, soulful state before consciousness emerged, the fabled Golden Age or Garden of Eden. Also, as just described, they have had to live in denial of the truth of the epidemic, the **'wholesale pathology'**, of human alienation. These denials, and many, many others that will be described in this book, are enormous lies. Humans have blocked out so much that is real and true. In this sense there is now, as Amos prophesied and Laing recognised, a **'famine of *hearing*'** anything that brings into focus humans' apparent lack of compliance with **'the Lord'**, which is the cooperative state.

In Homer's Greek legend, *The Odyssey* (composed sometime in the 9th century BC), the prophet Teiresias predicted that Odysseus (or 'Ulysses' in the later Roman version), on his return to Ithaca after the Trojan War, would have to undertake one final journey into a desperately barren land. Odysseus told his wife Penelope, **'Teiresias bade me travel far and wide, carrying an oar, till I came to a country where the people have never heard of the sea and do not even mix salt with their food.'** The sea is a metaphor for humans' original innocent instinctive self or soul that, as is explained in *Beyond* and referred to in essays in this book, became repressed with the emergence of our conscious self. This instinctive self or soul, which psychoanalyst Carl Jung referred to as our species' **'collective unconscious'** and for which the sea is such a good symbol, now resides deep in humans' subconscious. From there it bubbles up in dreams and on other occasions when the conscious self is subdued. As Jung wrote, **'The dream is a little hidden door in the innermost and most secret recesses of the psyche** [soul]**, opening into that cosmic night which was psyche long before there was any ego consciousness'** (*Civilization in Transition,* The Collected Works of C.G. Jung, Vol.10, 1945). Our psyche or soul ('psyche' in the dictionary means 'soul') has immense sensitivity, it has access to all the beauty and magic of life, the 'salt' or full flavour of existence—unlike the numb, seared, flavourless, 'saltless', alienated state. It follows that this **'country where the people have never heard of the sea and do not even mix salt with their food'** is the soul-destroyed, alienated world that R.D. Laing described as, **'the present age.'** The **'oar'**, Teiresias explained to Odysseus, is actually a **'winnowing shovel'**. This other journey, that Teiresias was predicting would eventually have to be undertaken, was into the centre of this alienated world we now live in, in order to winnow from its denial-compliant, evasively presented scientific discoveries the truth about the human condition. It is that other journey that has at last been completed.

It should be mentioned that Sir Laurens van der Post has written an essay, titled *The Other Journey*, that explains 'the other journey' component of the Teiresian prophecy in greater detail than I have here. It can be found in Sir Laurens' 1993 book, *The Voice of the Thunder.* I should point out that Sir Laurens titled that book **'The Voice of the Thunder'** in anticipation of the arrival of the deafening truth about the human condition. For Sir Laurens van der Post to have been able to make such penetrating insights into mythology and the human situation he must clearly have been one of those rare individuals who was sufficiently sound to confront and think freely about the human condition, someone living outside Plato's **'cave'** of denial in the truthful world of the illuminating **'sun'**, one of those people traditionally termed a prophet. Indeed in his full-page obituary in the *London Times* he was recognised as **'a prophet'** (20 Dec. 1996). Sir Laurens van der Post's extraordinary ability to reveal the confronting truth about humans has led some people to seek to destroy his credibility. More is said about the posthumous persecution of Sir Laurens in the *Plato* essay.

The point being made is that the extreme levels of alienation in humans now means that their 'deafness' to any discussion of the human condition could hardly be greater. In Plato's imagery, humans have retreated to the deepest, darkest corners of their **'cave'** world of denial, to a life that finds any light at all **'painful'** and **'hurtful'**.

The two most dreaded terms in the English language are 'human condition' and 'alienation', and yet they are the most-used terms in this book. This gives an immediate measure for the reader of how difficult it will be for their mind not to be, in the words of the cancer sufferer quoted earlier, **'in a state of shock'**, where the words will just seem **'to flood over'** them, and they will not **'hear'** what is said and will **'remember nothing from what** [is] **said in the initial consultation** [the first reading]'.

This book, *A Species In Denial*, published in 2003, is the third book I have written. The first, *Free*, was published in 1988 and *Beyond* was published in 1991. These two earlier books were written as concise and thus, it was hoped, easily accessed and understood presentations of the explanation of the human condition. However, far from being easily accessed and understood, the overwhelming reaction has been that people found they could not 'hear' what was being said in these books.

With only this reaction to judge by, people would be entitled to be sceptical and think that the lack of response was because the books were badly written, or the ideas lacked veracity, but consider the following responses to *Free* and *Beyond:* **'The words in your books have in my experience brought up emotional reactions in people and they reject the information, not able to get behind them and experience the profundity of where you are coming from...Your insights are so head on as to cripple some people'** (FHA Supporter G. Clark, letter to author, Jan. 1993). Similarly, **'From the reactions of people who have borrowed my copies of *Free* and *Beyond* I have started to wonder if the complete holistic picture presented may be too much to accept and absorb in one hit'** (New Zealand FHA Supporter P. Sadler, letter to author, 8 Nov. 1995). Years after he had finally overcome his mental state of denial of the books' subject matter one reader confided, **'When I first read your books all I saw were a lot of black**

marks on white paper' (comment by FHA Supporter G. Plecko, Mar. 2000).

In those first two books, the logic that led me to the strategy of presenting the direct explanation of the human condition, the concise account of the biological reason for why humans have been divisively behaved, was that since humans' denial of the issue of the human condition occurred because they were not able to understand why they have not been able to be ideally behaved, the key to getting them over their denial would be to give them the explanation of the human condition. What I had not sufficiently appreciated and factored in was the extent of the deaf effect. I was solving the problem of the human condition but the reader was in denial of the existence of the problem. The strategy with this book has been to reconnect the reader with the problem of the human condition because only when this is done can they appreciate and begin to take in the explanation. The deaf effect has been the stumbling block.

It can be seen from the comments in the preceding paragraph that Plato was right when he said that when the cave dweller **'emerged into the light his eyes would be so overwhelmed by the brightness of it that he wouldn't be able to see a single one of the things he was now told were real.'** Isaiah was also right when he said this about human condition-confronting information: **'"You will be ever hearing, but never understanding; you will be ever seeing, but never perceiving." This people's heart has become calloused** [alienated]**; they hardly hear with their ears, and they have closed their eyes'**.

When people begin reading direct description and analysis of the human condition the reality is that they 'hear' very little. The mind is literally in shock at hearing the subject even admitted, let alone talked about and talked about so openly and freely. The mind quickly brings up the blocks it has needed to protect itself from the condemning and thus depressing implications that thinking about the human condition has led to in the past. It may seem amazing, but people can read a page of those earlier two books over and over again and still not comprehend what is written there. In a typical description of this phenomenon, a married couple said: **'We have tried very hard to read *Beyond;* in fact my wife and I would sit in bed and read a page together, and then re-read it a number of times, but still we couldn't understand what was written there'** (FHA Supporter H. Saunders reporting a friend's comment, Oct. 1998).

It should be pointed out that the almost instantaneous triggering of the deaf effect is witness to the fact that a person's mind only has to become aware of the smallest amount of information relating to the human condition to know where the discussion is heading. It is an indication of just how intuitively aware humans are of the truths they are blocking out. The deaf effect response to this information is a measure of just how penetrating it is; in fact it authenticates the integrity of the material being presented.

Unaware of the 'deaf effect' readers blame the presentation

The major problem has been that people who have difficulty reading about the human condition—virtually all adults—cannot know that this deaf effect is occurring. If people knew they were alienated they would not be alienated; if we can recognise that we have blocked something out then necessarily we have not blocked it out. People living in denial are not aware that they are. The human mind, finding itself unable to absorb or 'hear' what is being said, and not knowing that the problem is the 'deaf effect', has to find some justification for the difficulty it is having. It usually decides the book must be badly written. It blames the presentation. It finds reading the material **'extremely heavy going'** and decides the concepts must be **'poorly expressed'**, or even that the text is **'too dense'** and the concepts **'too intellectual'** to understand. To the denial-maintaining, evasive, alienated brain the information appears **'obscure'** and **'impenetrable'**. These terms are comments that have been made by people who have attempted to read my earlier books.

Since I am dealing with the human condition, rather than living in denial of it, I do not have to merely allude to the truth, or find clever, coded, evasive, intellectual, sophisticated, esoteric ways to talk about it, as people living in denial have to. My writing only needs to be simple, plain and direct, but that is the problem, people who are living in denial do not want the simple truth, they can only tolerate the truth being cleverly alluded to. A mind that is living in denial does not allow itself, and cannot hear, the simple truth. What appears dense and impenetrable to the mind that is living in denial is simple and clear to a mind that is not in denial.

In the *Resignation* essay it will be explained that people are not born mentally in a state of denial of the human condition, rather it is a psychological adjustment they make during their adolescence. In that essay it will be explained that at about 12 years of age humans begin to attempt to understand the dilemma of the human condition; however with humanity unable—until now—to explain this deepest of issues, adolescents eventually, at about the age of 15, learnt they had no choice but to resign themselves to a life of denial of the extremely depressing subject. Each generation of humans has adopted the historic state of denial in their mid adolescence. The following unsolicited letter to the FHA from a 16-year-old student, Lisa Tassone, illustrates how young adolescents who have not yet adopted an attitude of denial of the issue of the human condition have been able to read and digest my books with the greatest of ease: **'Before stumbling upon *Free: The End Of The Human Condition* that was discreetly shoved in the back of the philosophy section, I was at the end of my road. I had experienced a year of complete and utter pain, confusion, anger and frustration. When I finally took the plunge to seek medical help (as I was suicidal), I was diagnosed with severe depression and put on medication. After reading your book (which I stayed up till 2am reading, I just couldn't put it down), I have been one of the fastest recovering depressants around. No wonder why. If everyone knew your insights, so much would be resolved. The purpose of this letter is to thank you for your courage in publishing your sure-to-be controversial work, and for basically recovering and saving this 16 year old. Not only is your work the absolute truth and has restored my faith in humanity, it has given me inspiration to help others. I may seem young to know what I'm talking about but, well, I do. I have tested all your work and others and yours always held up'** (Brisbane, 4 Oct. 1999).

The difficulty older people have reading about the human condition is not due to poor presentation of the explanations, it is because they are living in denial. The problem is alienation.

Sir Laurens van der Post's writings have been of the greatest influence in my own thinking. *Beyond* is dedicated to him and the reader will already have seen how much value is placed on his writings to support or illustrate the points being made in my books. Sir Laurens' writings are as 'terrifyingly honest' as those of William Blake. He was an exceptionally unevasive, honest thinker, and, as a result, many people have found his writings to be obscure. For example, in the obituary where he was described as **'a prophet'**, Sir Laurens' writings were variously described as **'not always clear'**, as **'imprecise'** and **'elusive'**. The fact is that Sir Laurens van der Post's writing is some of the clearest in recorded history, as his many quotes in this book attest. What is not clear are the minds of humans blocked with the historic denial of the human condition.

A publisher once said to me after reading *Beyond,* **'I find the concepts fascinating, but I also find the arguments elusively receding from my mind as soon as I stop reading them'** (FHA records, 3 June 1993). This comment raises another point that even when people begin to comprehend these understandings of the human condition it is hard for their mind to retain those understandings. So practiced at denying the issue of the human condition has the human mind become that it even has difficulty retaining argument about the subject when it has grasped it.

While people regularly say they find my writing obscure and impenetrable it is usually not difficult to demonstrate to someone that the text is not too obscure, dense or intellectual to understand. If a reader is taken carefully through the text, paragraph by paragraph, they usually cannot identify any step in the logic that is not clear. The problem is not the presentation, it is the implication of what is being said coming through in their mind subliminally that triggers their evasive defences and leads readers to decide they do not want to hear what is being said.

The deaf effect also causes readers to believe **'the text is too repetitive'**. In order to explain this response it first needs to be appreciated that, at best, the mind of nearly all humans can only tolerate the subject of the human condition being alluded to remotely and briefly. Acceptable glancing references to the human condition include: **'the meaning of humans and their place in the world'**, **'our human predicament or situation'**, **'our troubled human state and nature'**, **'what it is to be human'**, **'the dark side of our nature'**, **'the riddle of life'**, **'why are we the way we are'** and **'the root of human conflict'**.

The biologist, Edward O. Wilson, uses the term 'human condition' more frequently than any other writer I have encountered—Wilson has even acknowledged the importance of studying the human condition, stating that **'The human condition is the most important frontier of the natural sciences'** (*Consilience,* 1998, p.298 of 374). Although Wilson frequently uses the term 'human condition' I have only been able to find two descriptions of what he means by the term. Firstly, there is a reference he made to **'the persisting enigmas of the human condition—What are we? Where did we come from? What do we wish to become?'** (Wilson's 1998 Phi Beta Kappa Oration, pub. in *Harvard Magazine,* 1998 addresses). The meanings of these terms, 'what are we?', 'where did we come from?' and 'what do we wish to become?' *are* elusive. However, if you think honestly about the question 'what are we?', the essential truth has to be that while we humans do have a potential for generosity and kindness on one hand we nevertheless have a capacity to be extraordinarily aggressive and selfish on the other. Similarly, if you think deeply about the question 'where did we come from?', the real questions that are being raised are, 'where did the dark side of our nature come from?', and 'have we always been so aggressive and selfish?' Also, if you think deeply about the question 'what do we wish to become?', the real question involved is 'how are we to free ourselves from our corrupted, "fallen" state?' Wilson is only daring to allude to the all-important question of humans' corrupted condition. The second reference breaks the convention of evasion and clearly defines the human condition, but notice it takes a direct question to Wilson to bring about the precise admission of what the issue of the human condition is. When an interviewer asked, **'How is it that ethics can coexist in the same species, in fact in the same human being, with hatred and aggression?'**, Wilson responded, **'That is the human condition'**, and added, **'That was our biological fate. And there is no way of reconciling these things except by law and treaty'** (Futurology, *Talking About Tomorrow,* The Wall Street Journal interviewing E.O. Wilson, 2000). Saying there is no way of reconciling good and evil is, I suggest, an allusion to the fact that it has been virtually impossible for humans to confront the issue of the human condition. I should mention that, even though Wilson has said here that there is no way of reconciling good and evil, he has said elsewhere that **'Biology is the key to human nature'** *(On Human Nature,* ch.1, 1978). This is a more accurate response because biology *could* one day make sense of human nature, it could explain the human condition, as I am suggesting it now has.

The point is, most humans can only talk about the human condition in code, in ways that only the initiated can understand. They limit themselves to esoteric inference and innuendo. They appeal to the shared intuitive awareness in humans of the need to evade the deeper confronting truths about human life. They talk of certain things being 'self-evident'. They intellectualise the truth, learn to live with it at arm's length; at base they find a way to safely live in denial of the truth. In Plato's imagery, they prefer the darkness of a cave existence.

Continued direct and open description and analysis of the human condition greatly affronts the mind of most humans. Their mind does not want to keep hearing description of the subject but it cannot admit this to itself without admitting it is practicing denial, without admitting and confronting its alienation, which, as has been pointed out, it cannot do. Something *is* occurring repeatedly but it is *not* repetition of the same particular concept or material, it is repeated raising of a subject the human mind has become deeply committed to blocking out, a continual elaboration and analysis of a long-forbidden, exiled subject.

Age and the 'deaf effect'

Age is another important factor in people's capacity to take in or 'hear' description and analysis of the human condition. Alienation, and with it the deaf effect, tends to increase with age. The longer a person lives in the corrupted, alienated world, the more encounters they will inevitably have with that corruption and alienation, and the more they will become corrupted and alienated themselves as a result of those encounters.

Inevitably, the older the person the more they will have become adapted and habituated to living in denial of the issue of the human condition, and thus the more they will find the new human condition-confronting information difficult to 'hear' and thus consider.

Older people have always found it difficult adjusting to a new device or a new way of doing something or a new concept—as the proverb attests, **'you can't teach an old dog new tricks'**—and no new concept is as revolutionary as the arrival of understanding of the human condition. In fact, what is being introduced is not just a new concept, it is a whole new way of viewing our world—a new paradigm—and one that could not be more different to, and more confronting of the old paradigm. Indeed, what is being introduced brings the *real* 'culture shock', 'future shock', 'sea change', 'brave new world', 'tectonic paradigm shift' humanity has long anticipated. The change that the arrival of understanding of the human condition brings is so great most older adults will find it extremely difficult even 'hearing' the new understanding, let alone appreciating it and then adjusting to it. In fact, if you cannot 'hear' the information it is not possible to appreciate it.

The great Australian educator, Sir James Darling, recognised the extreme difficulty older people have adjusting to a renaissance when he wrote: **'At every time when there has been great activity and great originality, there has been opposition and tenacity from the old. Those who have grown up in another age...are terribly afraid of newness of life...they cannot adapt themselves to the new life...The mind of most men is not adaptable after a certain age and the onrush of a Renaissance is very rapid'** (*The Education of a Civilized Man,* 1962, p.53 of 223).

Historically it has always tended to be the young who take up new ideas, and it has to be expected that this will be especially so with a paradigm shift as monumental as the one that is now being introduced. Not being as wedded as older minds to the denial-maintaining paradigm, younger minds are more able to 'hear', consider, appreciate and adjust to this new denial-free, human condition-confronting-and-explaining paradigm.

The difficulty older minds have reading about the human condition is such that it is almost worth putting this notice at the beginning of this book: 'Warning: unless you are under 21 you will have difficulty reading and understanding the content of this book.'

Significantly, while older adults suffer the most from the deaf effect they are also the most aware of, and on their guard against, any information that brings the human condition into focus. Both the deaf effect and sensitivity to any human condition-confronting information increases with alienation, the more innocent having less to fear from information that brings the human condition into focus. This combination that accompanies alienation, of loss of 'hearing' but an acute 'radar' for any information relating to the human condition, is an extremely dangerous mix when understanding of the human condition arrives.

Historically any information and truth that brings the human condition into focus has been rightly feared as dangerous—because the apparent condemnation of humans' corrupted, divisive nature could lead to suicidal depression. However when the full, dignifying, ameliorating understanding of the human condition arrives it ends the unjust condemnation and thus the danger of depression. The full truth about humans dignifies them, it lifts the 'burden of guilt' from humans, it explains and ameliorates human nature. While human condition-confronting information has historically been dangerous and to be avoided, the human condition-confronting information that presents the full truth about humans is the direct opposite; it is liberating and to be embraced. The great danger arises from the fact that, as they suffer most from the deaf effect, the more alienated—who tend to be older adults—will not be able to 'hear' and thus appreciate that the new human condition-confronting information is presenting the safe, full truth about humans. Not being able to 'hear' and thus discover that the information is the safe liberating truth that humanity has for 2 million years searched for, but being the most aware that the information is bringing the human condition into focus, and the most afraid of that confrontation, older, alienated adults can set out to repress and even destroy the information without realising that they are destroying humanity's chance of liberation from the human condition.

Democracy, which allows for freedom of expression, is the mechanism that has been established to ward against the dangerous possibility of prejudice shutting down the human journey from ignorance to enlightenment. Tragically however, where information relating to the human condition is concerned, the more alienated can be tempted to abandon the democratic principle of freedom of expression and set out to destroy by any means available what they fear and cannot understand. In so doing they can commit the worst possible crime, that of destroying humanity's only chance for freedom. For someone to avoid the issue of the human condition is one thing, to seek to stop anyone else looking at it is quite another.

The danger posed by the intolerance of older adults towards human condition-confronting information will be elaborated upon later in the book.

It should be stressed that no one is precluded from accessing the liberating understanding of the human condition. For the older and/or more alienated it simply requires more patience and perseverance to access it than those who are young and/or less alienated. This important point will be examined shortly.

Intelligence and success also increase the 'deaf effect'

One of the obstacles this material faces is the arrogance of the human mind or, more precisely, the extent of its delusion that it is not alienated. People naturally feel it is an insult to their intelligence to be told their mind is incapable of comprehending something. People simply do not believe their mind could be as deaf as a post to well reasoned and explained analysis of a subject, but that is exactly what the human mind has been to the subject of the human condition. The letter from the 16-year-old girl Lisa Tassone described how she read and digested *Free* in one sitting in one night. To most people who have had to struggle mightily to read my books this is astonishing but in fact it is merely a measure of their alienation, a measure of how deeply they have blocked out the subject of the human condition. Christ described the alienated situation of adults perfectly when he said **'you have hidden these things from the wise and learned, and revealed them to little children'** (Matt. 11:25). As was mentioned, the *Resignation* essay explains why young people haven't yet adopted a strategy of denial of the issue of the human condition.

Very intelligent and successful people are often the most deaf to discussion of the human condition. After reading material about the human condition they frequently do not recognise that they cannot understand it and yet I know they have not understood it because if they had they would be moved by the accountability of the explanations, or at least unnerved by its penetrating truthfulness. Time and again I have seen people with obviously high IQs and backgrounds of exceptional academic and public achievement try and scan-read material about the human condition, certain that they can ascertain what is being said and take in anything of value, but basically they absorb nothing and do not begin to know what has been said. Life in Plato's cave became a life unto itself. Its residents lived in almost total darkness, and yet they were oblivious of their extremely limited view and behaved as if it were boundless, the only view. They acted with sublime confidence, incredible arrogance, in fact, extreme delusion. Christ highlighted the arrogant delusion that people living in denial have been capable of when he said, **'They like to walk round in flowing robes and love to be greeted in the market-place and have the most important seats in the synagogues and the places of honour at banquets. They devour widows' houses and for a show make lengthy prayers. Such men will be punished most severely'** (Luke 20:46). Delusion/alienation/denial will not be 'punished' with the arrival of the understanding of the human condition, but it will be made visible and will thankfully become unnecessary in human life.

Adult humans are so deluded, so unaware of being alienated, that while reading about the deaf effect they very often think that, while it may be true of some people, it does not apply to them and that they are quite capable of taking in any analysis or argument. In general, the more successful and intelligent the reader, the more confident they are that they do not suffer from the deaf effect. The truth is that when it comes to denial-free, human condition-confronting information, the more intelligent and successful are often the people who have been most able to master and refine the art of living in denial—hence their comfort and competence in the corrupt world of denial—and are thus the people most 'deaf' to the truth. Once again the *Bible* provides the most succinct description of the limitations of the sophisticated, clever, alienated mind in accessing presentation and analysis of the human condition: **'Do not deceive yourselves. If any one of you thinks he is wise by the standards of this age, he should become a "fool" so that he may become wise. For the wisdom of this world is foolishness in God's sight. As it is written: "He catches the wise in their craftiness"; and again, "The Lord knows that the thoughts of the wise are futile"'** (I Cor. 3:18–20 & Psalm 94:11).

In her introduction to the 1986 book, *Simone Weil, An Anthology,* Siân Miles recorded Simone Weil's account of the limitation of high intelligence: **'Simone Weil completed her last work in 1943 when at the age of thirty-four she died in an English sanatorium. Two weeks earlier she had written to her parents: "When I saw** [Shakespeare's] ***Lear* here, I asked myself how it was possible that the unbearably tragic character of these fools had not been obvious long ago to everyone, including myself. The tragedy is not the sentimental one it is sometimes thought to be; it is this: There is a class of people in this world who have fallen into the lowest degree of humiliation, far below beggary, and who are deprived not only of all social consideration but also, in everybody's opinion, of the specific human dignity, reason itself—and these are the only people who, in fact, are able to tell the truth. All the others lie. In *Lear* it is striking. Even Kent and Cordelia attenuate, mitigate, soften, and veil the truth; and unless they are forced to choose between telling it and telling a downright lie, they manoeuvre to evade it...Darling M., do you feel the affinity, the essential analogy between these fools and me—in spite of the Ecole and the examination successes and the eulogies of my 'intelligence'...** [which] **are positively intended to evade the question: 'Is what she says true?' And my reputation for 'intelligence' is practically equivalent to the label of 'fool' for those fools. How much I would prefer their label." Since then she** [Simone Weil] **has become known as one of the foremost thinkers of modern times, a writer of extraordinary lucidity and a woman of outstanding**

moral courage' (pp.1,2 of 310).

What Simone Weil has said here is reminiscent of Christ's description of the intellectual **'teachers'** of his day as **'blind fools'** (see Matt. 23:17). It has to be remembered that to live in denial of the issue of the human condition was not itself foolish because to confront the human condition, and the many truths that brought the human condition into focus, could lead to madness, even suicidal depression.

Clever people find it most difficult to make any headway with the truth, their intelligence having allowed them to hide themselves from it most effectively. Intellectualism is the opposite of instinctualism in that the former has been concerned with the art of denial while the latter is concerned with God or cooperative meaning-confronting, soulful truth. To date human intelligence has largely been concerned with the art of denial, not with truthful thinking. Artful, sophisticated, evasive, esoteric, cryptic, intellectual cleverness was needed to establish, defend and maintain the safe, non-confronting, escapist, alienated state. Universities selected for cleverness—you have not been able to attend a university unless you pass exams that test for mental aptitude—not because cleverness was best at thinking and learning, as those living in denial have evasively maintained, but because 'dumbness' or lack of intelligence could not be trusted to maintain the denial. The truth is the average IQ or intelligent quotient of humans is quite adequate for understanding. Why a high IQ or 'cleverness' was needed was to deny and evade the truth. That was the real art. Universities have high IQ entrance requirements because they have been the custodians of denial, keepers of 'the great lie'. A student had to be able to investigate the truth and talk about the truth without confronting it or admitting it, which is a very difficult, IQ-demanding undertaking. In her 1980 book, *War Within and Without,* Anne Lindbergh, wife of the renowned aviator Charles Lindbergh, recalled a conversation with the author, philosopher and denial-free thinker or prophet, Antoine de Saint-Exupéry: **'The three greatest human beings he has met in his life are three illiterates, he says, two Brittany fishermen and a farmer in Savoy. "Yes", I say, "it has nothing to do with speech—quick brilliant speech—though one *thinks* it has when one is young." "Oh, yes," he says, "mistrust always the quick and brilliant mind"'** (p.30 of 471). In his 1976 book, *Jung and the Story of Our Time,* Sir Laurens van der Post records that: **'Once asked then which people he** [the psychoanalyst Carl Jung] **had found most difficult to heal, he had answered instantly, "Habitual liars and intellectuals"...Jung maintained that**

the intellectualist was also, by constant deeds of omission, a kind of habitual liar' (p.133 of 275).

Now that the human condition is thankfully resolved, human intelligence can end its life of deluded denial, its lying arrogance, and take up a truthful, humble existence. As the Old Testament prophet, Isaiah said about the arrival of the human condition-ameliorated new world: **'I will put an end to the arrogance of the haughty'**, and **'all your pomp has been brought down'** (Isa. 13:11 & 14:11).

Age and alienation are important factors in determining how much someone will suffer from the deaf effect. To those factors should now be added the factors of IQ and the often-associated success in managing life in the corrupt world.

What has been said about success in the world of denial makes clear Christ's comment that **'it is easier for a camel to go through the eye of a needle than for a rich man to enter the kingdom of God'** (Matt. 19:24, Mark 10:25).

To illustrate how those who have been successful in the world of denial are especially prone to the deaf effect we can look at the situation of someone who finds themselves experiencing the opposite of success, someone whose life is in crisis. Such people can find they are able to hear human condition-confronting information because what occurs is that a crisis can leave a person so emotionally bereft that they abandon any pretence of denial and at that point they become open to the truth again. As Sir Laurens van der Post wrote, **'Until war or sudden social and individual disaster penetrate our conscious defences we see only the glittering civilised and defended scene lying below and before us'** (*The Dark Eye in Africa,* 1955, p.99 of 159). A reviewer of my first book, *Free,* obviously had no trouble accessing the book's content when she wrote: **'Was Jeremy Griffith struck by lightning on the road to Damascus...Such was my cynicism reading the summary...Then whack! Wham! Reading on I was increasingly impressed and then converted by his erudite explanation for society's competitive and self-destructive behaviour. His is not a band-aid cure for mankind's sickness but a profound thinking through to the biological cause of the illness'** (*Executive Woman's Report,* 12 May 1988). When I wrote to the reviewer commenting on her ability to comprehend my book this was her response: **'Our 12 year old son was killed in a bicycle accident eight years ago and God literally hammered on my soul and senses. Miracles, signs, phenomena were granted me because I kept open the communication lines and begged God to help me.'**

Despair can shatter the carefully constructed facade of denial and make it seem pointless and immaterial. In Olive Schreiner's extraordinarily honest 1883 book, *The Story of an African Farm,* there is a marvellous description of how despair can lead to the abandonment of the facades of life and, when that happens, how all the beauty—and truth—of life is suddenly accessible: **'There are only rare times when a man's soul can see Nature. So long as any passion holds its revel there, the eyes are holden that they should not see her...For Nature, ever, like the old Hebrew God, cries out, "Thou shalt have no other gods before me." Only then when there comes a pause, a blank in your life, when the old idol is broken, when the old hope is dead, when the old desire is crushed, then the Divine compensation of Nature is made manifest. She shows herself to you. So near she draws you, that the blood seems to flow from her to you, through a still uncut cord: you feel the throb of her life. When that day comes, that you sit down broken, without one human creature to whom you cling, with your loves the dead and the living-dead; when the very thirst for knowledge through long-continued thwarting has grown dull; when in the present there is no craving and in the future no hope, then, oh, with a beneficent tenderness, Nature enfolds you...And yellow-legged bees as they hum make a dreamy lyric; and the light on the brown stone wall is a great work of art; and the glitter through the leaves makes the pulses beat'** (p.298 of 300).

When the Australian bushranger and folk hero, Ned Kelly, was taken from his jail cell to the gallows and all hope was gone, he was suddenly able to see beauty, for **'he remarked how lovely were the flowers'** that he passed in the courtyard (*Ned Kelly: Australian Son,* Max Brown, 1948, p.223 of 262).

The obvious fact is that the more secure a person is in their state of denial, the more difficult it is for them to 'hear' denial-free information.

The non-falsifiable situation

This is an appropriate point to address the problem of the 'non-falsifiable situation'. There has been an inference that those who oppose this information are suffering from denial. In fact I will say that reading my books amounts to an alienation test; the more alienated the reader, the more their mind will resist the truth that is being presented. In support of this I cite Christ, who said, **'everyone who**

does evil [has become corrupted] **hates the light** [the unevasive, denial-free truth], **and will not come into the light for fear that his deeds will be exposed'** (John 3:20), and Plato, who has already been quoted as saying: **'if he** [the cave prisoner] **were forcibly dragged up the steep and rocky ascent** [out of the cave]**...into the sunlight...he would much object...his eyes would be so overwhelmed by the brightness...he would turn back and take refuge in the things which he could see'** (*Plato The Republic,* tr. H.D.P. Lee, 1955, p.280 of 405). Once you propose that alienation is almost universal then the situation exists where people who disagree with or oppose what you are putting forward can feel they are being dismissed as being alienated, evasive and 'in denial', leaving them no way to disprove or falsify the explanation being put forward.

The first point to consider is that I did not create the dilemma of the human condition that produced alienation in humans and this conundrum. It is not a ploy to defeat criticism as some have implied.

Secondly, and more importantly, the problem only really exists at the superficial level because the ideas being put forward can be tested as true or otherwise. These are not untestable hypotheses that must be accepted on faith. For example, the existence of denial of the issue of the human condition can easily be established by scientific investigation. I have already quoted many references to and descriptions of denial as initial evidence of its occurrence. In fact, since humans are the subject of this particular study, each person can experience and thus know the truth or otherwise of what is being put forward. Once the explanations are presented and applied—as is done in my books—you will discover they are able to make such sense of human behaviour that your own and everyone else's becomes transparent. As the *Encarta* summary of Plato's allegory states, once free of the cave, the prisoners recognise that **'the only things they have seen heretofore are shadows and appearances'**, not **'the real world'** at all. This new-found transparency confirms that this understanding is the long-sought explanation of the human condition.

Overcoming the 'deaf effect'

Of course presentation and analysis of the issue of the human condition produces responses other than the deaf effect. For instance, there is the response of trying to maintain the false arguments that have historically been used to deny and evade the issue of the human condition—such as the arguments used to deny that there is a cooperative, integrative purpose to existence, or that humanity once lived instinctively in a state of cooperation. These and many other historic denials are referred to in subsequent essays in this book, and examined in full in *Beyond.*

The other significant response has been one of extreme anger towards the exposing information. When confronted with description of the human condition, the human mind, especially the more alienated, cannot help demanding that the denial be restored. It says, in effect, 'either you present what you have to say in a way that I don't find confronting, or I will reject it'. That rejection can be extreme. It was mentioned that older, alienated adults, being the most aware that the information is bringing the human condition into focus, and being the most afraid of that confrontation, can be tempted to defy the democratic principle of freedom of expression, and set out to repress and even destroy the information. Indeed, people can have such an angry response to the information that they set out to attack the heresy by any means available, including persecuting its supporters. This response will be examined later in this book. ('Heresy' is defined as opinion contrary to orthodox thought, in this case the almost universal practice of denial.)

The two responses of trying to maintain the false arguments that have been used to maintain the denial and of becoming extremely angry towards this information are certainly major problems to be overcome. However, of more serious concern is the problem of the deaf effect. The deaf effect can stop any support for the crucial understanding of the human condition from developing. There is a saying that **'you can knock on a deaf man's door forever'**, and it is true that unless the deaf effect is overcome people will never be able to understand the human condition. Alienation protected humans from going mad, even suicidal self-destruction, but it also has the potential to deny humans their freedom from their dishonest, corrupted existence, and, with alienation continuing to increase, the potential to eventually bring about the destruction of the human race.

The fact is that when description and analysis of the human condition is presented, the words, as the cancer sufferer described it, **'just seem to flood over'** the reader and their mind takes in almost nothing. Tragically the minds of most people are initially unable to get past the denial to make the realisation that the subject of the human condition has at last been compassionately explained and that it is now both safe and necessary to confront the subject. The task now is to abandon the historic, immensely dishonest, false world of evasion and denial. Humanity is now able to 'come out of' the biggest 'closet' of all, that of denial of the human condition. Humans can be honest now and honesty is the only basis for psychological therapy and rehabilitation. As Christ said, **'the truth** [when it comes] **will set you free'** (John 8:32).

While it is now safe to look at the human condition, the problem is the reader's mind has to absorb what is being said for it to discover that it is indeed safe—but how is their mind to take in what is being said when it is effectively 'deaf' to any description and analysis of the human condition? How is the mind to learn that it is now safe to look into the human condition? If Genghis Khan's army has been invading our country for decades, and we have sensibly taken to a life of hiding in the forests, it is going to be a brave person who first ventures into the open when it is announced that the tyrant has finally been vanquished.

The wonderful news is that the Genghis Khan of the human mind, the 'fire-breathing dragon' that crops up in all mythologies, *is* slain. Why humans have been the way they have been, divisively rather than cooperatively behaved, has been biologically explained. With this compassionate understanding of human's divisive nature the insecurity that produces such behaviour subsides and the behaviour disappears. It is safe to come out of hiding now—and absolutely necessary if we are to end all the terrible destruction caused by humans' dishonest existence.

With the dignifying, compassionate understanding of the human condition at last found, the need for denial of the subject has been removed, but the problem remains of how are those humans who have been living in deep denial of the subject to discover that it is now safe to look into the human condition? How is the mind to learn that it is now safe when it cannot 'hear' what is being explained. The reader is in a catch-22 situation, apparently trapped by alienation.

Firstly, to help the reader trust that it is now safe to confront the issue of the human condition it has to be emphasised again that the human condition could not possibly be being discussed here so openly and freely if the subject had not been successfully penetrated and understood. It was not possible to talk clearly and directly about the human condition while it had not been explained. To be discussing it so openly means it must have been explained. Similarly, people could not be actively supporting and developing these understandings of the human condition, as those associated with the FHA are doing, if the understandings were not enabling them to safely confront the human condition. Humans could not be wandering happily around a minefield if the mines had not been defused. *The dilemma of the human condition was such that the subject held its own safeguard that it could not ever, and would not ever, be naively, trivially talked about.* Wittgenstein's comment, mentioned earlier, **'About that which we cannot speak, we must remain silent'**, confirms this point. We can only speak about the subject of the human condition when it has been rendered safe to speak of.

This book is written from the freedom and safety of outside the cave prison of denial where humans have had to live for some 2 million years. Jim Morrison, of the famed musicians, The Doors, sang about having to **'break on through to the other side'**; that has now occurred—humanity has finally broken through the wall of necessary denial to freedom from the human condition.

The strategy that has been used in this *Introduction* for the problem of the denial blocking the mind from hearing description and analysis of the human condition, has been to as quickly as possible resurrect the issue of the human condition and outline how it can now be coped with—to move from the beginning of the story to the other end and safety before the rejection reaction has time to assert itself.

Once the reader gains at least an intimation that the human condition does exist and that it has been safely explained, then they have gained the foothold needed to begin to overcome their denial of the subject.

The need for patience and perseverance

Once the reader has gained the initial foothold of awareness of the existence of the human condition and that it can now be overcome, then the main requirements to more fully overcoming the denial and its deaf effect are patience and perseverance.

It was mentioned earlier that in his cave allegory Plato wrote that **'if he** [a person] **were made to look directly at the light of the fire** [the issue of the human condition], **it would hurt his eyes and he would turn back and take refuge in the things which he could see, which he would think really far clearer than the things being shown him. And if he were forcibly dragged up the steep and rocky ascent** [out of the cave] **and not let go till he had been dragged out into the sunlight** [shown the reconciling explanation of the human condition], **the process would be a painful one, to which he would much object, and when he emerged into the light his eyes would be so overwhelmed by the brightness of it that he wouldn't be able to see a single one of the things he was now told were real'** (*Plato The Republic,* tr. H.D.P. Lee, 1955, p.280 of 405).

It is now relevant to add what immediately follows the above dialogue. Plato continues: **'Certainly not at first, because he would need to grow accustomed to the light before he could see things in the world outside the cave. First he would find it easiest to look at shadows, next at the reflections of men and other objects in water, and later on at the objects themselves. After that he would find it easier to observe the heavenly bodies and the sky at night than by day, and to look at the light of the moon and stars, rather than at the sun and its light. The thing he would be able to do last would be to look directly at the sun, and observe its nature without using reflections in water or any other medium, but just as it is. Later on he would come to the conclusion that it is the sun that produces the changing seasons and years and controls everything in the visible world, and is in a sense responsible for everything that he and his fellow-prisoners used to see. And when he thought of his first home and what passed for wisdom there and of his fellow-prisoners, don't you think he would congratulate himself on his good fortune and be sorry for them?** [what was considered wise in the world of denial was for the most part intellectualised blindness]**'** (ibid. p.280). The main point is that it takes time to adjust to living outside the cave of denial in the presence of the truth about the human condition.

In the example of the cancer sufferer's denial, the patient said, **'it took another two weeks before reality began to sink in.'** Readers have to be prepared to accept that their mind needs time to get over the shock of being confronted with the issue of the human condition. With time the historic denial *will* gradually be eroded and the mind *will* discover that it is at last safe to confront the issue of the human condition.

As already mentioned, the FHA is an organisation that has been established to promote and develop this understanding of the human condition. We have learnt that people's historic resistance to description and analysis of the human condition *can* be overcome with patience and perseverance. The mind simply needs time to ride out the initial shock of having the subject of the human condition broached and all manner of mystery unlocked. Once you break into the human condition, a veritable avalanche of answers is unleashed, and an attempt to absorb so many new insights quickly can become overwhelming. It must be acknowledged that a whole new paradigm is being introduced, and humans are notorious for taking time to adjust to even a small change, let alone a revolution to an entirely new way of viewing their world and ultimately of living. Alvin Toffler was anticipating the shock that the arrival of understanding of the human condition would cause when he wrote in his 1970 book *Future Shock*, **'Future shock...** [is] **the shattering stress and disorientation that we induce in individuals by subjecting them to too much change in too short a time'** (p.4).

In the case of accessing the explanations through this book and other FHA material, our experience is that you need to be prepared to persevere with re-reading the information. With repeated readings, the concepts *do* gradually become clearer, the deaf effect *is* eroded, and when that happens you will realise that the deaf effect is very real. With that crucial realisation, you will be encouraged to be even more patient until eventually you are able to access the full, compassionate, liberating, unevasive, denial-free explanation and understanding of ourselves. You enter a positive feedback loop. Plato was referring to this positive feedback loop when he said **'And when he thought of his first home and what passed for wisdom there and of his fellow-prisoners, don't you think he would congratulate himself on his good fortune and be sorry for them?'**

Reference was made earlier to a couple who could not access the information despite repeated reading of a certain page. The couple's experience may seem to contradict the assertion that re-reading the information helps overcome the deaf effect. However, their focus was too narrow; it is more a case of re-reading whole sections so that there is a chance for the compassionate big picture to gradually emerge and erode the mind's need for the denial or block-out. The whole truth about humans dignifies them but it is built from many partial truths that on their own appear to condemn humans. For example, a partial truth is that humans were once innocent. On its own this truth condemns humans' present corrupted state. However, once the reason for humans' corrupted state begins to become apparent, the truth of our past innocent state seems less condemning. You cannot punch an isolated hole in mist or fog, the hole will just close over; to see the entire view all the mist must lift.

Our experience with introductory talks about these ideas provides a good example of how the deaf effect can be eroded. During the early 1990s I regularly gave a standard talk introducing the concepts in my books. While we have moved to presenting the ideas through the Internet—as many university and other educational institutions are doing—our experience with this introductory talk showed that people needed to attend or listen to it at least three times to start to hear the information. All those who have overcome the deaf effect through listening to the introductory talk attest to the need to hear it a number of times. In fact, it became a point of amusement within the FHA how often people who, having attended a second or third introductory talk, said, **'That was a much better presentation this time than last time, the explanations and descriptions were so much easier to follow.'** The comment was amusing because the talks were virtually identical. What had dramatically improved was the listener's ability to 'hear' what was being said. Similarly, people who have read *Free* and *Beyond* more than once find their understanding improves on each reading.

Summary of the old and new highways of thought in response to analysis of the human condition

The habituated, historic response

↓

Analysis of the human condition

↓

Resurrects humans' state of contradiction with the cooperative ideals

↓

A sense of fear, shame, exposure, guilt and confrontation

↓

Depression

Reaction to the implied criticism for being 'bad' or 'evil'.
Alienation/block-out/denial of the analysis. Anger at
the implied criticism. Contrived excuses to try and
repel and undermine the logic of the analysis.

↓

The deaf effect asserts itself

Life in darkness

The response that is now possible

Analysis of the human condition

Resurrects humans' state of contradiction
with the cooperative ideals

Rational biological explanation which dignifies humans
by explaining how humans are 'good' despite all the
appearances that they are 'bad'. The biological reason
why humans are angry, alienated and egocentric.

The naked but compassionate truth ameliorates
humans' sense of fear, shame and guilt

Allows honesty and acceptance of humans' corrupt reality

Allows the human condition to subside

Liberation

Life in the sun

This move to a new human condition-ameliorated world has no relationship to the New Age Movement, or to a religious movement

It was mentioned at the beginning of this *Introduction* that the benefit of persevering with the journey into the human condition is to arrive at a state of extraordinary freedom. It was emphasised there that this freedom is not the sort of indoctrinated feel-good freedom that New Age and 'self-improvement' gurus achieved with self-affirming mantras and other motivational exercises. There is a fundamental insecurity in humans arising from the human condition and denial has been their main means of coping with it. When this denial began to wear thin, and the underlying insecurity began to re-emerge, people sought to reinforce their denial through the so-called 'power of positive thought'. Basically they were retreading their denial and adding fresh layers to it. The truth is real freedom from the human condition lay in confronting and solving the human condition, not in trying to further deny and escape it. Humans had to confront and find understanding of their human condition-tortured, guilt-ridden, God-fearing, insecure, upset angry, egocentric and alienated state. Evading, escaping and transcending the condition was not a real solution. This means that the think-positive, human-potential-and-self-esteem-promoting, self-improvement, motivational, feel-good—basically human condition-avoiding—pseudo-idealistic and artificially-utopian so-called New Age industry was being led by charlatans—false prophets. Those who proclaimed that the way for people to relieve and even heal their condition was to learn better ways to block-out, escape and transcend it were ultimately only leading humanity to a state of even greater denial/dishonesty/alienation. As the philosopher Thomas Nagel said, **'The capacity for transcendence brings with it a liability to alienation, and the wish to escape this condition...can lead to even greater absurdity'** (*The View From Nowhere,* 1986).

With regard to indoctrination, some might unfairly say that re-reading suggests that a form of indoctrination is involved in what is being advocated for this book. Such a comment would be unfair because the truth is the complete opposite; repeated reading is necessary to *dismantle* indoctrination. This is because living in denial of the human condition has been a state of self-indoctrination. As will be explained in the *Resignation* essay, humans have been indoctrinating themselves with the denial, brainwashing their mind of any form of deeper thinking since the emergence of consciousness and with it the dilemma of the human condition. They trained their mind not to think about certain subjects that they found they could not bear to think about. As Rod Quantock said, **'Thinking can get you into terrible downwards spirals of doubt.'** What is needed now to free humans from their highly programmed state of denial is a process of *de*-programming; people have to unlearn or dismantle this self-imposed indoctrination, their mind-controlling practice of living in denial of the human condition. The way to achieve that is to re-read the information until the practiced denial is eroded. As many readers would know, repeated reading is accepted practice for anyone learning a new subject.

It was also emphasised at the beginning of this *Introduction* that the state of extraordinary freedom that can be arrived at from persevering with this journey into the human condition has no correlation with the limited reinforcement humans have long derived from religious faith and belief. What is being offered is *real* reinforcement and thus *real* freedom, achieved through digesting the ameliorating *understanding* of what it actually is to be human. Unlike the limited reinforcement provided by religion and other forms of dogma, what is possible now is real reinforcement of self because it is first principle-based reconciling and thus dignifying biological explanation of human nature. There is absolutely no dogma, nothing that has to be blindly accepted, no leaps of logic and, most significantly, no leaps to faith. What is being presented is the very opposite of faith and belief; it is rational explanation. In fact, the explanation demystifies dogma and mysticism and religious metaphysics. It actually brings such fundamental freedom to the human situation that it takes humans beyond the state where they need dogmatic forms of reinforcement and religious faith and belief.

Importantly, any 'hold' the information has over the mind is due to its accountability, its ability to make sense of experience. What is being presented is first principle, rational, arguable, investigable, testable, verifiable biological explanation. The challenge is to think, question and understand, a process that is the very opposite of abandoning thought and deferring to some form of dogma, faith or belief.

It should also be said that people who have had the patience to persevere until they start to understand the human condition frequently develop an enthusiasm for the explanations—and for life in general. There are two points to make about this enthusiasm. Firstly, it is not a response to some seductive form of artificial 'idealism'. While understanding of the human condition brings about an idealistic world that is truly inspirational, the liberation is a result of knowledge. This real and enduring freedom is the opposite of the unreal, transitory 'freedom' offered by the mindless idealistic dogma that, in the past, has been attractive to some people, especially to some naive young people. Humanity has long held a vision of its ideal potential. This is being achieved now as a result of understanding reality, not from escaping reality and deferring to some dogmatic empty promise of ideality. Secondly, the enthusiasm should not be mistaken for the fervour that is sometimes associated with fanatical religious expression and mindless cults. Again the real situation is the complete opposite. What is occurring is not fanatical, *mindless* fervour, it is the profound satisfaction of *mindful* understanding. This information is brain nourishment, not brain anaesthetic.

It is not surprising that once people have overcome the deaf effect they hold on to these ideas that explain the human condition and defend them with tenacity and enthusiasm. The human mind has sought understanding of the human condition since the onset of rational thought. It is entirely to be expected that once humans find it they hold on to it dearly. As Plato said in his cave allegory, referring to those who learn to confront the **'sun'** (that is, overcome the deaf effect and understand the human condition): **'it won't be surprising if those who get so far are unwilling to return to mundane affairs, and if their minds long to remain among higher things'** *(Plato The Republic,* tr. H.D.P. Lee, 1955, p.282 of 405).

Anticipations of the arrival of the human condition-ameliorated new world

As mentioned, humanity has long held a hope and vision of the arrival of a liberated, upset-free, integrated, peaceful world. While the

New Age Movement, and the many other pseudo-idealistic movements, deluded themselves that the idealistic, peaceful, utopian state could be achieved through people artificially restraining, sublimating and even transcending their upset, corrupted condition, the truth is that state could only be achieved through confronting and solving the human condition. The New Age Movement, and other pseudo-idealistic movements adopted the following expressions of hope of the arrival of an idealistic world as being in support of their vision and message, when in fact what they were doing was teaching people to adopt even greater levels of denial. They were not leading the world to a new age of freedom, as they proclaimed, but *away* from it. Much more will be said about the extreme delusion and danger of pseudo-idealism, in an essay entitled *Death by Dogma,* that is to be published shortly in another book I have written.

While pseudo-idealistic movements have adopted the following expressions of awareness of the possibility of a peaceful, corruption-free world, and by so doing have tarnished and discredited them, we are now able to see that these expressions were actually remarkably accurate intuitive anticipations of the arrival of a human condition-ameliorated world.

A quote has already been included from the 1966 song *Break on Through,* written and sung by Jim Morrison of The Doors rock band, however having explained the problem of the human condition, all of the lyrics of this song can now be clearly appreciated: **'You know day destroys night** [truth destroys the denial or lies]/**Night divides** [us from] **the day/Tried to run/Tried to hide** [tried to live in the cave state of denial, but ultimately humanity had to]/[chorus repeated] **Break on through to the other side** [free ourselves from the bondage of the human condition]/**We chased our pleasures here/Dug our treasures there** [tried to find satisfaction through materialism]/**But can you still recall/The time we cried**/[chorus repeated] **Break on through to the other side/Yeah!/C'mon, yeah/Everybody loves my baby/She get high/I found an island in your arms/Country in your eyes/Arms that chain us/Eyes that lie** [as is explained in the essay in this book, *Bringing Peace To The War Between The Sexes,* women's neotenous image of innocence has inspired men to dream of a pure, human condition-free world but the truth is women are necessarily as psychologically corrupted as men, and thus women's inspiration of 'heaven', of a human condition-free, idyllic world, has only been an illusion and thus transitory]/[chorus repeated] **Break on through to the other side/Oh, yeah!/Made the scene/Week to week/**

Day to day/Hour to hour [tried to go along with the escapist, deluded, artificial, superficial world of denial]/**The gate is straight/Deep and wide** [the real path to freedom lay in trying to plumb the depths of the human condition, which had to be done if we were to]/**Break on through to the other side.'**

Three other prophetic composers created songs that contain the exact same sentiments as *Break on Through.*

Firstly, Mark Seymour, of the Australian rock band Hunters and Collectors, composed a prophetic anthem with his 1993 song *Holy Grail:* **'Woke up this morning from the strangest dream/I was in the biggest army the world had ever seen/We were marching as one on the road to the Holy Grail//Started out seeking fortune and glory** [tried to find satisfaction through materialism]/**It's a short song but it's a hell of a story/When you spend your lifetime trying to get your hands/on the Holy Grail** [the 'Holy Grail' of the human journey was to find understanding of the human condition]//**Well have you heard about the Great Crusade?/We ran into millions but nobody got paid** [once the fire of condemnation blocking the exit from the cave of denial is doused—as it now has been—humans will finally be free to selflessly take liberating understanding to all humans]/**Yeah we razed four corners of the globe for the Holy Grail//All the locals scattered...they were hiding in the snow** [those still living in the old cave prison of denial were taken aback by the enthusiasm of those who were liberated]/**We were so far from home...so how were we to know/There'd be nothing left to plunder** [humans' greedy, selfish efforts to try and satisfy their insecurity through materialism had almost destroyed the Earth]/**When we stumbled on the Holy Grail?//We were so full of beans but we were dying like flies** [humans were pretending to be happy but in truth they were all but dead with alienation]/**And those big black birds...they were circling in the sky/And you know what they say, yeah nobody deserves to die//Oh but I've been searching for an easy way/To escape the cold light of day** [I have tried to live in the cave state of denial]/**I've been high and I've been low** [I have oscillated from being able to block out my reality enough to feel some relief, to being unable to block it out]/**But I've got nowhere else to go** [trying to live through denial had run its course]/**There's nowhere else to go!//I followed orders** [I have tried to live through deferment to laws, rules and faith], **God knows where I've been/but I woke up alone, all my wounds were clean** [with understanding of the human condition found, humans' debilitating sense of guilt, with all its resulting physical sickness, is removed]/**I'm still here, I'm still a fool for the Holy Grail/I'm a fool for the Holy**

Grail [I still live in hope and faith that we will find the reconciling understanding of the human condition that will bring us this new, human condition-free world].'

Secondly, the 1987 song of rock band U2, *I Still Haven't Found What I'm Looking For,* lyrics by vocalist Bono: **'I have climbed the highest mountain/I have run through the fields/Only to be with you/I have run, I have crawled/I have scaled these city walls/Only to be with you** [I have tried to live off the inspiration of women]**/But I still haven't found what I'm looking for** [but I still haven't found any real satisfaction]**/I have kissed honey lips/Felt the healing in her fingertips/It burned like fire/This burning desire** [lived off the inspiration of women]**/I have spoke with the tongue of angels/I have held the hand of a devil/It was warm in the night/I was cold as a stone** [I have delved into mysticism and indulged superstitions but they ultimately left me as destitute as ever]**/But I still haven't found what I'm looking for/I believe in the kingdom come/Then all the colours will bleed into one/Well, yes, I'm still running** [I still believe in a time when humans will be free of the alienating state of the human condition and will live as one]**/You broke the bonds and you/Loosed the chains/Carried the cross/And all my shame/You know I believe it** [I have lived through faith in Christ, but ultimately we still had to find understanding of ourselves]**/But I still haven't found what I'm looking for.'**

Thirdly, John Lennon's 1971 song *Imagine,* a song that surveys report is regarded by many as the best ever written: **'Imagine there's no heaven/It's easy if you try/No hell below us/Above us only sky** [imagine the end of the duality of good and evil, the reconciliation and amelioration of the human condition]**/Imagine all the people/Living for today/Imagine there's no countries/It isn't hard to do/Nothing to kill or die for/And no religion too** [imagine the world free of the human condition-produced insecurities that necessitated either religious faith and obedience or the egocentric compensations of power, fame, fortune and glory]**/Imagine all the people/Living life in peace/Imagine no possessions/I wonder if you can/No need for greed or hunger/A brotherhood of man/Imagine all the people/Sharing all the world/You may say I'm a dreamer/But I'm not the only one/I hope some day you'll join us/And the world will be as one** [imagine a world free of the human condition and all the resulting alienation]**.'**

Regarding the ultimate futility of the escapist, materialistic existence referred to in the songs above, another major survey released at the end of the 20th century voted the 1965 Rolling Stones song *(I Can't Get No) Satisfaction* the most popular song of the 20th century.

Written by Mick Jagger and Keith Richard, the song is an anthem to the ultimate futility of trying to live an escapist, materialistic life: **'I can't get no satisfaction, I can't get no satisfaction/'Cause I try and I try and I try and I try/I can't get no, I can't get no/When I'm drivin' in my car, and that man comes on the radio/And he's tellin' me more and more about some useless** [human condition-denying, superficial] **information/ Supposed to fire my imagination/**[chorus repeated] **I can't get no satisfaction/When I'm watchin' my TV, and that man comes on to tell me/How white my shirts can be/Well, he can't be a man/'Cause he doesn't smoke the same cigarettes as me/**[chorus repeated] **I can't get no satisfaction/ 'Cause I try and I try/When I'm ridin' 'round the world, and I'm doin' this and I'm signin' that/And I'm tryin' to make some girl. Who tells me, baby/ Better come back later next week, 'cause you see I'm on a losing streak/** [chorus repeated] **I can't get no satisfaction/That's what I say.'**

I might mention that people have said 'why do you refer to songs, poetry, comic strips, cartoons, euphemisms, mythology, and the like; shouldn't you be drawing on academic studies to support your arguments?' My answer is that I have drawn on science, and in fact have been totally dependent on it to assemble the biological explanation of the human condition. However, when it comes to needing denial-free, truthful description of our world and what is really happening in it I take it where I find it, and that has been in various forms of expression and creativity where the truth has emerged through cracks in the heavy concrete slab of denial that has been laid over it. In a carpark flowers only grow around the edges and in the odd crack in the paving.

The benefit of an indirect approach

It was explained that an initial foothold of appreciation of the existence of the human condition and a trust that the human condition can now be overcome is required in order to consider description and analysis of the human condition. Once that foothold was gained, patience and perseverance was then needed to further overcome the deaf effect and more fully access the information being presented about the human condition. What we have also learnt from experience is that it helps the reader if the analysis of the human condition is not presented too directly.

Our experience at the FHA has taught us that the direct, concise explanation of the human condition that is presented in *Free* and *Beyond* is, to use the comment included earlier, **'so head on as to cripple some people.'** Considering denial of the issue of the human condition occurred because humans were not able to understand why they have not been able to be ideally behaved, it was reasonable to assume that what was needed to assist people to break their denial was to give them the explanation of the human condition. What has been learnt since the first two books were published is that people first had to be reconnected with the issue of the human condition and then given a less 'head on' presentation of the subject matter. After being reconnected with the subject the reader needed to be able to become gradually familiar with the whole subject of the human condition before delving into the specific explanation of it.

Direct explanation of the human condition is too overwhelming a step for most people, they are defeated by the deaf effect. We have learnt that the best way to gradually introduce people to the topic is by choosing interesting aspects of the human condition—such as religion or mythology or politics or the relationship between men and women or the stages of psychological maturation that humans go through—and explain these aspects in terms of what we are now able to understand about the human condition.

This book, *A Species In Denial*, is designed to provide this indirect, gradual approach to the subject of the human condition. Four highly relevant and, in most cases, extremely topical subjects have been selected for explanation. The subjects and their explanation are presented in essays.

The essays of 'A Species In Denial'

With the human condition understood it is now possible to explain all manner of mystery. It is the key understanding needed to unlock virtually everything humans ever wanted to know. If this statement is true, and you will come to see that it is, then choosing a previously inexplicable or unresolved issue and explaining it should be enthralling and hold the reader's attention. As emphasised, this strategy also allows people to become familiar with the subject of the human condition without having to overly confront it.

For the reasons that have now been explained, it is recommended that the reader read these familiarisation essays in this book before reading the direct, concise explanation of the human condition that is presented in *Beyond The Human Condition,* copies of which can be ordered from the FHA, or read on the website <www.human condition.info>.

The first essay, *Deciphering Plato's Cave Allegory: And in the Process Explaining How The Human Condition is Resolved,* analyses Plato's cave allegory in greater detail than has been done in this *Introduction* and shows how biological understanding of the human condition liberates humanity from its human condition-afflicted, cave-like existence.

The second essay, *Resignation,* looks at the most important psychological event in human life. It is an event that has dictated the very nature of existence for adult humans, yet it is an event that has not previously even been acknowledged. If humans are living in a state of deep psychological denial then the question arises, are they born with this denial, and if not, when and how do they adopt it? This essay explains that at about 12 years of age humans begin trying to understand the dilemma of the human condition; however, with humanity unable—until now—to explain this deepest of issues, adolescents eventually, at about the age of 15, learnt they had no choice but to *resign* themselves to a life of denial of the depressing subject. While humanity was unable to acknowledge the issue of the human condition and the resulting need to resign to a life of denial of it, it was not possible to describe and explain the unresigned life of children and adolescents. Their world has been a mystery to adults but once you read *Resignation* that mystery will evaporate.

The third essay is titled, *Bringing Peace To The War Between The Sexes and The Denial-Free History Of The Human Race.* Some of the deepest wounds in human life have been caused by the lack of understanding in the relationship between men and women. The bitterness, heartache, suffering and the hurt to children has been immense. There are many questions about the relationship between men and women that need answering. With understanding of the human condition it is now at last possible to answer *all* these questions and bring peace to the 'war' between the sexes. This understanding also makes it possible to present a denial-free history of the human race.

The fourth essay, titled *The Demystification Of Religion,* is an impacting demonstration of how understanding the phenomenon of resignation demystifies some previously impenetrable aspects of human life, in particular, the world of religious metaphysics and dogma. *All* the riddles, parables, abstract terms and mystical concepts in religion suddenly become explicable with understanding of the human condition. This essay is a demonstration of how science can and should be a winnower of mystery and superstition.

The fifth essay, titled *The Foundation for Humanity's Adulthood* is a brief profile of the FHA written by FHA Vice-President Tim Macartney-Snape AM.

It should be explained that since each of the first four essays has been written as a discrete essay about a particular subject, rather than as a chapter of a book, each essay can, to a degree, be understood without having read the preceding essays and *Introduction.* It is important to note however that what is being introduced is an entirely new paradigm or way of viewing our world, and unless the reader already has some familiarity with the new paradigm, reading the essays in order will be necessary.

The reaction of someone who read the draft of the *Resignation* essay indicates that this indirect access to the issue of the human condition is effective. Peter Landahl, a teacher at a prominent Sydney private school and father of an FHA Member, said after reading the draft, **'Of all the FHA literature I have read I found *Resignation* to be the best to understand. I didn't get stumped anywhere, it flowed and I understood the whole way through.'** He said, **'It was good to discuss a specific topic'**, adding that **'as a teacher, I found the explanations remarkably relevant and insightful.'** He added **'I read it in a day'**, and asked, **'could I read the draft of the sequel essay, *The Demystification Of Religion?*'** (FHA records, Oct. 2001). Given all that has been said, these words might seem scripted but they are in fact an accurate and genuine response.

The feedback from within the FHA has been even more positive. FHA Member, James West, wrote on 19 February 2002 that, **'In terms of the "give up"** [ie give up trying to maintain the denial] **document you have always striven for, this new book achieves this. When I read these essays I kept getting surprised by the way the issues have been handled in a new way, and the surprise is at their increased power to hold my attention. For example, I'm reading and thinking from past experience with the subject, "here comes three paragraphs that might be overwhelming** [ie overwhelmingly confronting] **for me", and I start to feel some "deaf" reaction kicking in** [the process of eroding deafness is an ongoing one, so that even people within the FHA who have had a long association with this information still have degrees of deafness/alienation to erode]**. Then**

I read the paragraphs and, bammo, the momentum picks up instead of drops off. So many times the writing hits it up the middle and talks about a heretical issue so directly, assuredly and interestingly that it wins the day on the issue. To me it's as though a committee or a university faculty has been researching the human condition for 50 years and this is the result, such is the plausibility and authority. It lifts the truth one step at a time out of the depths of unconsciousness into concrete reality. If communicating understanding of the human condition is chopping down a tree then these essays are like a chainsaw in their effectiveness (chainsaw doesn't reconcile very well with the sensitivity that's also present, but you know what I mean, sensitive and tough). In terms of the macro objective, it's truly fantastic to have got to this stage with these essays of having the measure of the incredibly difficult task of communicating understanding of the human condition despite the deaf effect' (FHA records, Feb. 2002).

We have also received feedback from other people outside the FHA similar to Peter Landahl's, indicating that this less direct presentation is an effective method of overcoming the deaf effect and by so doing will enable people to finally understand human nature; understand themselves.

Deciphering Plato's Cave Allegory

And in the Process Explaining How The Human Condition is Resolved

Written in approximately 360 BC as part of his great work, *The Republic,* Plato's cave allegory is the clearest description I have come across of the entire situation associated with the human condition. Using the narrative of bound prisoners confined to a life in a cave, Plato provides an insightful description of humans' life in denial of the human condition, and of the difficulties and the intellectual processes involved in solving and liberating ourselves from that life. A short interpretation of the allegory was included in the *Introduction,* but a more detailed analysis will help confirm and explain what exactly the human condition is.

Often the earliest work undertaken in a field of study is the most honest and therefore the most penetrating. Over time, the human mind perceives confronting implications in the work that it did not see initially, and duly puts in place protective denials. The result is that while the studies are made safe from condemning implications, they tend to become sophisticated in their evasiveness; intellectualised, sanitised—less honest and less penetrating. As will be described later, humanity has recently arrived at the point where many who go to university, our 'centres of learning', are taught deconstructionist, **'postmodern theory** [which elevates]**...lying to the status of an art and neutralise**[s] **untruth'** (Jeremy Campbell, *The Liar's Tale: A History of Falsehood,* 2001).

This essay will explain that alienation has grown exponentially with humanity progressing from a situation of no knowledge but total honesty some 2 million years ago, to a situation now of immense knowledge but no honesty. Plato's cave allegory is a marvellous example of the greater honesty of earlier work. In it, he let all the truth about the human condition 'out of the bag', as it were, truth that humans have been trying to put back in the 'bag' ever since, starting with Aristotle, who began reductionist, mechanistic science's narrow emphasis on objectivity and exclusion of subjectivity (this emphasis will be explained in full later). Now that the human condition is explained, the truth that Plato so clearly revealed can further verify the explanation.

The renowned 20th century philosopher Alfred North Whitehead once described the history of philosophy as merely **'a series of footnotes to Plato'**. The issue of the human condition is the core issue in human life, the subject that philosophy—which is **'the study of the truths underlying all reality'** *(Macquarie Dict.* 3rd edn, 1998)—has tried most to grapple with. Given the difficulties humans have had even alluding to the issue of the human condition, such a clear description of the whole situation associated with it was a remarkable achievement on Plato's part. The time of the early Athenian city state—2,300 years ago—was certainly a golden age in human history, a brief period of extraordinary clarity, when Plato and his courageous teacher and mentor, Socrates, together with a few other exceptional individuals, contributed almost all the alignment humanity needed for its journey from ignorance to eventual enlightenment of the human condition.

Plato wrote *The Republic* as a series of dialogues. In the case of the cave allegory, while it is ostensibly an imaginary conversation between Socrates and Glaucon, Plato's brother, it is believed that it substantially reflects Plato's own thinking as inspired by Socrates. The allegory begins with Socrates making a direct reference to the human condition: **'I want you to go on to picture the enlightenment or ignorance of our human conditions somewhat as follows. Imagine an underground chamber, like a cave with an entrance open to the daylight and running a long way underground. In this chamber are men who have been prisoners there'** *(Plato The Republic,* tr. H.D.P. Lee, 1955, p.278 of 405). Interpretations of the original Greek differ, with a number of other translations stating 'our nature' rather than 'our human conditions', however there is no doubt that the material that follows is specifically about the human condition, in what must be one of the earliest, if not the earliest, mention of the concept.

This analysis of the cave allegory relies largely upon H.D.P. Lee's 1955 translation of *The Republic,* and unless otherwise stated all quotes will be from this translation. However the following summary, taken from the 1996 *Encarta Encyclopedia* entry on Plato which was referred to extensively in the *Introduction,* provides a useful summary of the cave allegory and will also be referred to in this essay. It states: **'The myth of the cave describes individuals chained deep within the recesses of a cave. Bound so that vision is restricted, they cannot see one another. The only thing visible is the wall of the cave upon which appear shadows cast by models or statues of animals and objects that are passed before a brightly burning fire. Breaking free, one of the individuals escapes from the cave into the light of day. With the aid of the sun, that person sees for the first time the real world and returns to the cave with the message that the only things they have seen heretofore are shadows and appearances and that the real world awaits them if they are willing to struggle free of their bonds. The shadowy environment of the cave symbolizes for Plato the physical world of appearances. Escape into the sun-filled setting outside the cave symbolizes the transition to the real world, the world of full and perfect being, the world of Forms, which is the proper object of knowledge.'**

The metaphor of the sun and fire, the meaning of existence and the demystification of God

Before explaining the reason for the cave existence we need to examine what the 'sun' and 'fire' represent. In *The Republic,* prior to presenting the allegory, Plato introduced his 'world of Forms', particularly the idea of what he termed the 'Form of the Good'. In explaining this term he said that **'the highest form of knowledge is knowledge of the essential nature of goodness'** (p.268 of 405). He proceeded to talk about the **'steps in the ascent** [of **reason**] **to the universal, self-sufficient first principle'** (p.277), also stating that **'the final thing to be perceived in the intelligible realm, and perceived only with difficulty, is the absolute form of Good'** (p.282), and that **'Good, then, is the end of all endeavour, the object on which every heart is set'** (p.269).

When Plato was pressed to explain **'the Good'** more directly than he was able to do in the above quotes he said, **'I'm afraid it's beyond**

me' (p.270), but agreed to do so by means of an allegory, in which **'the Good'** was compared to **'the sun'**. He said that **'the Good...gives the objects of knowledge their truth and the mind the power of knowing...** [just as] **the sun...makes the things we see visible...The Good therefore may be said to be the source not only of the intelligibility of the objects of knowledge, but also of their existence and reality; yet it is not itself identical with reality, but is beyond reality, and superior to it in dignity and power'** (p.273). He talked of **'objects** [being] **illuminated by daylight...** [just as they are] **by truth and reality'** (p.273), adding that **'the sun...controls everything in the visible world, and is in a sense responsible for everything'** (p.280).

With the benefit of the discoveries of science it is now *possible to explain*—and with the understanding of the human condition available, it is now *safe to admit*—that the **'absolute form of Good'**, the **'highest form of knowledge'**, the **'universal, self-sufficient first principle'**, is what was referred to in the *Introduction* as the cooperative, loving, selfless ideals of life. While these are the ideals **'on which every heart is set'**, the ideals that every human aspires to live by, they are also the ideals that humans' competitive, aggressive and selfish reality has been so at odds with, and which humans therefore found so condemning. The truth of the ideals of life is a truth that stands above the fact of humans' divisive reality, it is a truth that, as Plato said, **'is not itself identical with reality, but is beyond reality, and superior to it in dignity and power.'**

It is necessary to explain the greater significance of the cooperative, selfless, loving ideals of life. It was very briefly mentioned in the *Introduction* that the theme of all existence, in fact the **'universal first principle'**, is the development of order or integration of matter. This integrative theme of existence is described in detail in my book *Beyond* in the chapter 'Science and Religion'. Negative entropy, like gravity, is one of the physical laws of existence. This law states that in an open system, such as Earth's, where energy can come in from outside the system, in Earth's case from the Sun, matter becomes ordered and more complex. Negative entropy causes matter to self-organise, order itself into larger and more stable wholes. It forces matter to integrate, develop order. Thus, due to the influence of negative entropy, atoms have organised themselves or come together or integrated to form molecules, molecules have then integrated to form compounds, compounds have integrated to form single-celled organisms, single-celled organisms have integrated to form multicellular organisms, and these in turn have integrated to form societies. Humans are surrounded by evidence of the development of order of matter.

The scientist philosopher Arthur Koestler acknowledged integrative meaning in his 1978 book, *Janus: A Summing Up,* in a chapter titled 'Strategies and Purpose in Evolution': **'One of the basic doctrines of the nineteenth-century mechanistic world-view was Clausius' famous "Second Law of Thermodynamics". It asserted that the universe was running down towards its final dissolution because its energy is being steadily, inexorably dissipated into the random motion of molecules, until it ends up as a single, amorphous bubble of gas with a uniform temperature just above absolute zero: cosmos dissolving into chaos. Only fairly recently did science begin to recover from the hypnotic effect of this gloomy vision, by realizing that the Second Law applies only in the special case of so-called "closed systems" (such as a gas enclosed in a perfectly insulated container), whereas all living organisms are "open systems" which maintain their complex structure and function by continuously drawing materials and energy from their environment...It was in fact a physicist, not a biologist, the Nobel laureate Erwin Schrödinger, who put an end to the tyranny of the Second Law with his celebrated dictum: "What an organism feeds on is negative entropy"...Schrödinger's revolutionary concept of negentropy, published in 1944...is a somewhat perverse way of referring to the power of living organisms to "build up" instead of running down, to create complex structures out of simpler elements, integrated patterns out of shapelessness, order out of disorder. The same irrepressible building-up tendency is manifested in the progress of evolution, the emergence of new levels of complexity in the organismic hierarchy and new methods of functional coordination'**. Koestler proceeded to talk of **'the active striving of living matter towards'** order, of **'a drive towards synthesis, towards growth, towards wholeness'**. He said **'the integrative tendency has the dual function of coordinating the constituent parts of a system in its existing state, *and* of generating new levels of organization in evolving hierarchies'** (pp.222–226 of 354).

Significantly, in terms of behaviour, **'the integrative tendency'** requires **'coordination'**, as Koestler said. It requires that the parts of the new whole *cooperate,* behave selflessly, place the maintenance of the whole above maintenance of self. Put simply, selfishness is divisive or disintegrative while selflessness is integrative.

The concept of 'holism' is an acknowledgment of integrative meaning. The 'alternative culture' has embraced the word on the superficial basis that it refers to the interconnectedness of all matter;

however the true, deeper, core meaning of holism is **'the tendency in nature to form wholes'** (*Concise Oxford Dict.* 5th edn, 1964). The concept of 'holism' was introduced by the statesman, philosopher and scientist Jan Smuts in his 1926 book *Holism and Evolution.* He conceived 'holism' as being **'the ultimate organising, regulative activity in the universe that accounts for all the structural groupings and syntheses in it, from the atom, and the physico-chemical structures, through the cell and organisms, through Mind in animals, to Personality in Man.'**

'Teleology', **'the belief that purpose and design are a part of nature'** (*Macquarie Dict.* 3rd edn, 1998), is another word that, like holism, has been used to describe the integrative, cooperative, loving, selfless purpose or meaning or theme or design in the universe.

'Holism' and 'teleology' acknowledge the **'universal first principle'**, as Plato referred to it, of the cooperative, integrative purpose or meaning of life and indeed of all existence.

As was also briefly mentioned in the *Introduction,* and is also explained in the 'Science and Religion' chapter in *Beyond,* this truth of the cooperative, integrative meaning of existence has been termed 'God' in the metaphysical, religious domain, such as in monotheistic Christian mythology. 'God' is the metaphysical term that has been used for integration, the **'universal first principle'** of life, the **'absolute form of Good'**.

With regard to God being negative entropy, the physicist Stephen Hawking said **'I would use the term God as the embodiment of the laws of physics'** (*Master of the Universe,* BBC, 1989). Hawking is the Lucasian Professor of Mathematics at Cambridge University, the position once held by Isaac Newton. Another leading physicist, Paul Davies, has similarly said that **'these laws of physics are the correct place to look for God or meaning or purpose'** (*God Only Knows, Compass,* ABC-TV, 23 Mar. 1997), and that **'humans came about as a result of the underlying laws of physics'** (*Paul Davies—More Big Questions: Are We Alone in the Universe?,* SBS-TV, 1999).

In a feature article titled *The Time of His Life* that appeared in the *Sydney Morning Herald* in 2002, Hawking elaborated on his comment about God being the laws of physics. Written by Gregory Benford, a professor of physics at the University of California, the article recorded his recent meeting with Stephen Hawking at the University of Cambridge. Benford reported that at one stage he commented that **'there is amazing structure we can see from inside** [the universe]**'**, with which Hawking agreed, saying, **'The overwhelming impression is of order. The more we discover about the universe, the more we find that it**

is governed by rational laws. If one liked, one could say that this order was the work of God. Einstein thought so…We could call order by the name of God, but it would be an impersonal God. There's not much personal about the laws of physics' (27–28 Apr. 2002). (Later, in *The Demystification Of Religion* essay, in the section, 'The demystification of God', some of the main problems humans have accepting the demystification of God will be examined, in particular the problem that interpreting God as the laws of physics seems to destroy the personal, spiritual dimension that humans have come to associate with God.)

'God' is the personification of the negative entropy-driven integrative cooperative, loving, selfless ideals, purpose and meaning of life. The old Christian word for love was **'caritas'**, which means charity or giving or selflessness (see the *Bible,* Col. 3:14, 1 Cor. 13:1–13, 10:24 & John 15:13), therefore 'God is love', or unconditional selflessness, or commitment to integration.

In another of his dialogues, *Timaeus,* Plato recognised the Godly, integrative theme of existence. He wrote, **'God desired that all things should be good and nothing bad, so far as this was attainable. Wherefore also finding the whole visible sphere not at rest, but moving in an irregular and disorderly fashion, out of disorder he brought order, considering that this was in every way better than the other'** (tr. Benjamin Jowett, 1877).

The burning effect of the sun and fire

It is worth reiterating that while there was no explanation for the dilemma of the human condition—for why humans have been competitive, aggressive and selfish when the ideals are to be cooperative, loving and selfless—humans had no choice other than to block the whole issue from their minds, for the condemnation they otherwise experienced from these ideals was dangerously—even suicidally—depressing.

Faced with the unjust criticism that the cooperative, loving, selfless ideals of life represented, humans eventually learnt they had no choice other than to simply accept their divisive reality as normal and deny the whole concept and truth of ideality. If there is no acknowledgment made of the existence of cooperative ideality there is no issue about human divisiveness, no dilemma of the human condition to become depressed about. The strategy of denial is one humans have employed in many diverse situations. For instance it was used early last century to resist the now-accepted concept of Continental Drift. Opponents of that concept simply maintained there were no plates in the Earth's crust, in which case there was nothing to drift.

As will be described there are many truths relating to an ideal state that humans learnt to live in denial of because those truths unjustly criticised humans' corrupt reality. Of those criticising truths the main one that humans learnt they had to live in denial of was the truth of the integrative, cooperative, loving, selfless meaning of existence. While integrative meaning is the most profound and thus important of all truths it is also the truth that has appeared to most condemn humans and, tragically, which humans have most feared and found most difficult to confront. As Plato said, the **'absolute Good'** was **'perceived only with difficulty'**.

Being divisively rather than integratively behaved and unable to explain the necessary reason for that state, humans have had no choice but to evade or deny the truth of integrative meaning. They have been a 'God-fearing' rather than a 'God-confronting' species. Integrative meaning is the universal truth, the one great ultimate truth, the truth that, as Plato said, **'controls everything'** and is **'responsible for everything'**, and is **'the source...of their** [everything's] **existence and reality; yet it is...superior to it** [humans' **reality**] **in dignity and power.'** It is superior to humans' reality, in fact it is a truth that humans have lived in mortal fear and thus denial of.

Even science was forced to comply with this crucial need to deny the truth of integrative meaning. As will be explained in detail later in this essay, in the section 'Science—the liberator', and as is fully explained in *Beyond,* science has been reductionist and mechanistic, not holistic; it has focused on the details and mechanisms of the workings of our world and avoided the dangerously depressing, whole, integrative meaning-confronting view. In place of the truth of integrative meaning, mechanistic, reductionist science has maintained that evolution is a meaningless, purposeless, random process of selfish genetic opportunism. Such emphasis on selfishness conveniently excused humans' selfish, divisive behaviour, it avoided the psychological issue of the human condition.

After his description of integrative meaning, quoted in the previous section, the denial-free thinker or prophet, Arthur Koestler proceeded to clearly describe this state of denial of the truth of integrative purpose in life, writing that **'although the facts** [of the integration of matter] **were there for everyone to see, orthodox evolutionists were reluctant to accept their theoretical implications. The idea that living organisms, in contrast to machines, were primarily *active*, and not merely *reactive;* that instead of passively adapting to their environment they were...creating...new patterns of structure...such ideas were profoundly distasteful to Darwinians, behaviourists and reductionists in general...Evolution has been compared to a journey from an unknown origin towards an unknown destination, a sailing along a vast ocean; but we can at least chart the route...and there is no denying that there is a wind which makes the sails move...the purposiveness of all vital processes...Causality and finality are complementary principles in the sciences of life; if you take out finality and purpose you have taken the life out of biology as well as psychology'** (*Janus: A Summing Up*, 1978, pp.222–226 of 354).

Despite the great danger of acknowledging integrative meaning without first explaining the human condition, there has, in recent times, been a movement by some scientists and science commentators to follow the brave examples of Schrödinger and Koestler and recognise the truth of holism or teleology or integrative meaning. The titles of the books written by these scientists and commentators offer evidence (particularly the words I have underlined) of this recent development. Professor David Bohm wrote *Wholeness and The Implicate Order* in 1980; professors Ilya Prigogine and Isabelle Stengers wrote *Order Out of Chaos* in 1984; Professor Paul Davies wrote *God and the New Physics* in 1983, *The Cosmic Blueprint* in 1987 and *The Mind of God: Science and the Search for Ultimate Meaning* in 1992; Professor Charles Birch wrote *Nature and God* in 1965, *On Purpose* in 1990 and *Biology and The Riddle of Life* in 1999; Roger Lewin wrote *Complexity: Life at the Edge of Chaos, the major new theory that unifies all sciences* in 1992; M. Mitchell Waldrop wrote *Complexity: The Emerging Science at the Edge of Order and Chaos* in 1992; and Professor Stuart Kauffman wrote *The Origins of Order: Self-Organization and Selection in Evolution* in 1993, *At Home in the Universe: The Search for the Laws of Self-Organization and Complexity* in 1995 and *Anti-chaos* in 1996.

Complexity results from the integration of matter, and as Roger Lewin wrote in his above-mentioned book, **'the study of complexity represents nothing less than a major revolution in science'** (p.10 of 208). Complexity/order/self-organisation/integrative meaning gained some recognition when the independent organisation, the Santa Fe Institute for the Study of Complexity was formed in America in 1984. Stuart

Kauffman is a biology professor at the Institute.

In an article titled *Science Friction* journalist Deidre Macken, summarised the resistance that scientists who have recognised order/complexity/teleology/holism/purpose have encountered from the orthodox scientific world. In it she spoke of a **'scientific revolution'** and a coming **'monumental paradigm shift'**, and said that the few scientists who have **'dared to take a holistic approach'** are seen by the scientific orthodoxy as committing **'scientific heresy'**. Macken went on to say that scientists taking the **'holistic approach'**, such as the Australian scientists whose books are mentioned above, **'physicist Paul Davies and biologist Charles Birch'**, are trying **'to cross the great divide between science and religion'**, and are **'not afraid of terms such as "purpose" and "meaning"'**, adding that **'Quite a number of biologists got upset** [about this new development] **because they don't want to open the gates to teleology—the idea that there is goal-directed change is an anaethema...The emerging clash of scientific thought has forced many of the new scientists on to the fringe. Some of the pioneers no longer have university positions, many publish their theories in popular books rather than journals, others have their work sponsored by independent organisations...Universities are not catering for the new paradigm'** (*Sydney Morning Herald, Good Weekend* mag. 16 Nov. 1991).

It is significant that both professors Birch and Davies have been awarded the prestigious and, at $US1 million, financially rewarding Templeton Prize, for **'increasing man's understanding of God'** (*The Templeton Prize,* Vol.3, 1988–1992, p.108 of 153).

In discussing a **'scientific revolution'** and a coming **'monumental paradigm shift'**, Deidre Macken was intimating that acknowledging holism or integrative meaning is becoming a trend, but the truth is, until understanding of the human condition was found, holism could not be accepted by humanity as a whole without the disastrous consequences of madness and suicidal depression on a global scale. In his 1987 book *The Cosmic Blueprint* Paul Davies wrote that: **'We seem to be on the verge of discovering not only wholly new laws of nature, but ways of thinking about nature that depart radically from traditional science...Way back in the primeval phase of the universe, gravity triggered a cascade of self-organizing processes—organization begets organization—that led, step by step, to the conscious individuals who now contemplate the history of the cosmos and wonder what it all means...There exists alongside the entropy arrow another arrow of time, equally fundamental and no less subtle in nature...I refer to the fact that the universe is *progressing*—through the**

steady growth of structure, organization and complexity—to ever more developed and elaborate states of matter and energy. This unidirectional advance we might call the optimistic arrow, as opposed to the pessimistic arrow of the second law...There has been a tendency for scientists to simply deny the existence of the optimistic arrow. One wonders why' (from Chapters 10,9,2 respectively). The reason **'why'** integrative meaning was denied was because it was too dangerous to acknowledge without the necessary understanding as to the cause of humans' divisive, apparently non-integrative condition. Evidently Davies now appreciates this danger. In the last 10 years the focus of his books appears to have shifted from integrative meaning to less confronting issues, in fact to 'escapist' issues that allow humans to distract themselves from the real issue before us as a species, namely the human condition. His recent books include *The Last Three Minutes* (1994), on the ultimate fate of the universe; *About Time* (1995), on the puzzles and paradoxes of time; *Are We Alone?* (1995), on the search for extraterrestrial life; *The Fifth Miracle* (1998), on the possibility of life emerging in the deep, hot subsurface of our planet; and *How to Build a Time Machine* (2001), on the threat of an asteroid strike destroying life on Earth.

Plato's use of the sun as a symbol for the truth of integrative meaning is an obvious metaphysical choice for such an all-pervading but totally unapproachable truth that will 'cremate' all those who dare to go near it. Interestingly, the first monotheistic religion, the first religion based on the worship of a single god, was introduced by the Egyptian pharaoh and denial-free thinker or prophet, Ikhnaton, in 1377 BC and, appropriately, its object of worship was the sun-god Aton. Of course, the Earthly equivalent of the sun is fire, which, with its burning heat and blinding glare, is the other obvious choice of metaphor for the unapproachable **'universal first principle'** and **'absolute form of Good'** of integrative meaning or God.

Thus in Plato's cave allegory, both the **'sun'** and the **'fire'** represent the cooperative ideals of life or integrative meaning, the truth of which so condemned humans and which they have understandably lived in such fear of. Plato made this association clear, saying, **'the light of the fire in the** [cave] **prison corresponds to the power of the sun'** (p.282).

The illuminating effect of the sun and fire

As metaphors for integrative meaning, the sun and fire represent not only the condemning, hurtful truth of integrative meaning, but also its illuminating truthfulness. Humans need to acknowledge the fundamental truth of integrative meaning if they want to illuminate the world, make it intelligible. As mentioned earlier, Plato said, **'the Good** [integrative meaning]**...gives the objects of knowledge their truth and the mind the power of knowing...** [just as] **the sun...makes the things we see visible...The Good therefore may be said to be the source not only of the intelligibility of the objects of knowledge, but also of their existence and reality...** [the point being that] **objects** [are] **illuminated by daylight...** [just as they are] **by truth and reality'**. Living in denial of the unbearably condemning **'universal first principle'** and **'absolute form of Good'** of integrative meaning saved humans from suicidally depressing criticism, but it meant they were in no position to think truthfully. All thought was coming off a dishonest, false, unreal base and was thus unsound in its inferences and conclusions. This was especially so when the truth they were denying was the most fundamental of all truths, the truth of integrative meaning. Instead of being able to illuminate, have insight into our world, humans were going about mentally blinkered.

While the heat and glare of the sun and the fire could 'burn' and 'blind' humans (ie the truth of integrative meaning could be unbearably condemning), the light they shed was needed to illuminate the world, make it intelligible.

The reason for the cave existence

Plato said that between the fully **'illuminating'**, **'intelligible'** sunlit world and humans' **'cave'** existence stands a **'brightly burning fire'** that prevents the prisoners from leaving the cave. Understanding what the sun and fire represent we can understand why humans were virtual prisoners in a cave. As we have seen, the **'sun'** and the **'brightly burning fire'** represent the cooperative, integrative ideals of life, the confronting heat and glare of which were so searing and so bright—so condemning, depressing, hurtful and blinding—that humans had to turn their back on them, live in denial of them. As Plato says, **'if he** [a prisoner in the cave] **were made to look directly at the light of the fire, it would hurt his eyes and he would turn back'** (p.280). The prisoner had to face away from the fire and look only at the shadows cast by the real world on the back wall of the cave.

Integrative meaning, the sun and fire, caused humans' insecure state, which is the human condition. So intense was **'the power'** of the exposing and condemning **'sun'/'fire'** that humans could not face it, let alone approach it, and they were so held in bondage by the human condition, so **'chained'** up, as even to be estranged—alienated—from each other, to the extent that they **'cannot see one another.'**

Fire has been used in a number of mythologies as a metaphor for the integrative ideals of life, the condemning implications of which prevented humanity's 'escape' from its **'restricted'**, alienated condition. In the Zoroastrian religion, **'Fire is the representative of God...His physical manifestation...Fire is bright, always points upward, is always pure'** (*Eastern Definitions,* Edward Rice, 1978, p.138 of 433). In Christian mythology, in the story of Genesis, there was **'a flaming sword flashing back and forth to guard the way to the tree of life'** (Genesis 3:24). The *Bible* also recorded that the Israelites said, **'Let us not hear the voice of the Lord our God nor see this great fire any more, or we will die'** (Deut. 18:16). In the biblical account, Job pleaded for relief from confrontation with the issue of the human condition when he lamented, **'Why then did you** [God] **bring me out of the womb?...Turn away from me so I can have a moment's joy before I go to the place of no return, to the land of gloom and deep shadow, to the land of deepest night'** (Job 10:18, 20-22). Job's **'land of gloom and deep shadow...land of deepest night'**, the state of deepest and darkest depression that resulted from trying to confront the issue of the human condition, equates perfectly with life in Plato's cave. Only by facing away from the sun/fire, living psychologically in denial of the integrative meaning of life, could humans avoid the terrible—even suicidal—depression.

Facing away from the fire, living in denial of the human condition and all the truths that related to it, may have saved humans from the glare, the condemning criticism of these truths, but such a strategy necessitated living in a false, unreal world. As was mentioned, all thinking in that world was coming off a false base, it was a flawed view of the world, a world of delusions and illusions, a dishonest, ugly, limited existence. Scottish psychiatrist and denial-free thinker or prophet, R.D. Laing, offers this honest description of just how fraudulent humans' alienated world of denial has been: '[In the world today] **there is little conjunction of truth and social "reality". Around us are pseudo-events, to which we adjust with a false consciousness adapted to see these events as true and real, and even as beautiful. In the society of men the truth resides now less in what things are than in what they are not. Our social realities are so ugly if seen in the light of exiled truth, and beauty is almost no longer possible if it is not a lie'** (*The Politics of Experience* and *The Bird of Paradise,* 1967, p.11 of 156).

Humans' world of denial, of **'exiled truth'**, was a world of lies and thus of delusions and illusions. Plato's shadows on the back wall of the cave symbolise this world of appearances. Bound in such a way that they face the back wall, Plato's **'prisoners'** cannot see anything **'except the shadows thrown by the fire on the wall of the cave opposite them...And so they would believe that the shadows of the objects...were in all respects real'** (p.279).

Liberation from the cave

In the cave allegory, Plato says that humans have been living in deep denial of integrative meaning and all the truths associated with it, with the result that they can only see a highly distorted representation of the world. The question then is, how were humans to free themselves from this awful world, how were they going to escape their life of bondage in the 'cave'? The next part of Plato's allegory deals with that question. It was summarised in the *Encarta* entry thus: **'Breaking free, one of the individuals escapes from the cave into the light of day. With the aid of the sun, that person sees for the first time the real world and returns to the cave with the message that the only things they have seen heretofore are shadows and appearances and that the real world awaits them if they are willing to struggle free of their bonds. The shadowy environment of the cave symbolizes for Plato the physical world of appearances. Escape into the sun-filled setting outside the cave symbolizes the transition to the real world, the world of full and perfect being, the world of Forms, which is the proper object of knowledge.'**

Plato is saying that someone has to be free of the cave existence of denial—has to be living in the illuminating light of the **'sun'**, that is, acknowledging the fundamental truth of integrative meaning and therefore thinking truthfully—to be able to find understanding of the human condition and by so doing see through the denial and illusion of the cave existence.

While Plato's cave allegory is a simplified account of how humanity is liberated from the human condition, how it **'escapes from the cave'**, it nevertheless contains all the elements involved in the process of becoming free. An analysis of these elements will confirm just how incredibly insightful Plato was.

There are three aspects to consider:

Firstly, in **ASPECT A**, we will examine the possibility of someone being sufficiently free of denial, that is sufficiently sound and secure in self, to be able to confront and think truthfully and effectively about the issue of the human condition—how it is possible for someone to **'escape from the cave into the light of day.'**

Secondly, in **ASPECT B**, we will examine the knowledge that a person would need to assemble the liberating understanding of the human condition, and by so doing end humanity's need to live in denial—how it is possible for someone to **'see for the first time the real world'**.

Thirdly, in **ASPECT C**, we need to look at the difficulty of introducing those who are living in denial of the human condition to the liberating understanding of the human condition. In Plato's metaphor we need to look at what happens when the person who has assembled the liberating understanding of the human condition **'returns to the cave** [prisoners] **with the message that the only things they have seen heretofore are shadows and appearances and that the real world awaits them'**.

These areas of examination have been divided into Aspects 'A', 'B' and 'C' to assist the reader with their comprehension of the information.

ASPECT A

The need to live in denial of the human condition has not been totally universal amongst humans. The fact that people are variously exposed to the corruption in the world and thus variously corrupted as a result of that exposure means that there has always been a spectrum of denial or alienation in humanity, with some individuals suffering little alienation and others extreme alienation.

In the lead-up to his cave allegory Plato recognises these different degrees of denial or alienation when he talks of **'four states of mind: to the top section Intelligence, to the second Reason, to the third Opinion, and to the fourth Illusion. And you may arrange them in a scale, and assume that they have degrees of clarity corresponding to the degree of truth and reality possessed by their subject-matter'** (p.278). **'Intelligence'**, the ability to confront integrative meaning and as a result think honestly, truthfully, effectively and thus intelligibly, is the least alienated state. To live in denial of integrative meaning, but not so in denial that your thinking was entirely alienated, deluded and dishonest, is the state where you could be trusted to be able to at least **'reason'**. Those more alienated but still able to think with some semblance of effectiveness were those who were capable of at least having an **'opinion'**. The most alienated were those people who were living with no **'truth'** and not facing any **'reality'** about their corrupted state, people who were mentally living a life of complete **'illusion'**.

It makes sense that at the relatively unalienated, uncorrupted end of the spectrum there would occasionally be individuals who were sufficiently sound and secure in self to be able to confront and think truthfully and effectively about the human condition.

This is a very brief summary of the answer to the question of how some people could be sound and secure enough in self to confront the issue of the human condition. However, this answer raises further issues that need to be addressed, and the purpose of the seven sections of ASPECT A is to respond to those questions.

Four particular questions that will be examined are:

1. **When humans are corrupted what is it that is corrupted?**

ASPECT A Section 1, '<u>The difficulty of acknowledging the human soul and the spectrum of alienation, or denial of soul</u>', deals with this question. This section will explain that it is our soul that is corrupted and look at the difficulty humans have had acknowledging the human soul and the spectrum of denial of it.

2. Since the answer to the first question is that it is our soul that is corrupted, the question then is: **What is our soul?**

A2 and A3 deal with this question. A2, '<u>Plato's acknowledgment of the nature of our soul</u>', presents Plato's acknowledgment of the nature of our soul. A3, 'Recognition of the soul's world in mythology', looks at the recognition mythologies have given to the existence and nature of our soul.

3. **How did we acquire our soul?**

A4 and A5 deal with this question. A4, 'The biological explanation for how we acquired our soul, our instinctive orientation to integrative, cooperative meaning', presents a summary of the biological explanation of how we acquired our soul. A5, 'The contrived excuses for humans' divisive nature', looks at the contrived excuses humans have used to avoid recognising the existence of integrative meaning and our soul's orientation to that meaning.

4. **What was the reason for this corruption of our soul that left humans variously alienated?**

A6 and A7 deal with this fourth question. A6, 'A brief description of how our soul became corrupted', explains how our soul was corrupted. A7, 'There were people who could live in the sun's light, but they were unbearably condemning for the cave prisoners', explains that there were degrees of corruption amongst humans, with some people sufficiently uncorrupted not to be condemned by the truth of integrative meaning and it is these people who have been able to safely confront the issue of the human condition.

A1: The difficulty of acknowledging the human soul and the spectrum of alienation, or denial of soul

The previous section described how humans are living in various states of corruption, but posed the question of what is it that is corrupted. The answer to this question is that it is the 'soul' of humans that is corrupted, but that again leaves other questions to be answered; what is our soul and how was it corrupted. To begin to answer these questions it is first necessary to appreciate that the subject of the human soul has been one of the many areas that humans have hitherto been unable to examine and have thus had to live in denial of.

It was pointed out that humans learnt that the only way to avoid condemning criticism of their corrupted angry, egocentric and alienated state was to deny the whole concept of cooperative ideality. If there is no ideal state there is no issue with human divisiveness, no dilemma of the human condition to become depressed about. It was also mentioned that there are many truths relating to an ideal state that humans have had to learn to live in denial of, the most profound and confronting of all being the truth of integrative, cooperative meaning.

It now needs to be explained that the second most important representation of ideality that humans have had to live in denial of has been the truth that they once lived in an utterly integrated, harmonious, cooperative, all-sensitive, loving state, the instinctive memory of which is what we have long referred to as our 'soul'. In their corrupted angry, egocentric and alienated state, the existence of this loving, original instinctive self or soul criticised humans unbearably, leaving them no choice other than to avoid recognising our soul's true meaning.

The extent of humans' denial of their integratively-orientated instinctive self or soul can be gauged from the fact that it has been pushed so far beyond conscious awareness, has been so psychologically repressed, that it now resides deep in humans' subconscious. From there this **'collective unconscious'** self (as psychoanalyst Carl Jung termed our shared-by-all instinctive self) emerges only in dreams and on other occasions when our conscious self is subdued. As Jung wrote, **'The dream is a little hidden door in the innermost and most secret recesses of the psyche** [soul]**, opening into that cosmic night which was psyche long before there was any ego consciousness'** (*Civilization in Transition,* The Collected Works of C.G. Jung, Vol.10, 1945). Both the truth of humans' current numb, seared, alienated state, and the truth of all the beauty and magic of life that our species' original instinctive self or soul is aware of, have been repressed and denied.

Compliant with the denial, science has denied the existence of our soul's world, the fabled state of harmony and enthralment that humans once lived in before they became corrupted and alienated. Far from acknowledging that our ancestors lived innocently, sensitively and cooperatively, scientists have maintained the view that our ancestors lived competitively and aggressively. As will be described shortly, biologists especially have maintained the view that humans' forebears were competitive, survival-of-the-fittest-driven, reproduce-your-own-genes-at-all-cost, selfish and aggressive beasts.

Not only has the true nature of our soul been denied but all truths relating to our soul's world have also been denied. The concepts of innocence (the absence of corruption of soul) and of alienation (the presence of corruption of soul and resulting denial of anything relating to ideality) bring the issue of the human condition into focus and are thus concepts that most humans have found unbearably difficult to acknowledge.

To acknowledge human corruption and the resulting varying degrees of alienation has until now been impossible. Any differentiation only produced prejudice, condemnation and the insinuation that some people are good and others bad, when, with understanding of why humans became corrupted we are able to appreciate that no human is fundamentally bad or 'evil'. Understanding the human condition, knowing the biological reason for humans' corrupted angry, egocentric and alienated state, dignifies humans, makes the concept of 'evil' obsolete and lifts the 'burden of guilt' from the human race. Only with that dignifying explanation is it at last safe to acknowledge the cooperative integrative meaning of existence, the fact that humanity once lived cooperatively in accordance with this meaning, and that humans have since become variously corrupted, and thus variously alienated from that harmonious original instinctive state.

A2: Plato's acknowledgment of the nature of our soul

An individual from the more innocent end of the alienation spectrum, Plato was of sufficient soundness (alienation-free) and security (unafraid of integrative meaning) to confront and describe the human condition, so it is not surprising that he was also able to acknowledge the existence of all aspects of ideality. As has already been described, he recognised in his world of **'perfect forms'**, in particular the **'form of absolute Good'**, the existence of integrative meaning. In his dialogue, the *Phaedo,* Plato also recognised that our soul is our species' memory of a time when we lived ideally, harmoniously, cooperatively, lovingly and all-sensitively.

Of the more than two dozen dialogues Plato composed, *The Republic* and the *Phaedo,* written during his inspired middle period are considered his greatest works. The dialogue in the *Phaedo* commences with the assertion that humans are born with the ability to recognise what is ideal and what is not, that humans have an innate ability to know when something **'falls short'** of, or **'inadequately resembles'**, or lacks **'equality'** with what is ideal (*Phaedo*, tr. H. Tredennick). Plato went on to say that if we obtained **'knowledge of these standards...these absolute realities, such as beauty and goodness...before our birth, and possessed it when we were born, we had knowledge, both before and at the moment of birth, not only of equality and relative magnitudes, but of all absolute standards. Our present argument applies no more to equality than it does to absolute beauty, goodness, uprightness, holiness, and, as I maintain, all those characteristics which we designate in our discussions by the term "absolute"'** (ibid). Plato was acknowledging that humans are born with not only what we now refer to as a 'conscience', an ability to recognise what is 'right' and 'wrong' behaviour, but with an awareness of what is beautiful and what is not.

Plato linked our innate awareness of **'these absolute realities, such as beauty and goodness'** with our soul, saying, **'it is logically just as certain that our souls exist before our birth as it is that these realities exist...** [and our] **soul is in every possible way more like the invariable** [absolute entities] **than the variable** [non-absolutes]**.'** In an unambiguous statement Plato said that the **'soul resembles the divine'** (ibid).

Interestingly, Plato also emphasised the immortality of the soul when he said that **'a man's soul...will exist after death no less than before birth'**, because **'that absolute reality...** [such as] **absolute equality** [with goodness] **or beauty** [that our soul is imbued with]**...remain always constant'** (ibid). With understanding of the human condition we can understand that humans' corrupted state is, like our integratively-orientated soul, also part of the purpose of existence and is thus also eternally meaningful and thus eternally enduring—immortal.

As he does in *The Republic*, Plato emphasises in the *Phaedo* that effective inquiry and learning involve the recovery of awareness of the **'absolute realities'** that our soul knows, stating, **'what we call learning will be the recovery of our own knowledge'** (ibid). Christ similarly saw the limitations to learning imposed by humans' cave state of living in denial when he said, **'you have hidden these things from the wise and learned, and revealed them to little children'** (Matt. 11:25). As will be explained in the *Resignation* essay, since humans did not adopt an attitude of denial until mid-adolescence, children are free of denial, are honest in their thinking. 'Great' thinkers are simply people capable of thinking honestly. There are many sayings to the effect that genius is the ability to think like a child; the Chinese philosopher Mencius said, **'The great man is he who does not lose his child's-heart'** (*Works*, 4–3 BC, 4, tr. C.A. Wong). In the following quote Plato described the corruption of humans' capacity to think truthfully: **'when the soul uses the instrumentality** [of denial] **of the body for any inquiry...it is drawn away by the body into the realm of the variable, and loses its way and becomes confused and dizzy, as though it were fuddled...But when it investigates by itself** [ie, free of the body's intellect's capacity for denial], **it passes into the realm of the pure and everlasting and immortal and changeless, and being of a kindred nature, when it is once independent and free from interference, consorts with it always and strays no longer, but remains, in that realm of the absolute, constant and invariable'** (*Phaedo*, tr. H. Tredennick).

Plato's acknowledgment that the **'soul resembles the divine'** is echoed in one of the greatest (most honest) poems ever written, William Wordsworth's *Intimations of Immortality from Recollections of Early Childhood* (1807), in the wonderful line **'But trailing clouds of glory do we come/From God, who is our home'**. It is worth including more of this poem because of its extraordinary acknowledgment of humans' past state of uncorrupted, alienation-free innocence. **'There was a time when meadow, grove, and streams/The earth, and every common sight/To me did seem/Apparelled in celestial light/The glory and the freshness of a dream/It is not now as it hath been of yore/Turn wheresoe'er I may/By night or day/The things which I have seen I now can see no more//The Rainbow comes and goes/And lovely is the Rose/The Moon doth with delight/Look round her when the heavens are bare/Waters on a starry night/Are beautiful and fair/The sunshine is a glorious birth/But yet I know, where'er I go/That there hath past away a glory from the earth.'** Wordsworth then described how nature and the innocence of youth reminded him of this lost paradise: **'Thou Child of Joy/Shout round me, let me hear thy shouts, thou happy Shepherd-boy!/Ye blessed Creatures, I have heard the call/Ye to each other make; I see/The heavens laugh with you in your jubilee/...While Earth herself is adorning/This sweet May-morning/And the Children are culling/On every side/In a thousand valleys far and wide'**. Wordsworth is then reminded of his loss of innocence and the alienation that has set in, adding: **'But there's a Tree, of many, one/A single Field which I have looked upon/Both of them speak of something that is gone/...Whither is fled the visionary gleam?/Where is it now, the glory and the dream?//Our birth is but a sleep and a forgetting/The Soul that rises with us, our life's Star/Hath had elsewhere its setting/And cometh from afar/Not in entire forgetfulness/And not in utter nakedness/**

But trailing clouds of glory do we come/From God, who is our home/ Heaven lies about us in our infancy!/Shades of the prison-house begin to close/Upon the growing Boy/...And by the vision splendid/Is on his way attended/At length the Man perceives it die away/And fade into the light of common day/...Forget the glories he hath known/And that imperial palace whence he came.' Wordsworth proceeded to say that it is only a denial-free thinker or a **'prophet'** who can plumb the depths of the forgotten realm: **'Thou best Philosopher, who yet dost keep/Thy heritage, thou Eye among the blind/That, deaf and silent, read'st the eternal deep/Haunted for ever by the eternal mind/Mighty Prophet! Seer blest!/ On whom those truths do rest/Which we are toiling all our lives to find/In darkness lost, the darkness of the grave'**. The **'darkness lost, the darkness of the grave'** perfectly equates with Plato's cave existence.

A3: Recognition of the soul's world in mythology

Wordsworth's acknowledgment of an innocent, cooperative past for humans is by no means unique. There is overwhelming evidence in all mythologies of the existence of a time in humans' past when we lived in uncorrupted innocence, harmoniously and cooperatively. For example there is the story of the **'Garden of Eden'** where humans lived before our, so-called, **'fall from grace'**. In Greek mythology there are references to a **'golden age'** in humanity's past, as seen in the following poem, *Theogony,* written by the famous 8th century BC Greek poet, Hesiod: **'When gods alike and mortals rose to birth/A golden race the immortals formed on earth/Of many-languaged men: they lived of old/ When Saturn reigned in heaven, an age of gold/Like gods they lived, with calm untroubled mind/Free from the toils and anguish of our kind/Nor e'er decrepit age misshaped their frame/The hand's, the foot's proportions still the same/Strangers to ill, their lives in feasts flowed by/Wealthy in flocks; dear to the blest on high/Dying they sank in sleep, nor seemed to die/Theirs was each good; the life-sustaining soil/Yielded its copious fruits, unbribed by toil/They with abundant goods 'midst quiet lands/All willing shared the gathering of their hands'** (tr. Elton).

In his best selling book of 1987, *The Songlines,* the explorer and philosopher Bruce Chatwin wrote: **'Every mythology remembers the innocence of the first state: Adam in the Garden, the peaceful Hyperboreans, the Uttarakurus or "the Men of Perfect Virtue" of the Taoists. Pessimists**

often interpret the story of the Golden Age as a tendency to turn our backs on the ills of the present, and sigh for the happiness of youth. But nothing in Hesiod's text exceeds the bounds of probability. The real or half-real tribes which hover on the fringe of ancient geographies—Atavantes, Fenni, Parrossits or the dancing Spermatophagi—have their modern equivalents in the Bushman, the Shoshonean, the Eskimo and the Aboriginal' (p.227 of 325).

In *The Heart of The Hunter,* Sir Laurens van der Post acknowledged that, **'There was indeed a cruelly denied and neglected first child of life, a Bushman in each of us'** (1961, p.126 of 233). D.H. Lawrence recognised that, **'In the dust, where we have buried/The silent races and their abominations/We have buried so much of the delicate magic of life'** *(Son of Woman: The Story of D.H. Lawrence,* D.H. Lawrence, 1931, p.227 of 402). Similarly Jean-Jacques Rousseau noted that **'nothing is more gentle than man in his primitive state'** *(The Social Contract and Discourses,* 1755; tr. G.D.H. Cole, pub. 1913, Book IV, *The Origin of Inequality,* p.198 of 269).

If the reader would like more references to our lost state of innocence I suggest Richard Heinberg's book *Memories & Visions of Paradise.* Considering the extent of denial in the world today it is an astonishing collection of evidence from many mythologies of belief in the existence of an integrated, cooperative past in humanity's journey and of humanity's current corrupted alienated state. The following is a sample from the 1990 edition of the book: **'Every religion begins with the recognition that human consciousness has been separated from the divine Source, that a former sense of oneness...has been lost...everywhere in religion and myth there is an acknowledgment that we have departed from an original...innocence...the cause of the Fall is described variously as disobedience, as the eating of a forbidden fruit, and as spiritual amnesia'** (pp.81–82 of 282).

With the human condition at last able to be explained it is finally safe to acknowledge the existence of our soul and the true world it is aware of. At last able to acknowledge our soul we are faced with the question of how we acquired it.

A4: The biological explanation for how we acquired our soul, our instinctive orientation to integrative, cooperative meaning

The question is, how did humans acquire their 'soul'? As Bishop John Shelby Spong says in his 1992 book, *Born Of A Woman,* **'If only human beings have souls** [and not other animals]**, as the church has taught, one must be able to say when humanity became human and was infused with its divine and eternal soul'** (p.34).

The detailed biological explanation for how humans acquired an instinctive self or soul that is orientated to behaving cooperatively and lovingly is given in *Beyond,* in the chapter headed 'How We Acquired Our Conscience'. What follows is a brief summary of that explanation.

As has been explained, the meaning or purpose of existence is the integration or development of order of matter. This development occurs as a result of the law of physics called negative entropy. In this process of developing larger and more stable wholes the parts of the developing whole must be able to develop the capacity for unconditional selflessness if the fully integrated whole is to develop. As Koestler said, in terms of behaviour, **'the integrative tendency'** requires **'coordination'**. It requires that the parts of the new whole *cooperate,* which means behave selflessly, place the maintenance of the whole above maintenance of self. Put simply, selfishness is divisive or disintegrative while selflessness is integrative.

The process we refer to as genetic reproduction (more properly described as the genetic learning or genetic information-processing system) has been one of the main tools for this development of integration or order of matter on Earth. While genetics has enabled the development of a great deal of ordered matter—it has made possible the emergence of the great variety of 'life' on Earth—as a tool for developing order it has one particular limitation, which is that it requires traits to always be selfish. If an unconditionally selfless trait appears, such as the inclination to sacrifice yourself in defence of your group, then over time that trait self-eliminates, it cannot become established. Only selfish traits carry on.

(Those seeking to deny the truth of integrative meaning have made much of this inability of the genetic learning system to develop unconditional selflessness. Indeed, it is necessary to dedicate the next section to briefly dealing with all the contrived excuses for humans' divisive nature. In that section it will be described how the fact that genes have to be selfish—even though they are committed to the task of integrating matter—has been used to infer that all life is essentially selfish, thus justifying human selfishness. It will be described how this particular contrived excuse—that genes are selfish because life is—has evolved through a number of stages of refinement, from 'Social Darwinism', to 'Sociobiology' and, most recently, to 'Evolutionary Psychology'.)

The particular limitation of the genetic information refining or learning system is that it normally cannot learn or refine or develop unconditionally selfless traits because such genetic traits self-eliminate and thus cannot become established in a species. However, there was a way to overcome this genetic limitation to developing the order or integration of matter on Earth—through nurturing. While nurturing is a selfish trait (by nurturing and fostering the next generation—which has the parent's nurturing trait—the nurturing trait is selfishly ensuring that it carries on from generation to generation), from an observer's point of view it *appears* to be selfless behaviour. The mother is giving her offspring food, warmth, shelter and protection for *apparently* nothing in return. This point is significant, because it means that from the infant's perspective, its mother is treating it with real love, which is unconditional selflessness. (With regard to unconditional selflessness being love, it was pointed out earlier that the old Christian word for love was **'caritas'**, which means charity or giving or selflessness.) The infant's brain is therefore being trained or conditioned or indoctrinated with selflessness, and with enough training in selflessness the infant will grow to be an adult that behaves selflessly.

The 'trick' in this 'love-indoctrination' process lies in the fact that nurturing is encouraged genetically because the better infants are cared for the greater are their chances of survival, but there is an integrative side effect, which is that the more infants are nurtured the more their brain is trained in unconditional selflessness. There are very few situations in biology where animals appear to behave selflessly towards other animals; they normally selfishly compete for food, shelter, space and mating opportunities. Maternalism, a mother's fostering of her infant, is one of the few situations where an animal appears to be behaving selflessly towards another animal.

It was this *appearance* of selflessness that exists in the maternal situation that provided the opportunity for the development of love-indoctrination.

To develop nurturing—this 'trick' for overcoming the genetic learning system's inability to develop unconditional selflessness—a species required the capacity to allow its offspring to remain in the infancy stage long enough for the infant's brain to become indoctrinated with unconditional selflessness or love. Primates are especially facilitated for leaving offspring in infancy and thus developing love-indoctrination. Being semi-upright as a result of their arboreal heritage, their arms were free to carry a helpless, dependent infant. Species that cannot carry and easily look after their infants cannot develop love-indoctrination. Upright walking and its result, bipedalism, in humans is a direct result of the love-indoctrination process, which means bipedalism must have occurred early on in the emergence of humans, as fossil records now reveal.

There is a limiting factor in the development of love-indoctrination, which is that while the nurturing of infants is strongly encouraged genetically because it ensures greater infant survival, the side effect of training infants to behave selflessly as adults is that the selflessly behaving and even self-sacrificing adults don't tend to reproduce their genes as successfully as selfishly behaved adults. The genes of exceptionally maternal mothers don't tend to endure because their offspring tend to be too selflessly behaved. While substantial unconditional selflessness is able to be developed through love-indoctrination due to the greater initial survival of infants who have been well cared for, there was a limit to how much it could be developed genetically. What fortunately developed at this point of limitation was <u>self-selection</u>. In the coming section 'A brief description of how our soul became corrupted', it will be briefly explained how the love-indoctrination process liberated consciousness in our ape ancestors. It was the emerging conscious intellect in our ape ancestors that began to support the development of selflessness. As our ape ancestors gradually became conscious they began to recognise the importance of selflessness and as a result began to actively select for it. (While the integrative, selfless theme and purpose of existence has been denied by humans suffering from the human condition, it is an obvious truth to a conscious being who is not living in denial of it.) They could do this by consciously seeking out love-indoctrinated mates. These were members of the group who had experienced a long infancy and were closer to their memory of infancy (that is, younger). The older individuals became, the more their infancy training in love wore off; our ape ancestors began to recognise that the younger an individual, the more integrative he or she was likely to be. Our ancestors began to idolise, foster and select youthfulness because of its association with cooperativeness or integrativeness. The effect, over many thousands of generations, was to retard our physical development so that we became infant-like in our appearance as adults. This explains how we came to regard neotenous (infant-like) features—large eyes, dome forehead, snub nose and hairless skin—as beautiful. The physical effect of neoteny was that we lost most of our body hair and became infant-looking compared with our adult ape ancestors.

Since males were preoccupied with competing for mating opportunities before love-indoctrination, females were first to self-select for integrativeness by favouring integrative rather than competitive and aggressive mates. This helped love-indoctrination subdue the males' divisive competitiveness. Without being aware of the process of love-indoctrination, primatologists have noted self-selection of integrativeness by females. **'Male** [baboon] **newcomers also were generally the most dominant while long-term residents were the most subordinate, the most easily cowed. Yet in winning the receptive females and special foods, the subordinate, unaggressive veterans got more than their fair share, the newcomers next to nothing. Socially inept and often aggressive, newcomers made a poor job of initiating friendships'** (Shirley Strum, *National Geographic* mag. Nov. 1987). **'The high frequencies of intersexual association, grooming, and food sharing together with the low level of male-female aggression in pygmy chimpanzees may be a factor in male reproductive strategies. Tutin (1980) has demonstrated that a high degree of reproductive success for male common chimpanzees was correlated with male-female affiliative behaviours. These included males spending more time with estrous females, grooming them, and sharing food with them'** (*The Pygmy Chimpanzee,* ed. Randall L. Susman, ch. 13 by Alison & Noel Badrian, 1984, p.343 of 435). The bonobo or pygmy chimpanzees have the most matriarchal, female-centric societies of all apes and they are also the most peaceful, cooperative and intelligent of all apes. In fact, as is explained in *Beyond,* bonobos are living examples of what life was like in humanity's 'Garden of Eden' stage, the time when humans lived utterly cooperatively and selflessly.

Historian Jacob Bronowski recognised the significant role played by self-selection when he stated: **'We have to explain the speed of human evolution over a matter of one, three, let us say five million years at most. That is terribly fast. Natural selection simply does not act as fast as that on animal species. We, the hominids, must have supplied a form of selection of our own; and the obvious choice is sexual selection'** (*The Ascent of Man*, 1973).

Eventually, with love-indoctrination occurring over many generations, selflessness became instinctive or innate. This occurred because once unconditionally selfless individuals were *continually* appearing the genes 'followed' the whole process involved reinforcing it, including reinforcing selflessness. Similarly, when the conscious mind fully emerged and went its own way—embarked on its course for knowledge—the genes followed reinforcing that development. Generations of humans whose genetic make-up in some way helped them cope with the human condition were selected naturally, making humans' alienated state somewhat instinctive in humans today. We have been 'bred' to survive the pressures of the human condition. The genes would always naturally follow and reinforce any development process—in this they were not selective. The difficulty was in getting the development of unconditional selflessness to occur, for, once it was occurring, it would naturally become instinctive over time.

It was through nurturing, the process of love-indoctrination and the accompanying self-selection of cooperativeness or selflessness, that humans were able to develop an instinctive orientation to behaving unconditionally selflessly and as a result become an utterly integrated cooperative, selfless, loving species. In our instinctive past, prior to becoming fully conscious and corrupted by the problem of the human condition, all humans behaved selflessly and considered the welfare of the group above their own. This instinctive memory of a loving, cooperative, alienation-free, all-sensitive past is what we term our soul, one expression of which is our conscience, the instinctive expectation within us that we behave selflessly, lovingly, cooperatively.

When Bishop Spong referred to **'the church'** teaching that **'only human beings have souls'**, he is almost certainly referring to the Genesis passage in the *Bible* which says that **'God created man in his own image'** (1:27). Since God is integrativeness, when humans became utterly integrated they *were* finally **'in God's image'**. Other animals that have not yet overcome the genetic limitation to developing unconditional selflessness and thus pure integration are not yet **'in God's image'**, they don't yet have an instinctive orientation to integrative meaning or God, such as that of the human soul.

It should be emphasised that although animals do not have an instinctive orientation to integrative meaning they are still a part of the Godly, integrative process, as are corrupted humans. There is such a thing as the 'animal condition', the stalled situation that most other animal species have found themselves in, of being unable to overcome the genetic limitation to developing order, of being unable to develop unconditional selflessness. They have to grow up and compete for mating opportunities, a horribly divisive activity. In the sense however that a 'soul' is that part of a being that is associated with, or belongs to, God, namely the integrative process, then all of life is soulful, meaningful and part of the integrative process.

The following wonderful quote, from Sir Laurens van der Post's 1953 book *The Face Beside the Fire,* clarifies how all of life has participated fully in the great journey to develop order. Note that Sir Laurens acknowledges that there were races whose innocence was such that they could not cope with the new levels of corruption that emerged in the world and who died out as a result; in doing so, he fearlessly acknowledges different levels of alienation amongst humans and recognises that those more innocent races who died out also participated fully in the heroic journey to develop order on Earth. Thankfully, with the human condition reconciled, the more innocent and the more corrupt can live harmoniously, and before long corruption too will disappear from the human situation.

'I was allowed to attend a victory parade, as it were, of all the life that has ever been. I saw all that has ever been come streaming through the long lanes and corridors of my blood, through their arch of admiralty, round the inner-square and then straight down past my own white lighted Hall. Out of the darkness that preceded Genesis and flood, it began with a glimmer and a worm of the unformed earth in love with the light to come. Yes! a worm with a lantern, a glow-worm with phosphorescent uniform, marched proudly at the head, and behind came great streams of being protozoic and pre-historic. Nothing was excluded and everything included, their small fires of being clearly lit, tended and well beloved. This, it was said, is the true, the noble heroic and unique crusade of the love of life. For look, among them not a brain but only matter tentatively and awkwardly assembled. Yet remark on their bearing and the trust with which they hurl themselves into the uncomprehended battle. Ah! tears of love and gratitude burned in my eyes at so urgently moving and life confiding a sight. To feel, at last, the burden that they carry for me in my own blood, to know at

every second several of these reflected in white corpuscle and scarlet cell are dying unflinchingly in battle for my all, to know that giant lizard and lion as well as unicorn came after, and were hurled too into similar struggle and defence of the totality of all. I was allowed, too, to see the first man and registered the seismographic thrill of the marching column at the appearance of so skilled and complex a champion. I was allowed to speak to him and I touched his skin riddled with snake bite, his shoulder pierced by mastodon's spike, his skull deep-scarred with sabre-tooth's claw. And as reverently and tenderly I took his hand shaking with marshy malarial fever, I was moved to pity him by the evidence of such dread and unending war. But he would have none of it. He looked me fearless in the eye and in a voice that boomed like a drum in his stomach said: "Brother, it was worth it. Whatever they tell you, add this, it was worth it."

I spoke to a Bushman half-eaten by a lion in the Kalahari, his only vessel a brittle ostrich egg with red and black triangles painted neatly on it, now broken and sand scattered. He looked in my grey eyes with the brown eyes of a people at dusk, slanted to bridge a chasm behind the face of a dying member of a dying and vanishing race. He too, my dying nomad brother, said: "Add, add quick before I go, 'it was worth it'." I spoke to an aborigine in the bight of the great gulf Tattooed with dung he said: "I vanish, but it was worth it." In New Guinea, I met a stone-age Papuan, his black skin sheened with green after centuries in the jungle between basin and fall of water and spurting volcano, and he too said: "Doubt it not, it was worth it." Everyone said, "Lovely gift of a life that we blindly trust burns with such loving fire in the dark that at any price, no matter how great, it is worth it."

Yes, they all agreed and utterly convinced me, so that I can never doubt again. I wept when the great procession came to an end, for one and all, great and small—I loved them all. Yes, even to the worm that brought up the rear, with shaded night light and a nurse's white, in its dress concealing a phial of the drug of the greater sleep made with a touch of the hand of God's great, good night. Yes...I love them all; I believe them; I am ready for battle; and to continue at their head the journey of them all to the end of the road in my blood. At last, purified and complete, I am ready to awaken and defend my love' (pp.292–294 of 312).

A5: The contrived excuses for humans' divisive nature

As emphasised, while humans couldn't truthfully explain their corrupted, divisive condition they had no choice other than to live in denial of the truth of their corrupted state. The main means for maintaining that denial was to deny there was an integrative, selfless purpose to existence. In fact humans deluded themselves that they were selfish because life was selfish. Rather than face the obvious truth of an order-developing, integrative, selfless, loving, cooperative purpose to existence, humans contrived excuses for their divisive selfish, aggressive and competitive behaviour.

The business of inventing or contriving excuses for humans' divisive behaviour has gone through many stages of refinement. Humans' original excuse for their competitive and aggressive behaviour was that 'it is only natural because, after all, we are only animals and animals are always fighting and killing each other; animals are "red in tooth and claw", so that's why we are'. With the emergence of science this original misrepresentation of the genetic learning system as a divisive, rather than integrative, process was given a biological basis. This was Social Darwinism, the misrepresentation of Charles Darwin's 1859 *The Origin of Species* theory of natural selection as being concerned with 'the survival of the fittest'. As emphasised, the real concern or objective of genetic refinement, or 'natural selection' as Darwin originally termed it, was the integration or development of order of matter on Earth. Order is what was being learnt or refined or developed. In fact the word 'development' should replace the word 'evolution' in biology because evolution can, and has been, misrepresented as meaning change is undirected, meaningless and random, when it is not.

It should be explained that it was Darwin's associates, Herbert Spencer and Alfred Russel Wallace, who persuaded Darwin to replace the term 'natural selection', that Darwin used in the first editions of his 1859 *The Origin of Species,* with the term 'survival of the fittest'. They said the term 'natural selection' could be interpreted as implying the involvement of a personal selector. Darwin's friend and great defender, Thomas Huxley called it an **'unlucky substitution'** (*Charles Darwin,* Sir Gavin de Beer, 1963, p.178 of 290), and it certainly was. While a personal, interventionist God was not involved, God in the form of an integrative purpose to existence *was*. While Darwin's idea of natural selection did not recognise the involvement of integrative purpose in change, the concept of natural selection did not preclude it. Natural selection simply recognised that some varieties of a species reproduced more than others. Whether those that reproduced more could be viewed as winners, as being 'fitter' or more worthwhile or 'better' than others, was not decided. Now that we can safely acknowledge integrative meaning, we can see that losing, in the sense of not reproducing, can be consistent with integration. Acts of unconditional selflessness, where an individual gives their life for the maintenance of the larger whole, and, as a result, does not reproduce, can be very meaningful, a fitter, 'better' way of behaving. As a denial-free thinker or prophet, Christ was able to acknowledge the truth of the importance of selflessness when he said, **'Greater love has no-one than this, that one lay down his life for his friends'** (John 15:13). Again, the old Christian word for love was **'caritas'**, which means charity or giving or selflessness. In this light, what Christ said can be interpreted as 'greater selflessness has no one than to behave selflessly'.

Social Darwinism became further refined with the publication of Edward O. Wilson's 1975 book *Sociobiology: The New Synthesis,* a text that claimed to be **'the systematic study of the biological basis of all social behavior'**, and asked readers in its final chapter to **'consider man in the free spirit of natural history, as though we were zoologists from another planet.'** Essentially, Wilson's theory of Sociobiology proposed that human selfishness is due to humans' need to perpetuate their genes.

The final refinement of this evasive, contrived excuse that 'genes are selfish and that's why we humans are' has been the emergence in recent years of the theory of 'Evolutionary Psychology'. This theory argues that even acts of altruism amongst humans can be explained in terms of genetic selfishness. In his 1994 book *The Moral Animal: Why We Are The Way We Are—The New Science of Evolutionary Psychology,* Robert Wright attributed a cooperative inclination in humans—our **'morals'**—to biological situations of reciprocity, to situations where animals cooperate for mutual benefit. He maintained that acts of selflessness amongst humans are really acts of biological selfishness, of our genes 'saying' 'I'll scratch your back on the condition you scratch mine'. With this theory, denial-compliant biologists finally found a means to misportray the cooperatively orientated and ideal-world-aware instinctive self or soul in humans as being nothing more than an expression of a subtle form of selfishness in the human make-up!

In his 1998 book, *Consilience: The Unity of Knowledge,* Edward O. Wilson took the art of denial to its absolute extreme point, suggesting that Evolutionary Psychology's supposed ability to explain the moral aspects of humans means that biology and philosophy, the sciences and the humanities, could be reconciled. He talked about **'the attempted linkage of the sciences and humanities...of consilience, literally a "jumping together" of knowledge...to create a common groundwork of explanation'** (p.6 of 374), and went so far as to say, **'The strongest appeal of consilience is...the value of understanding the human condition with a higher degree of certainty'** (ibid. p.7). An extract from *Consilience,* published in the prestigious journal *The Atlantic Monthly* (Apr. 1998), and boldly titled **'The Biological Basis of Morality'**, was introduced thus: **'Philosophers and theologians have almost always conceived of moral instincts as being transcendent or God-given. Is it possible, though, that ethical reasoning derives not from outside but from our very nature as evolving material creatures?'** To illustrate just how bold Wilson was in his claims to have made sense of the philosophical aspect of human life using biology, one of the headings used in the extract was **'The Origins of Religion'**. Religions have been the custodians—albeit using abstract, metaphysical terms—of the truth of the existence of the Godly ideals of life, and all the other great truths associated with those ideals, in particular the truth that within humans there exists a soul and spirit embued with awareness of the Godly, integrative, ideal state. These truths *can* be explained biologically, and **'the human condition'**, the dilemma of the existence of good and evil in the human make-up, *can also* be explained biologically, *but to use biological lies to explain them is an outrage;* the ultimate example of deceit, dishonesty, delusion and alienation from the truth.

Having dismissed the human soul as merely a subtle form of selfishness, Wilson brazenly summarised by saying **'Rousseau claimed,** [that humanity] **was originally a race of noble savages in a peaceful state of nature, who were later corrupted...** [but what] **Rousseau invented** [was] **a stunningly inaccurate form of anthropology'** (*Consilience,* 1998, p.37 of 374). The 'stunningly inaccurate form of anthropology' is in fact Evolutionary Psychology. It is Jean-Jacques Rousseau who was stunningly accurate in his belief that humans have a pure, altruistic instinctive awareness within them. As emphasised, it is the ultimate lie to claim that the magic (integratively-orientated, non-alienated and thus all-sensitive) 'child' within us (our original instinctive self) is nothing more than a conniving, selfish, Machiavellian manipulator. Wilson, who is a professor at Harvard University, has twice won the Pulitzer Prize, basically for being the lord of lying, the supremo of anti-soul—and, since Christ was the living expression of our soul, also a leading expression of the Antichrist. In the following quote, Randolph Nesse, Professor of Psychiatry and Psychology at the University of Michigan, expresses a justified feeling of alarm and revulsion towards the theory of Evolutionary Psychology: **'The discovery that tendencies to altruism are shaped by benefits to genes is one of the most disturbing in the history of science. When I first grasped it, I slept badly for many nights, trying to find some alternative that did not so roughly challenge my sense of good and evil. Understanding this discovery can undermine commitment to morality—it seems silly to restrain oneself if moral behavior is just another strategy for advancing the interests of one's genes. Some students, I am embarrassed to say, have left my courses with a naïve notion of the selfish-gene theory that seemed to them to justify selfish behavior, despite my best efforts to explain the naturalistic fallacy'** (*The Origins of Virtue,* Matt Ridley, 1996, p.126 of 295). The position of humans under the duress of the human condition has been that they would rather destroy **'commitment to morality'** than face the truth of integrative meaning. Such has been the extent of human insecurity, and such has been the degree of alienation or lying or denial that has developed on Earth.

To summarise, genetic reproduction is an *integrative* process, a way of developing the order of matter on Earth. It is *not* a 'survival of the fittest', divisive process, as Social Darwinism, Sociobiology and Evolutionary Psychology have taught. As a tool for integrating matter, the *limitation* of genetics is that it normally cannot develop unconditionally selfless traits. Unconditional selflessness—the ability to consider the good of the whole above the good of self—is the ultimate integrative trait for parts of a whole to have, and the inability to develop it is a serious limitation in the development of larger wholes.

The reason the genetic tool for developing order cannot normally develop unconditional selflessness is that for a genetic trait to carry on from generation to generation it has to be selfish. For example, if an animal is born with a genetic inclination to unconditionally selflessly give its life whenever its group is threatened, then, when it does give its life, that unconditionally selfless trait disappears. Unconditionally selfless traits normally cannot become established genetically. (The exception was the development of unconditional selflessness through love-indoctrination.) Selflessness can only develop genetically up to the situation of reciprocity, because reciprocity is basically a selfish trait. It is a *limitation* of genetic refinement that unconditional selflessness cannot normally develop. This inability *does not* mean that the meaning of existence is to be selfish, as these contrived excuses maintain.

A more comprehensive list of the contrived excuses that have been used for humans' divisive behaviour include the original excuse that 'animals are red in tooth and claw and that's why humans are'; Social Darwinism, the need to compete for survival; B.F. Skinner's Operant Conditioning Theory which argued that man is a slave to reward and punishment; Konrad Lorenz's Theory which excused humans' divisive behaviour by saying it is stereotyped and the product of past experiences—that it is instinctive; Robert Ardrey's Theory which stated human competitiveness was due to an imperative need to defend their territory; Edward O. Wilson's Sociobiology Theory, which argued that our selfishness is due to humans' need to perpetuate their genes; Chaos Theory, with its emphasis on the world being chaotic rather than ordered; and Evolutionary Psychology, with its use of reciprocity to account for any acts of altruism in human behaviour.

A6: A brief description of how our soul became corrupted

The manner in which humans' conscious search for knowledge corrupted their original, innocent, loving, cooperative instinctive soulful state, creating their corrupted angry, egocentric and alienated behaviour, is the central explanation presented in *Beyond*. What follows is a brief summary of that explanation.

Understanding how humans became corrupted—explaining the so-called 'origin of sin'—depends on appreciating the difference between the gene-based and nerve-based learning systems.

What distinguishes humans from other animals is that we are fully conscious, able to understand the relationship between cause and effect. Consciousness is a product of the nerve-based learning system's ability to remember. It is memory which allows understanding of cause and effect to develop. Once you can remember past events you can compare them with current events and identify common or regularly occurring patterns. This knowledge of, or 'insight' into, what has commonly occurred in the past enables you to make predictions about what is likely to occur in the future—and with feedback or 'experience' these predictions can be refined. Sufficiently developed, this capacity to understand the relationship of events that occur through time gives rise to the ability to self-adjust. Being aware or 'conscious' of how experiences are related puts the intellect in a position to manage events to its own chosen ends. It can wrest management of life from the instincts.

Humans became conscious some 2 million years ago. How this happened is explained in *Beyond* in the chapter 'How We Acquired Consciousness'. Basically it was the development of a truthful mind, a mind trained in unconditional selflessness, which occurred following the development of love-indoctrination, that liberated consciousness in humans. In the explanation of 'How we acquired consciousness' in *Beyond,* what is revealed is that the human mind has been alienated from truthful, effective thinking—kept ignorant—twice in its history. The first time was when we were like other animals, genetically distanced from thinking selflessly. Normally unconditionally selfless behaviour (that is, outside the nurturing, love-indoctrination situation) led to the elimination of the individual who behaved that way, with the result that over time only individuals who were not inclined to think that way, individuals who had a genetic 'block' against such selfless thinking, survived. As soon as a mind begins to become conscious, able to understand cause and effect, it begins to realise that selflessness is a meaningful way of behaving, but such thinking becomes discouraged through natural selection, with the result that full consciousness can never develop. The human mind was kept from truthful and effective thought a second time when, approximately 1.5 million years ago, the duress of living with the human condition forced adults to psychologically live in denial of selfless-meaning acknowledging thought. This essay is primarily concerned with explaining how alienation has (deliberately as it turns out) kept the human mind ignorant, unable to think properly. With the advent of denial of the human condition, the human mind has retreated from consciousness into virtual unconsciousness. We are now nearly as mentally incognisant as animals. In fact there is an animated cartoon called *Wallace & Gromit* that plays on this state of affairs. Wallace is a lonely, sad—alienated—human figure whose dog

Gromit is very much on an intellectual par with him in his world. They both have the same blank, stupefied expression and muddle their way through life's adventures together.

Wallace & Gromit is cowritten by Nick Park & Bob Baker & produced by Aardman Animations

Incidentally, the suggestion that it was 2 million years ago that the ancestral human mind was liberated from the situation that other animals are presently in, where they have a block in their mind stopping truthful, effective thinking, comes from many indicators. For instance, the spread of our human ancestors from their ancestral home in central Africa began approximately 2 million years ago. This ability to wander and, no doubt, wonder 'what lay over the hill' strongly suggests consciousness, the ability to self-manage, had emerged. There would also have been an emerging need for self-distraction from the human condition which new realms would have helped supply. Also, while brain size is only an indicator of intelligence, the sudden and rapid increase in brain volume of humans and their ancestors from 500 cc to 1400 cc where it is today, began around 2 million years ago. Interestingly anthropologists have long recognised that the average brain volume of humans has not significantly increased in half a million years, the rapid increase in brain volume having ceased at 1400 cc at this time. Anthropologists have not been able to explain this phenomenon, but with understanding of the human condition and an appreciation of the fact that the more intelligent a person is, the sooner and more defiantly do they challenge their instinctive self, we can now see that eventually a balance had to be struck between answer-finding but corrupting cleverness and soul-obedient soundness. That particular level of IQ or intelligence quotient seemingly sets a limit on how much brain power and therefore brain volume humans have been able to develop.

To return to the issue of how our soul became corrupted: prior to becoming fully conscious and able to self-manage—consciously decide how to behave—humans were controlled by and obedient to their instincts, as other animals still are. As novelist Aldous Huxley pointed out: **'Non-rational creatures do not look before or after, but live in the animal eternity of a perpetual present; instinct is their animal grace and constant inspiration; and they are never tempted to live otherwise than in accord with their own...immanent law'** (*The Perennial Philosophy,* 1946).

Unlike the nerve-based learning system, the gene-based learning system is not insightful and cannot become conscious of the relationship of events that occur through time. Genetic selection gives animals adaptions or orientations—instinctive programming—for managing their lives, but those genetic orientations, those instincts, are not understandings. When our conscious mind emerged it was not enough for it to be oriented by instincts. It *had to* find understanding to operate effectively and fulfil its great potential to manage life. The problem is that when it began to experiment in the management of life from a basis of understanding in the presence of already established instinctive behavioural orientations, a battle broke out between the two.

Our intellect began to experiment in understanding as the only way of finding out the correct and incorrect understandings for managing existence, but the instincts, being in effect 'unaware' or 'ignorant' of the intellect's need to carry out these experiments, 'opposed' any understanding-produced deviations from the established instinctive orientations. The instincts in effect 'criticised' and 'tried to stop' the conscious mind's necessary search for knowledge. Unable to understand and thus explain why these experiments in self-adjustment were necessary, the intellect was unable to refute this implicit criticism from the instincts. The unjust criticism from the instincts 'upset' the intellect and left the intellect no choice but simply to defy the 'opposition' from the instincts.

The intellect's defiance expressed itself in a number of ways. It attacked the instincts' *unjust criticism,* tried to deny or block from its mind the instincts' *unjust criticism,* and tried to prove the instincts' *unjust criticism* wrong. Humans' upset angry, alienated and egocentric state—precisely the divisive condition we humans suffer from—

appeared. (Note, the dictionary defines 'ego' as 'conscious thinking self', so ego is another word for the intellect. The word 'egocentric' then means that the intellect became centred or focused on trying to prove the instincts' criticism wrong; it became focused on trying to prove its worth, prove that it was good and not bad.)

This conflict was then greatly compounded by the fact that the angry and aggressive behaviour was completely at odds with humans' particular instinctive orientation, which was to behave lovingly and cooperatively. From an initial state of upset humans then had to contend with a sense of guilt which greatly compounded their insecurity and frustrations, making them even more angry, egocentric and alienated. This escalating situation could only be ended by the dignifying, relieving understanding of why we became upset in the first place—an understanding that depended on the arrival of science and the ability to explain the differing natures of the gene-based and nerve-based learning systems.

A portion of Wordsworth's astonishingly truthful poem, *Intimations of Immortality*, was included earlier. After describing the soul's world and humans' loss of it, Wordsworth intimated at the reason for our loss of innocence, writing, **'High instincts before which our mortal Nature/Did tremble like a guilty thing surprised'**.

Eugène Marais, who was the first to study primates in their natural habitat, described the emergence of the conflict between instincts and intellect in his remarkable 1930s book, *The Soul of the Ape*: **'The great frontier between the two types of mentality is the line which separates non-primate mammals from apes and monkeys. On one side of that line behaviour is dominated by hereditary memory, and on the other by individual causal memory...The phyletic history of the primate soul can clearly be traced in the mental evolution of the human child. The highest primate, man, is born an instinctive animal. All its behaviour for a long period after birth is dominated by the instinctive mentality...As the...individual memory slowly emerges, the instinctive soul becomes just as slowly submerged...For a time it is almost as though there were a struggle between the two'** (written in 1930s, first pub. 1969, pp.77–79 of 170).

The biblical story of the Garden of Eden contains an acknowledgment that it was the emergence of consciousness and the resulting necessary search for knowledge that led to humans' corrupted, 'fallen' state. In Genesis it says, **'God did say, "you must not eat fruit from the tree that is in the middle of the garden...or you will die...** [although it was also pointed out that] **God knows that when you eat of it your eyes**

will be opened, and you will be like God, knowing good and evil...the fruit of the tree was...desirable for gaining wisdom"' (3:3,5,6). Conscious humans had to defy their integrative meaning-orientated, Godly instinctive self—take the forbidden fruit from the tree of knowledge—in order to eventually become knowing or understanding of the Godly, integrative world.

In Zen and Buddhism there is acknowledgment that the corruption of humans' innocent instinctive state was caused by the interference of the conscious mind. In Zen and Buddhism the loss of innocence is called **'the affective contamination (*klesha*)'** or **'the interference of the conscious mind predominated by intellection (*vijnana*)'** (*Zen Buddhism & Psychoanalysis,* D.J Suzuki, Erich Fromm, Richard Demartino, 1960, p.20).

This extraordinarily honest quote from the writings of Sir Laurens van der Post reveals the truth of just how corrupted humans have become, and of the estrangement they now feel: **'This shrill, brittle, self-important life of today is by comparison a graveyard where the living are dead and the dead are alive and talking** [through our soul] **in the still, small, clear voice of a love and trust in life that we have for the moment lost...** [there was a time when] **All on earth and in the universe were still members and family of the early race seeking comfort and warmth through the long, cold night before the dawning of individual consciousness in a togetherness which still gnaws like an unappeasable homesickness at the base of the human heart'** (*Testament to the Bushmen,* 1984, pp.127–128 of 176).

In a quote from the writings of Carl Jung that was included earlier, Jung alluded to how much of the magic, true world humans lost access to when they became conscious. He wrote of **'that cosmic night which was psyche** [soul] **long before there was any ego consciousness'** (*Civilization in Transition,* The Collected Works of C.G. Jung, Vol.10, 1945).

In *Beyond* there are numerous quotes from literature that recognise that the emergence of consciousness and the conflict that ensued with humans' already established instinctive self was the cause of humans' corrupted condition.

It is this biological explanation of humans' corrupted condition that liberates humans from the sense of guilt that has plagued them for 2 million years, and caused them to have to live in a cave-like state of denial and alienation. The historic 'burden of guilt' has been lifted from the human race. We can at last understand that there was a sound (ie integrative) biological reason for why humans became divisive and corrupted.

Again, as was emphasised in the *Introduction,* the finding of understanding of humans' non-ideal, upset, corrupted, divisive behaviour does not condone such behaviour, it does not sanction 'evil'; rather, through bringing compassion to the situation, it allows the insecurity that produces such behaviour to subside, and the behaviour to disappear.

A7: There were people who could live in the sun's light, but they were unbearably condemning for the cave prisoners

As explained, the corruption of the human soul began some 2 million years ago with the emergence of consciousness, yet we can expect it to only have become well entrenched roughly one million years ago. This means that the great majority of our species' ancestry, from approximately 8 million years ago (when primates had become sufficiently adapted to bipedal, arms-free walking to be able to hold a helpless infant and thus develop love-indoctrination) to one million years ago, was spent living in a predominantly cooperative, loving state. Therefore our species' instinctive heritage, and thus instinctive expectation, is essentially one of encountering a cooperative, loving world, a world that behaviourally is almost totally at odds with the world humans are now born into.

The corruption that began with the emergence of consciousness has been accumulating ever since. Each new generation of humans has entered a world that was already corrupt and their instinctive self or soul, as a result, has been compromised and corrupted by the encounter. The level or extent of corruption in the world has been increasing as each generation added to the existing corruption the corruption it incurred from its own conscious search for knowledge. While self-restraint could, to a degree, contain the corruption and its propagation, it was ultimately, only the reconciling, dignifying and ameliorating understanding of the reason for this corrupt behaviour that could end the cumulative process.

What this means is that all adults today are going to be corrupted to various degrees from encounters with the corruption already present in the world. It follows that those who have had less encounters in their upbringing with the corrupted angry, egocentric and alienated human condition-afflicted behaviour of humans are going to be less corrupted than those who have experienced more. People who have been more sheltered from the corruption in the world during their upbringing, more nurtured in their infancy and childhood, are less soul-damaged, less departed from our species' original instinctive orientation to the cooperative, integrative ideals of life and, as a result have had to use less denial of those ideals to cope with them.

With some people less corrupted than others, it also follows that amongst the millions of people on Earth, at any one time there were always going to be a few people who were sufficiently uncorrupted—sufficiently free of insecure competitive egocentricity, anger and alienation—not to be condemned by the integrative, cooperative, loving, selfless ideals of life, who thus did not have to live in denial of those ideals. If you are not at odds with the ideals you do not have to deny them. Plato hypothesised such innocence thus, **'But suppose...that such natures were cut loose, when they were still children, from the dead weight of worldliness, fastened on them by sensual indulgences like gluttony, which distorts their minds' vision to lower things, and suppose that when so freed they were turned towards the truth, then the same faculty in them would have as keen a vision of truth as it has of the objects on which it is at present turned'** (p.284).

Plato's reference to **'four states'** of effectiveness in thinking of the human mind was mentioned earlier when explaining the spectrum of denial in the world. These various states of denial or alienation and resulting effectiveness in thinking are a result of humans' different encounters during their upbringing with humanity's corrupting struggle with the human condition. Variously in denial and thus variously separated or alienated from the truth, humans have therefore been living in various states of dishonesty and delusion.

Those who were exceptionally fortunate in escaping encounters with the **'worldly'** products of humanity's necessary battle with the human condition, those who had few encounters with the anger, egocentricity and alienation that exists to various degrees within people, the very few who retained exceptional innocence, could face the **'sun/fire'**—face the truth of the cooperative, integrative ideals of life, or God, and the issue that arose immediately, of 'why aren't humans innocent?'—without feeling confronted by their own lack of innocence. Only innocent people have been able to confront the subject of the human condition with impunity. Given the extent of the corruption in humans now as a result of humanity's long struggle with the human condition, the chances of someone receiving an upbringing that avoids encounter with that corruption are extremely slim, so the occurrence of exceptional innocence now is extremely rare.

In Plato's symbolism, these people who could confront the integrative ideals were those who could live outside the cave in the illuminating sunlight. Historically individuals who were uncorrupted enough, sound enough, to confront the truth of integrative meaning and look into and talk truthfully about the human condition have been termed 'prophets'. As Wordsworth so eloquently said: **'Thou best Philosopher, who yet dost keep/Thy heritage, thou Eye among the blind/That, deaf and silent, read'st the eternal deep/Haunted for ever by the eternal mind/Mighty Prophet! Seer blest!/On whom those truths do rest/Which we are toiling all our lives to find/In darkness lost, the darkness of the grave'**. Without the ability to recognise the issue of the human condition, and all the vastly differing states of corruption and its outcome, denial and alienation, it was not possible to demystify prophets; but now we can, and must. They are not supernatural beings, simply people at one end of the spectrum of alienation that existed in a world struggling with the dilemma of the human condition, with their own particular contribution to make in the human journey.

Moses was an innocent, unevasive, denial-free thinker or prophet, and, as it says in the *Bible*, **'no prophet has risen in Israel like Moses, whom the Lord knew face to face'** (Deut. 34:10). Unlike most people, Moses was, to quote Isaiah, a person able to **'delight in the fear of the Lord'** (Isa. 11:3). Jacob was another exceptionally innocent person and marvelled that he was able to confront the truth of integrative meaning and the resulting dilemma of the human condition and survive, saying, **'I have seen God face to face and yet I am still alive'** (Genesis 32:30). Moses described how **'The Lord spoke to you** [the Israelite nation] **face to face out of the fire on the mountain.** [This was possible only because] **At that time I stood between the Lord and you to declare to you the word of the Lord, because you were afraid of the fire'** (Deut. 5:4,5). (Note the use of the fire metaphor again here for the condemning cooperative, integrative ideals, or God.)

Thus, there have always been rare individuals who could confront the truth of integrative meaning or God, individuals who could live outside the cave of denial in **'the light of day.'** While some of them, such as Moses, lived among people who could tolerate them, their truthfulness was normally too confronting and condemning for those around them who were having to live in denial of the human condition. As a result, prophets were often persecuted. As it says in the *Bible* **'was there ever a prophet your fathers did not persecute?'** (Acts 7:52).

Plato's *Republic* is concerned with the questions of **'what is a just state?'**, and **'who is a just individual?'** *(Encarta)*. In addressing these questions Plato argued that the achievement of a just society required an ideal type of leader, whom he referred to variously as **'philosopher rulers'** or **'philosopher kings'** or **'philosopher princes'** or **'philosopher guardians'**, by which he meant **'the true philosophers, those whose passion is to see the truth'** (p.238), or, to quote from the *Encarta* entry, individuals whose **'minds have been so developed that they are able to grasp the Forms and, therefore, to make the wisest decisions'**—individuals who are referred to in this and my earlier books as sound, denial-free, unevasive thinkers. The truth is the ideal leader is someone who can explain situations sufficiently clearly for people to be able to manage situations from their own understanding. The ideal leader is a guide and educator. The *Encarta* text continues, **'Indeed, Plato's ideal education system is primarily structured so as to produce philosopher-kings.'**

While the ideal *is* to be able to live in the presence of the **'sun'**, or the **'Form of the Good'**, or the truth of integrative meaning—look at **'God face to face'**, and be able to think truthfully, effectively and thus intelligibly, the reality is that such people **'whose passion is to see the truth'** have been very rare and their denial-free, unevasive truthfulness far too confronting for 'normal', human condition-afflicted, soul-corrupted, angry, egocentric and alienated people. The fact is that most people have had the exact opposite of a **'passion to see the truth'**, their passion has been to avoid and deny it.

Plato argues that, **'If you get, in public affairs, men who are so morally impoverished that they have nothing they can contribute themselves, but who hope to snatch some compensation for their own inadequacy from a political career, there can never be good government. They start fighting for power...** [whereas those who pursue a life] **of true philosophy which looks down on political power...** [should be] **the only men to get power...men who do not love it...rulers** [who] **come to their duties with least enthusiasm.'** Plato says that people who have returned from seeing clearly in the sun and become **'used to seeing in the dark'** are greatly advantaged because **'once you get used to it you will see a thousand times better than they** [the cave prisoners] **do and will recognize the various shadows, and know what they are shadows of, because you have seen the**

truth about things right and just and good' (pp.286,285).

Plato's idea, that he summarised when he said, **'isn't it obvious whether it's better for a blind man or a clear-sighted one to keep an eye on anything?'** (p.244), is all very well, but it has been largely unworkable due to its extreme idealism. Ideally society would select for **'philosopher rulers or guardians'**, for prophets, for people who could confront **'the Good'** and thus think truthfully, free of distortion. However, without the explanation that ameliorates the human condition and explains and dignifies corrupted humans, differentiating in society according to people's soundness or level of alienation only led to prejudice, to the more corrupted and alienated being treated and made to feel that they were evil, bad, worthless and inferior. Only with understanding of the human condition, only with the biological explanation for why humans unavoidably became corrupted, does it become safe to acknowledge the different degrees of corruption and alienation among humans. To even acknowledge the existence of alienation while humanity lacked the biological reason for it, was unjustly and thus unbearably condemning of most people. It was almost impossible to mention anything to do with the human condition until it could be explained. Almost all humans *had to* live in a **'dark cave'** of denial.

The problem with the concept of having **'philosopher rulers or guardians'**—prophets—as leaders of society or, in terms of education, of deliberately cultivating such leaders, was that it was *only* an ideal. It could not easily work in practice. The soundness of truthful thinkers or prophets was so confronting for everyone around them that as a consequence, instead of cultivating them through the education system and letting them lead society, they were normally not cultivated in education and, as will shortly be described, even eliminated from society.

Indeed, how much innocence and its soundness an individual could tolerate being confronted with, or a society could tolerate in its midst depended on that individual or society's own level of soundness. The more corrupted and alienated they were, the more hurtfully condemning innocence was and the less they could afford to acknowledge innocence. Therefore, a relatively innocent and sound society could tolerate being led by prophets. The early Athenian society must have been composed of relatively innocent people because it elected only uncorrupted, innocent shepherds to run its society, people whose lifestyle kept them isolated from the corrupting world and kept them close to nature, the home of our instinctive self or soul, and thus reinforcing of that soul. Indeed the prophet Mohammed observed **'that every prophet was a shepherd in his youth'** (*Eastern Definitions*, Edward Rice, 1978, p.260 of 433). Sir Laurens van der Post notes that in the turbulent period of Plato's time Pericles, a close friend of Plato's stepfather, **'urged the Athenians therefore to go back to their ancient rule of choosing men who lived on and off the land and were reluctant to spend their lives in towns, and prepared to serve them purely out of sense of public duty and not like their present rulers who did so uniquely for personal power and advancement'** (Foreword to *Progress Without Loss of Soul*, by Theodor Abt, 1983, p.xii of 389).

Of course not being able to differentiate according to soundness or alienation has not stopped societies differentiating according to cleverness and egocentricity. The more clever and the more egocentric have almost always dominated innocence and its soundness. In fact innocence in all its forms everywhere has been ruthlessly, and often brutally, oppressed and repressed. The reality is that 'might has ruled over right'. Countries and industries are typically run by the most powerful, not by the most sound. In education you cannot enter tertiary education facilities without passing exams that essentially test your level of mental cleverness or IQ. We do not have exams to test a person's level of soundness or lack of alienation. (Incidentally, to say we do not know who is alienated and who is innocent is untrue. For example, to ignore, deny, repress and in the extreme persecute to the point even of crucifying innocence, as humans have done, they first had to recognise it. It would be as easy, indeed, probably much easier, to design exams that test for a person's level of alienation as it has been to design exams that test for people's level of IQ. The problem is not the ability to do so but the danger of doing so.)

The practice of allowing the clever and the powerful to rule has been permissible because to be made to feel inferior for not being clever or powerful was not anywhere near as psychologically devastating as having your alienation pointed out and exposed. Now that the human condition has been explained and we can at last safely acknowledge soundness without condemning those who are corrupted, soundness can, should and must come to the fore. As Christ anticipated, **'the meek...inherit the earth'** (Matt. 5:5); **'many who are first will be last, and many who are last will be first'** (Matt. 19:30, 20:16; Mark 10:31; Luke 13:30). When Christ said, **'The stone the builders rejected has become**

the capstone' (Ps. 118:22; Matt. 21:42; Mark 12:10; Luke 20:17; Acts 4:11; 1 Pet. 2:7) he also was referring to this time when innocence with all its soundness would come to the fore from its previously repressed and denied position. Sir Laurens van der Post referred to this biblical analogy when he anticipated this new situation, writing that, **'It is part of the great secret which Christ tried to pass on to us when He spoke of the "stone which the builders rejected" becoming the cornerstone of the building to come. The cornerstone of this new building of a war-less, non-racial world, too, I believe, must be...those aspects of life which we have despised and rejected for so long'** (*The Dark Eye in Africa,* 1955, p.155 of 159). It should be emphasised that this coming situation does not mean that the less innocent are now going to be oppressed as the innocent have been. With understanding of the human condition now available, the corrupt anger, egocentricity and alienation produced by the battle with the human condition will subside and eventually disappear from the human make-up forever. While this is happening the innocent will be able to understand and thus be compassionate towards those who are more embattled, those whose instinctive self or soul has been upset from encounters with humanity's necessary battle with the human condition.

With understanding of the human condition found, Plato's ideal of society being led by soundness can finally come to fruition. In the new human condition-understood, genuinely compassionate world, everyone will have a role based on the acknowledgment, rather than denial of each individual's level of soundness. (The different roles in this new world are depicted in the final cartoon in *Beyond* under the heading 'The Activities of the New World'. This cartoon is also reproduced in this book towards the end of the *Resignation* essay.)

While placing sound people in positions of leadership would seem the sensible thing to do, prior to finding understanding of the human condition the tragic reality was that, instead of leading society, innocence had to a degree to be repressed because of its unjust condemnation of those who were no longer innocent. The danger of excessive repression of innocence was that it led to an overly corrupt society, for while exceptionally innocent, truth-confronting thinkers or prophets threatened the world of denial, and thus were often persecuted and even murdered, they contributed the most soundness to society. They were in fact the strongest balancing influence a society could have.

Over the years, an effective means of reducing the condemning criticism of prophets was to assign to them a divine status and regard them as beings from some remote, ethereal realm, separate from the human world. If they were divine and not Earthly then humans could avoid hurtful comparisons with their own alienated selves, and could easily treat the prophets as figures to be revered, even worshipped. In fact, in pre-scientific times, religions were sometimes founded around the truthful words and lives of prophets. People deferred to the prophet's truthful, sound state and were thus 'born-again' to ideality. Religions offered people an immensely valuable way of living meaningful lives despite their alienated state.

Even prophets who became revered as deities after their death were often subjected to persecution and martyrdom during their life because of their exposing truthfulness. As mentioned, the *Bible* refers to this process when it asks, **'was there ever a prophet your fathers did not persecute?'** Like Plato, Christ used the metaphors of light and darkness to describe people's hate of the exposing truth, when he said: **'the light shines in the darkness but...everyone who does evil hates the light, and will not come into the light for fear that his deeds will be exposed'** (John 1:5, 3:20). Christ also knew how persecuted prophets could be—before being murdered himself, he lamented, **'O Jerusalem, Jerusalem, you who kill the prophets'** (Matt. 23:37).

In summary, there have always been a few exceptionally innocent people who could confront integrative meaning without being condemned by it, people we have historically referred to as prophets. While they could confront and think about the human condition—using Plato's symbolism, they could think with **'the aid of the sun'**—their truthfulness has been extremely confronting and condemning for the rest of humanity.

ASPECT B

In analysing how humanity liberates itself from the human condition it was established that there was a need for someone to be sufficiently free of corruption of soul and its associated denial to be able to confront and think truthfully and effectively about the issue of the human condition. In Plato's metaphor, it needed to be explained how someone **'escapes from the cave into the light of day'** *(Encarta)*. In answering that question there were three aspects to look at.

ASPECT A involved explaining that it was the human soul that was being corrupted, then explaining how we acquired our soul, how it became corrupted and the spectrum of alienation that resulted from that corruption.

We now need to look at ASPECT B, which is the particular knowledge that someone sound enough to confront the issue of the human condition would need if they were to assemble the liberating understanding of the human condition. In Plato's metaphor of the cave, we need to look at how someone, **'sees for the first time the real world'**.

ASPECT B has two sections. B1, 'Science—the liberator', explains how mechanistic science had to find all the details and mechanisms of the workings of our world without confronting the truth of integrative meaning. B2, 'Soul—the synthesiser', explains how a denial-free thinker or prophet has to take the hard-won insights about the workings of our world that mechanistic science found, and from these insights synthesise the liberating explanation of the human condition.

B1: Science—the liberator

We have established that there always existed a few people who could confront and look into the issue of the human condition, however this ability was obviously not enough to solve the human condition otherwise it would have been achieved long ago.

The other component required to solve the human condition, and **'see for the first time the real world'**, was knowledge.

In *The Republic,* Plato frequently emphasises that the two ingredients needed for wisdom are nurture and education. For instance he talks about **'the one great thing,—a thing, however, which I would rather call, not great, but sufficient for our purpose...**[is] **education and nurture...for good nurture and good education implant good constitutions'**. (This quote is from Benjamin Jowett's 1877 translation of *The Republic.* Lee, in his translation of these passages, which appear on page 169, uses 'well brought up' and 'sound character', instead of 'nurture'.) All humans were once innocent—living outside the cave—and when the dilemma of the human condition emerged with consciousness some 2 million years ago they could have avoided having to live in denial of it—having to hide in the cave—if it were not for the fact that they did not have the knowledge with which to explain the human condition. Nurtured soundness alone was not enough, knowledge was also required. The overall journey that humanity has been traversing has been the journey to find sufficient knowledge to be able to explain the human condition, clarify the issue of whether humans are fundamentally good or evil, find the biological *reason* for humans' divisive behaviour. It has been a journey from ignorance to understanding, a journey in search of knowledge.

An innocent individual today is in the same predicament as the entire human race was in when all its members were innocent, in that no one could explain the human condition until there was sufficient knowledge available from which to assemble that explanation. Before an innocent individual could synthesise the liberating explanation of the human condition, humanity first had to find sufficient knowledge to make that synthesis possible. That was the plan, although it could never be acknowledged, for doing so would require admission of the existence of humans' non-ideal state and confrontation with the problem of the human condition—an impossible ask prior to solving the human condition.

Therefore, the primary task facing humanity was to find the knowledge that would make it possible to explain the human condition—of course the difficulty with this problem was how were you to find knowledge when you could not face the truth of integrative meaning and thus think truthfully and effectively? It was a catch-22.

In the words of the *Encarta* summary of the cave allegory, Plato emphasised that **'the proper object of knowledge'** was to achieve **'the transition to the real world'**; that is, end the alienated state of denial that humans have had to live in, get out of the cave, solve the human condition. Historian Jacob Bronowski stressed this objective of having to find understanding of ourselves, in his 1973 television series and book, *The Ascent of Man,* saying, **'We are nature's unique experiment to make the rational intelligence prove itself sounder than the reflex** [instinct]**. Knowledge is our destiny. Self-knowledge, at last bringing together the experience of the arts and the explanations of science, waits ahead of us'** (p.437 of 448).

While religions played a crucial role in sustaining humans who were living under the duress of the human condition, they could not lift the burden of guilt from humanity, find the dignifying biological understanding of human nature. The poet AlexanderPope said in his 1733 *Essay on Man,* **'Know then thyself, presume not God to scan/The proper study of Mankind is Man'**. Pope's admonition that we should not leave it to **'God to scan'** made the point that faith was not going to be sufficient. While religious assurances such as 'God loves you' could comfort us, ultimately we had to *understand why* we were lovable. There had to be a biological explanation for humans' divisive behaviour and our responsibility as conscious animals was to find that explanation. The ancients were right to have emblazoned across their temples the phrase, **'Man, know thyself'**. The biologist, Edward O. Wilson, was more specific when he said, **'The human condition is the most important frontier of the natural sciences'** (*Consilience,* 1998, p.298 of 374), and **'Biology is the key to human nature'** (*On Human Nature,* ch. 1, 1978).

Humanity's overriding responsibility and task *was* to liberate itself from the human condition, find the dignifying biological explanation for humans' divisive nature and by doing so end the historic state of denial—leave the cave. Solving the human condition *was* the objective; however, the difficulty to be overcome was *how* to go about achieving that objective while almost the entire human race was unable to confront the sun/fire, unable to confront the condemning truth of integrative meaning, which was what gave rise to the issue of the human condition in the first place.

Clearly, what was required was a way of investigating reality without confronting the truth of integrative meaning and the suicidally depressing issue of the human condition that followed. As is fully explained in *Beyond* in the chapter 'Science and Religion', the means humanity developed for this purpose was *mechanistic science.* Science has been mechanistic, *not* holistic, in its approach to inquiry. It has focused on the details and mechanisms of the workings of our world while all the time carefully avoiding the holistic view of the confronting truth of the cooperative or integrative meaning of existence.

Science *had to* comply with humans' need to live in denial of the fundamental truth of integrative meaning, it had to be mechanistic in its approach to inquiry, because humans could not be exposed to the potentially suicidally depressing truth of integrative meaning until the compassionate, dignifying, ameliorating understanding of the human condition was found.

It has been pointed out that having to investigate reality without confronting the human reality—the human condition—was an extremely difficult task, and a loathsome one because you could not admit and talk about your plan. You had to live a life of lying, never able to explain why you were lying. You were living as if there were no meaning to life or to your endeavour, when in fact there was a crucially important direction to life and to everything both scientists and humanity in general were doing.

The final episode of *Evolution,* a television series produced in 2001, examined the controversy in American schools and universities over the teaching of 'natural selection' as a godless, meaningless, blind process. The program's title, *What about God?,* questioned why 'God' is left out of science's interpretation of existence? The answer is that integrative meaning was left out for humans' own good, it saved them from suicidal depression. Leaving the concept of God abstract, and undefined in scientific terms, saved humans from direct confrontation with the truth of integrative meaning, a confrontation they could not survive until understanding of humans' divisive nature was found. Now this understanding has been found, science can acknowledge that 'natural selection' is in fact dedicated to ordering matter into larger and more stable wholes, that 'evolution' is an integrative, directed, meaningful, Godly process. The French scientist-prophet Pierre Teilhard de Chardin anticipated the time when science could stop practicing denial, when writing in 1938: **'I can see a direction and a line of progress for life, a line and a direction which are in fact so well marked that I am convinced their reality will be universally admitted by the science of tomorrow'** (*The Phenomenon of Man,* 1955; tr. Bernard Wall, 1959, p.142 of 320).

In recent times the concept of 'Intelligent Design' in the universe has been introduced by those attempting to combat mechanistic science's godless, meaningless view of the universe. Again, until the dilemma of the human condition was solved, it was dangerous and irresponsible to demystify 'God', as the Intelligent Design movement has been attempting to do. The immense value of mechanistic science was that it allowed humanity to approach the issue of the human condition without confronting the dangerous truth of the integrative, cooperative meaning/design/purpose/theme/direction in existence. It allowed humanity to find all the details and mechanisms about the workings of our world without confronting humans with the unbearably depressing truth of integrative meaning. Once we had enough knowledge about the details of existence, clarifying explanation for humans' divisive behaviour could be synthesised and the dilemma of the human condition eliminated from human life—the 'fire' could be extinguished and humans could leave their 'cave' world of denial.

Of course the difficulty in that final task of assembling the truth about the human condition from all of mechanistic science's hard-won insights into the workings of our world, was that all this knowledge was presented in a manner that avoided any condemnation of humans. It was all evasive and denial-compliant in its orientation.

The following examples illustrate just some of the evasions or denials that humanity developed during this mechanistic process: Psychiatrists acknowledged that psychosis was a part of the human make-up but avoided acknowledging the human condition as the source of the psychosis. Mathematicians and scientists identified many of the laws of physics but avoided acknowledging that there was any purpose or theme to the physics of the universe. Biologists acknowledged that there was natural selection of organisms but avoided acknowledging that there was an integrative purpose behind the emergence of the variety of life on Earth. Anthropologists postulated that the prime mover in the emergence of humans from our ape ancestor was, variously, the advent of language, or upright walking, or tool use, or the mastery of fire, etc, etc, when the real prime mover was nurturing or loving of our infants, a truth they had to evade because it confronted humans with their inability to nurture their infants, an inability that arose from their necessary preoccupation with the battle with the human condition. (The extent of humans' insecurity about their inability to nurture their offspring is evident in this quote, **'The biggest crime you can commit in our society is to be a failure as a parent...people would rather admit to being an axe murderer than being a bad father or mother'** [*Sun-Herald, Sunday Life* mag. 7 July 2002].) Primatologists emphasised an aggressive past for humans and evaded the truth of our cooperative, loving and selfless past because of its unjust condemnation of humans' competitive, aggressive and selfish present. Society avoided acknowledging the immense differences in alienation between individuals, sexes, generations, races and cultures—because it would have led to prejudice—and instead maintained the 'politically correct' lie that there were no significant differences in alienation between people. Individual humans lived in denial of their corrupted state because they could not face the truth of it.

Humanity accumulated a great deal of knowledge but all in a way that was safely considerate of humans' delicate state of insecurity. It constructed a sophisticated world of artful denial, a cave-like existence where humans were only able to see shadows of the real world.

B2: Soul—the synthesiser

The strategy of investigating reality while avoiding the truth of integrative meaning was all very necessary but at some stage that truth would have to be confronted if understanding of the human condition was to be found.

Mechanistic science, as the prime vehicle for humanity's inquiry into reality, complied with humans' need to live in denial of the issue of the human condition and all the truths associated with it. While it portrayed itself as rigorously objective and free of personal, subjective bias, mechanistic science was in truth an extremely subjective discipline. The real 'discipline' of mechanistic science was at all times to maintain the 'great lie' or denial. Investigating in this evasive manner, mechanistic science gradually found all the necessary details and mechanisms of the workings of our world, and subsequently presented them all in a safely evasive way. The problem then was, how could the truth about the human condition be synthesised from all this evasively presented information? Humanity was faced with another impasse, which was that you cannot assemble the truth from lies.

What was required was someone who could collect all the hard-won but evasively presented insights and while defying the evasive components, take the truthful components and from them synthesise the truth about the human condition, find the biological explanation for humans' divisive nature.

As was mentioned in the *Introduction,* in Homer's Greek legend, *The Odyssey,* the prophet Teiresias predicted that on returning home to Ithaca after the Trojan War, Odysseus (or 'Ulysses' in the later Roman version) would have to undertake one final journey, this time into a desperately barren land. Odysseus told his wife Penelope, **'Teiresias bade me travel far and wide, carrying an oar, till I came to a country where the people have never heard of the sea and do not even mix salt with their food'**. The sea is a metaphor for humans' innocent instinctive self or soul that became repressed because of its unjust condemnation of humans' divisive state. The sea's salt is our innocent instinctive self or soul's immense sensitivity, its access to all the beauty and magic of life—unlike the numb, seared, flavourless, 'saltless', alienated state. Therefore this **'country where the people have never**

heard of the sea and do not even mix salt with their food' is the soul-destroyed, alienated world of denial where humans currently reside. In Plato's terms it is humans' dark, imprisoned cave existence. The **'oar'**, Teiresias explained to Odysseus, is actually a **'winnowing shovel'**. This other journey then, that Teiresias was predicting humanity would eventually have to undertake, was into the centre of this alienated cave world humans now live in, in order to winnow from its evasive, mechanistic scientific insights the unevasive, holistic scientific truth about the human condition.

While mechanistic science had to arduously find all the details and mechanisms of the workings of our world experiment by experiment, clue by clue—all the while making sure not to confront humans with unjust condemnation—at the very end of that process someone living unevasively in the presence of the illuminating sunlight, had to appear and **'winnow'** out the truth about the human condition.

Mechanistic science had to do all the ground work—find, as it were, all the pieces of the jigsaw of explanation—but because it could not look at those pieces picture-side-up or truthfully, it could not put the jigsaw together to reveal the true picture or story about ourselves. Only someone sufficiently innocent not to need to be living in denial—a denial-free, unevasive thinker or prophet—could look at the jigsaw pieces picture-side-up and thus be in a position to assemble the jigsaw, explain the human condition, and by so doing break free of, expose and end the historic denial. In Plato's cave allegory, you could not assemble the truth about the human condition in the dark cave, that had to be achieved outside the cave in the illuminating sunlight. While for most people the heat of the sun and the fire was searing (ie the truth of integrative meaning was unbearably condemning), the light from the sun/fire was required if you were to illuminate the world and make it intelligible.

If we recall the key scientific explanations given earlier—of negative entropy being the meaning of existence; of how nurturing's 'love-indoctrination' gave humans their instinctive self or soul that was orientated to behaving cooperatively; of how consciousness in humans was able to develop once love-indoctrination overcame the genetic block to thinking selflessly and thus effectively; and how the conflict between the already established instinctive orientation to behaving cooperatively and the intellect's search for knowledge led to humans' corrupted state—we can see that all these breakthrough ideas depended on being able to recognise truths that humans normally practice denying, in particular the truth of integrative meaning and the significance of nurturing. These key understandings could not be reached living in the cave state of denial.

It can be seen that, in terms of solving the human condition, prophets had a crucial, albeit minuscule, concluding role to play. In *Beyond* I describe the situation using the metaphor of a game of gridiron: 'In many ways prophets only got in the way while we were searching for understanding because they confronted us with truths that depressed us and which we therefore had to evade. Exceptional innocence played an important but minuscule concluding role in our search for knowledge. In gridiron football the team as a whole (with one exception) does all the hard work gaining yardage down the field. Finally when the side gets within kicking distance of the goal posts, a specialist kicker, who until then has played no part, is brought onto the field. While he—in his unsoiled attire—kicks the winning goal, the win clearly belongs to the exhausted players who did all the hard work' (p.163 of 203).

While historically honest thinkers or prophets 'got in the way', overly confronted and condemned humans, and as a result were repressed and even removed, or deified through the formation of religions based upon their sound world, the situation for prophets in the modern, contemporary, scientific age is very different. With the development of science, unevasive, truthful thinkers have a role, but it is the very opposite to that of condemning humans or of creating a religion. Their task now is to bring dignifying, ameliorating, rational understanding to humans, and in the process demystify religions and make the need for deferment of self to a faith obsolete. The role now for truthful thinkers is not to confront and condemn humans or lead them, but to give them understanding so they can at last understand themselves and become effective self-managers.

By definition, to have been able to grapple with the human condition and synthesise the understanding of it, I must have had the soundness required. I must be a contemporary prophet. There are many modern-day or contemporary prophets, people who think unevasively, holistically, who confront the truth of integrative meaning and bring attention to the issue of the human condition. They include in my experience such thinkers as Jean-Jacques Rousseau, William Blake, William Wordsworth, Arthur Schopenhauer, Charles Darwin, Søren Kierkegaard, Friedrich Nietzsche, Olive Schreiner,

Sigmund Freud, Eugène Marais, Nikolai Berdyaev, Carl Jung, Pierre Teilhard de Chardin, Kahlil Gibran, D.W. Winnicott, Antoine de Saint-Exupéry, Louis Leakey, Joseph Campbell, Erich Neumann, Arthur Koestler, Sir Laurens van der Post, Simone Weil, Albert Camus, Ilya Prigogine, Robert A. Johnson, R.D. Laing, Dian Fossey, Stuart Kauffman, and, in New Zealand's case, the biologist and theologian John Morton. In Australia's case, the great educator Sir James Darling and the writer A.B. 'Banjo' Paterson were able to think unevasively, holistically, while the Templeton Prize-winning Australian scientists, physicist Paul Davies and biologist Charles Birch, have both been described as, and are, prophets. Davies has been described as a **'latter day prophet'** (ABC-TV *Compass, God Only Knows,* 23 Mar. 1997), and Birch as a **'scientist-prophet'** *(Sydney Morning Herald,* 30 May 2000). Importantly, none of these people have been concerned with confronting and condemning humans or with creating a religion. They have been concerned only with an unevasive holistic approach to inquiry.

The human journey had arrived at the situation where mechanistic science had completed its role of finding sufficient understanding of the details and mechanisms of the workings of our world to make clarification of the human condition possible. In fact, the journey had stalled, it had achieved as much progress in inquiry as was possible from an evasive perspective. Being so fundamentally flawed in its orientation there was a limit to how much worthwhile knowledge mechanistic science could achieve. General Omar Bradley summarised the final imbalance inherent in the mechanistic approach when he said, **'The world has achieved brilliance...without conscience. Ours is a world of nuclear giants and ethical infants'** (Armistice Day Address, 10 Nov. 1948, *Collected Writings of General Omar N. Bradley,* Vol.1). Charles Birch also emphasised the limits of the integrative meaning-denying and soul-oppressing reductionist, mechanistic approach when he wrote that: **'Reductionism or Mechanism...is the dominant mode of science and is particularly applicable to biology as it is taught today...** [it is] **A view or model of livingness that leaves out feelings and consciousness...** [and] **I believe it has grave consequences...In the name of scientific objectivity we have been given an emasculated vision of the world and all that is in it. The wave of anti-science...is an extreme reaction to this malaise...I believe biologists and naturalists have a special responsibility to put another image before the world that does justice to the unity of life and all its manifestations of experience—aesthetic, religious and moral as well as intellectual and rational'** *(Two Ways of Interpreting Nature, Australian Natural History,* Vol.21 No.2, 1983).

On the subject of inquiry having ground to a halt from its own accumulated dishonesty, Charles Birch, during his FHA 1993 Open Day address in Sydney, commented that **'there is a problem about that** [mechanism], **it can't deal with certain questions...every individual entity, be it cell or an atom, and certainly human beings...is different by virtue of the relationships that they have with the whole that they belong to. Now that is the most important thing I think that one can begin to think about, the** [integrative, holistic] **nature of the world, the universe...and I think this is the sort of exploratory area which could transform a lot of thinking. In other words, science can't deal with subjectivity...This** [subjective, holistic, cooperative-meaning, human condition-confronting aspect] **is something that is very difficult to get your teeth into and yet it is the most important thing in the world...what we were all taught in universities for decades is really recognised now as pretty much a dead end'** (*FHA Newsletter 26*, 1993).

Biologist Mary E. Clark expressed a similar sentiment in her 1989 book *Ariadne's Thread: The Search for New Modes of Thinking*: **'Formal learning has become a meaningless vaccination process, and the information transmitted is next to useless for properly understanding the world.'**

Paul Davies recognised that a holistic approach to inquiry was necessary if humans were to re-integrate themselves with what Plato saw as, **'the real world, the world of full and perfect being'**, when he wrote: **'But there is a deeper reason for the wide-spread antipathy. It is connected with the underlying philosophy of science itself. For 300 years science has been dominated by extremely mechanistic thinking. According to this view of the world all physical systems are regarded as basically machines...I have little doubt that much of the alienation and demoralisation that people feel in our so-called scientific age stems from the bleak sterility of mechanistic thought...Mechanistic thought has undoubtedly had a stifling effect on the human spirit. Liberation from this centuries-old straight jacket** [ie the adoption of a holistic approach] **will enable human beings to re-integrate themselves and the physical world of which they are a part'** (*Living in a non-material world—the new scientific consciousness, The Australian*, 9 Oct. 1991).

In his 1999 book, *Biology and the Riddle of Life*, Charles Birch emphasised that **'the onus is now on biologists to demonstrate the importance of self-organisation in biological evolution'** (p.110 of 158). 'Self-organisation' is a reference to negative entropy's ordering or integration of matter, to the integrative, cooperative, teleological, holistic purpose or meaning of life that mechanistic science has evaded because of its unjust criticism of humans' divisive state. What

Birch is saying is that the responsibility of biologists in recent times has been to finally face the truth of integrative meaning and, in so doing, confront the issue of the human condition—and as such it *was* a predicament for biologists to solve because the human condition is about the behaviour of the human animal. The title of Birch's book, *Biology and the Riddle of Life,* acknowledges that a holistic, biological approach to the human condition was required.

Humanity has been stalled, waiting for the human condition to be addressed and resolved. Someone had to appear who, using the insights of mechanistic science and benefiting from the collective work of all the other unevasive, holism-acknowledging, truthful thinkers, could resolve the human condition. As has been mentioned, living in denial of its corrupted state, humanity could not acknowledge that this was what it was waiting for. It could not even acknowledge that it was waiting. Therefore it could not even cultivate the innocence required. It simply had to have 'hope and faith' that, sooner or later, before the destructive effects of excessive alienation destroyed our world, someone would appear sound enough to assemble the answers. The biblical story of David and Goliath is a metaphorical description of this situation. The whole of humanity's army was, as it were, arrayed at the edge of a great battlefield, stalled, unable to venture onto it and tackle the monster Goliath, who symbolised the unconfrontable issue of the human condition. It had to wait for a boy, symbolising innocence, to appear who was capable of going out into the dangerous battlefield and solving the human condition, slaying Goliath. Elsewhere in the *Bible,* in Isaiah 11, this truth is more clearly spelt out where Isaiah describes how **'a child will lead them'** to the state where the corrupt and the innocent will be reconciled; to where, he says, the **'wolf will live with the lamb'**. This same mythology occurs in Hans Christian Andersen's 1837 story, *The Emperor's New Clothes,* where it takes a child (innocence) to break the spell of the denial of the human condition and disclose the truth. In Australian literature our most celebrated poem is Banjo Paterson's 1895 poem, *The Man From Snowy River.* The poem describes how a **'stripling'** boy (the embodiment of innocence) goes beyond where the alienated adults dare go, down the **'terrible descent'** of the mountain side where **'any slip was death'** to confront the issue of the human condition and retrieve the truth about ourselves as symbolised by the thoroughbred horse that has escaped into the impenetrable mountains.

Only an approach that was free of denial could assemble the truth from science's hard-won but evasively presented insights and liberate humanity from the human condition. In Plato's imagery someone had to be living outside the cave, living in the illuminating sunlit world, in order to liberate those within the cave, **'release** [them] **from their bonds and cure** [them] **of their delusions'** (p.279), explain the human condition and break through the historic denial. To repeat the relevant section from the *Encarta* summary, it says that, **'Breaking free, one of the individuals escapes from the cave into the light of day. With the aid of the sun, that person sees for the first time the real world and returns to the cave with the message that the only things they have seen heretofore are shadows and appearances and that the real world awaits them if they are willing to struggle free of their bonds.'**

Some comment is required in regard to the phrase **'breaking free'**. While the person who is able to confront the human condition is not somebody who has 'broken free' from the cave—rather he is someone who retained his innocence and never adopted the bondage of denial, never adopted a 'cave existence'—at some stage in his thinking he nevertheless has to **'break free'**, or **'escape'** the influence of the denial that necessarily surrounds him. With everyone else living in denial it will be a struggle for him to decide whether the way he views the world is right or whether the view that is being presented to him is right. If he is sufficiently sound, however, and holds on to his view long enough to find reconciling understanding, or at least appreciation, of the world of denial, he will 'break free' of that codependent situation and learn to trust himself rather than what he is being told. Christ, for example, was an exceptionally sound thinker, and when he said **'I have overcome the world'** (John 16:33) he was saying that he had overcome or broken free of the coercion from the dishonest world of denial, it no longer had any sway over him.

The human condition has been solved and the way it was solved was exactly as Plato predicted. The evidence that it has been solved is that the denial has been exposed, someone has **'returned to the cave with the message that the only things they have seen heretofore are shadows and appearances and that the real world awaits them if they are willing to struggle free of their bonds.'** When you are living in denial, living in the cave, you cannot be free of it, you cannot expose the denial. Alienation cannot expose alienation, it is a contradiction in terms. As Plato foresaw, only someone outside the cave could bring **'the message'** (the understanding that exposes the cave prisoners' denial and makes it obsolete) into the cave.

Humanity has at last found the dignifying and thus liberating biological explanation for why humans have not been ideally behaved. The explanation was summarised early in this essay, in the section 'A brief description of how our soul became corrupted', and is presented in full in my earlier books, *Free* and *Beyond*. It is humanity as a whole that has found this explanation of the human condition because it is only as a result of the discoveries of science, being the peak expression of *all* human intellectual effort, that it has been possible for me to synthesise the biological explanation of the human condition. In fact it is 'on the shoulders' of eons of human effort that our species' freedom has finally been won. As has been explained, there have always been a very small number of people in society who were sufficiently innocent to confront the issue of the human condition, but until mechanistic science had completed the hard task of finding understandings of the mechanisms and details of the workings of our world, no liberation from the human condition was possible. Innocence—uncorrupted guidance from our cooperatively orientated instinctive self or soul—was the synthesiser, but science was the liberator, the so-called 'messiah' of the human race.

Unevasive, holistic, subjective introspection, and evasive, mechanistic, objective science, both played crucial roles in the liberation of humanity from ignorance. Einstein was expressing this truth when he said, **'Science without religion is lame, religion without science is blind'** (*Out of My Later Years,* 1950). Obviously it was not religious faith that synthesised the explanation of the human condition, but it was the denial-free aspect or orientation of the world of religions that was required.

ASPECT C

Having looked at the nature of the human soul and how it became corrupted in ASPECT A, and what knowledge would be needed to synthesise the liberating understanding of the human condition in ASPECT B, we now need to look at ASPECT C.

ASPECT C is concerned with the difficulty of introducing the liberating understanding of the human condition to people who are living in denial of the human condition. Using Plato's metaphor, we are going to look at what happens when the person who has assembled the liberating understanding of the human condition **'returns to the cave** [prisoners] **with the message that the only things they have seen heretofore are shadows and appearances and that the real world awaits them'**.

ASPECT C has three sections. C1, 'The difficulty of taking the truth back into the cave', will explain the difficulty of introducing the denial-free truth about the human condition to people who are living in denial of the human condition. C2, 'The cave allegory was the theory, this is what happened in practice', will explain what actually happened when the denial-free truth was introduced to the 'cave dwellers'. C3, 'Humanity's departure from the cave to life in the sun', will explain how humanity is finally liberated from its tortured state of denial.

C1: The difficulty of taking the truth back into the cave

The *Encarta* entry says that the prisoners in the cave have to be **'willing to struggle free of their bonds'**. These words bring us to the final impasse that is encountered in the human journey from ignorance to self-understanding. When the explanation of the human condition is finally assembled, how is it to be presented to people who have been living in the cave in fear and denial of the whole issue of the human condition? Having lived this way so long, how will humans tolerate someone entering their cave of denial to tell them that the human condition has been solved?

As has already been described in some detail in the *Introduction,* at first mention of the human condition a mind that has been living in denial blocks all further words from conscious recognition and effectively becomes 'deaf' to any further comment. The analogy used earlier was of a country raided for decades by Genghis Khan to the extent that its people had resorted to a life hidden in the forests and caves. It will be a brave person who first ventures beyond their recognised safety zone once it is announced that Genghis Khan has been destroyed and it is now safe to come out of hiding. It was explained and emphasised in the *Introduction* that overcoming humans' habituated fear of the subject of the human condition takes patience and perseverance, but the benefit is to finally see the world truthfully, in its full magnificence. As it is described in the *Encarta* summary, **'Escape into the sun-filled setting outside the cave symbolizes the transition to the real world, the world of full and perfect being'**.

In *The Republic* Plato described the resistance that humans will have to leaving the cave thus: **'if he** [the cave prisoner] **were made to look directly at the light of the fire** [the truth of integrative meaning and the issue of the human condition it gives rise to]**, it would hurt his eyes and he would turn back and take refuge in the things which he could see, which he would think really far clearer than the things being shown him. And if he were forcibly dragged up the steep and rocky ascent** [out of the cave] **and not let go till he had been dragged out into the sunlight** [shown the reconciling explanation of the human condition]**, the process would be a painful one, to which he would much object, and when he emerged into the light his eyes would be so overwhelmed by the brightness of it that he wouldn't be able to see a single one of the things he was now told were real'** (p.280).

In addition to the 'deaf effect' or, as Plato referred to it, 'blindness' (**'his eyes would be so overwhelmed by the brightness of it that he wouldn't be able to see a single one of the things he was now told were real'**), there have been other responses to description and analysis of the human condition.

There has been the response of trying to maintain the denials and false arguments that historically have been used to evade the issue of the human condition. Plato was referring to this response when he said **'he** [the prisoner from the cave] **would turn back and take refuge in the things which he could see, which he would think really far clearer than the things being shown him'**. Some of the denials that have been used to avoid confronting the human condition were mentioned earlier. They include the denial that there is an integrative purpose to existence; that nurturing played the crucial role in the maturation of our species; that humanity once lived in a state of instinctive cooperation; and that humans are extremely alienated and vary greatly in their degree of alienation. These and many other historic denials and evasions are also referred to in subsequent essays in this book, and addressed in full in *Beyond*.

Another significant response has been one of extreme anger towards the human condition-confronting information. The following comment by Professor Iain Davidson, who at the time was head of the archaeology and palaeoanthropology department of the University of New England in NSW, Australia, illustrates the fury that extreme holism can induce in mechanistic scientists. In 1995, on marking an essay by UNE student and FHA Member Lee Jones, the professor wrote beside Lee's references to *Beyond:* **'There is absolutely nothing in this book that has the slightest value. Do not waste your time or mine with references to it'**. He then recommended his own book, saying, **'I hope you will find it more coherent than Griffith's rantings'**.

In fact people can have such an angry response to the information that they attack the heresy by any means available, including persecuting its supporters—an historic response to unevasive, denial-free truth. In the quote above from Plato's allegory he was referring to this response when he said **'if he** [the cave prisoner] **were forcibly dragged up the steep and rocky ascent** [out of the cave] **and not let go till he had been dragged out into the sunlight, the process would be a painful one, to which he would much object'**. Plato later elaborates on the extent of that objection when he specifically states that any person who came back from successfully confronting the 'sun', confronting the human condition, would be accused of being mad, would be told that trying to escape the cave was futile, and if that person tried to lead people out of the cave they would try to kill him—**'they would say that his visit to the upper world had ruined his sight, and that the ascent was not worth even attempting. And if anyone tried to release them and lead them up, they would kill him if they could lay hands on him'** (p.281).

Plato knew too well about the murderous response to anyone who challenged society's established view of the world for he had witnessed the murder of Socrates on a charge of **'impiety** [lack of reverence for the gods of the day] **and corrupting the young'** (p.11). The truth is, Socrates was one of the most honest thinkers in history. Plato described him as **'the most upright man then living'** (p.12). In the entry on Plato the *Encarta Encyclopedia* states that Socrates was charged with **'atheism and corrupting Athenian youth'**, while a *National Geographic* article written on the US Library of Congress, mentioned that **'Socrates, condemned to death, is charged with introducing strange gods and corrupting the young. His questioning and reasoning shattered too many illusions. Most Athenians, it seemed, preferred an unexamined life'** (Nov. 1975). The mention of an 'unexamined life' is a reference to Socrates' famous statement that **'the unexamined life is not worth living'**. The truth is, humans have *had to* live an 'unexamined life', had to live in a dark cave, they have had to avoid confrontation with the issue about themselves of the human condition—all of which is the reason that Socrates was persecuted.

A charge of impiety, of lack of respect for the religious customs of the day, was also made against the supporters of Christianity. When the apostle Stephen was seized, **'They produced false witnesses, who testified, "This fellow never stops speaking against the holy place and against the law. For we have heard him say that this Jesus of Nazareth will destroy this place and change the customs Moses handed down to us"'** (Acts 6:13-14). To say that Christ was trying to change the customs of Moses was a misrepresentation charged by **'false witnesses'**, because Christ merely elaborated upon the denial-free honesty of Moses. Like Moses, what Christ essentially did was challenge the practice of denial, and that was the real reason he and his supporters were persecuted.

Challenging the status quo has led to resistance, persecution and even outright physical attack, due to the fact that humans do not like change, in fact they find any new concept difficult to accept. A new device, a new method of doing something or a new idea or way of thinking will typically be resisted. People see change as a challenge to themselves and/or their world, and in an effort to defend themselves against the perceived challenge they mount a resistance to it, sometimes even resorting to fabrication and misrepresentation. For example, when younger, more adaptable minds take up new ways, adherents of the old ways sometimes falsely claim the younger minds have been corrupted.

Historically all new ideas have had to endure initial opposition, and even persecution, from the established order. Humans have been *extremely* insecure as a result of the human condition, to the extent that *any* challenge to their particular behaviour or view of the world is perceived as a form of criticism and thus something to be resisted.

The German philosopher Arthur Schopenhauer summarised the journey that new ideas in science have historically had to undergo when he **'said that the reception of any successful new scientific hypothesis goes through predictable phases before being accepted'**. First, **'it is ridiculed'** and **'violently opposed'**. Second, after support begins to accumulate **'it is stated that it may be true but it's not particularly relevant'**. Third, **'after it has clearly influenced the field it is admitted to be true and relevant but the same critics assert that the idea is not original'**. Finally, **'it is accepted as being self-evident'** (compiled from two references to Schopenhauer's work—*New Scientist*, 15 Nov. 1984 & *PlanetHood*, Ferencz & Keyes, 1988). Note that each stage of recognition is achieved in a way that protects the ego of the onlookers. The extent of insecurity in the human make-up is very apparent.

The adjustment that accompanies the introduction of something new has also been difficult for humans because of the problem of habituation. The more humans practice a pattern of thinking and behaving, the more established and automatic it becomes and thus the more difficult it is to alter. On the subject of overcoming a practiced way of thinking and behaving the best book I have come across is *Courage to Heal,* by Laura Davis and Ellen Bass (1988). On patterns of behaviour it states, **'A pattern is any habitual way of behaving. By its nature it is deeply entrenched, set by repetition...Patterns have a life of their own, and their will to live is very strong. They fight back with a vengeance when faced with annihilation'** (p.175 of 495). In Hans Christian Andersen's 1837 fable, *The Emperor's New Clothes,* where the child breaks the spell of deception that the emperor is beautifully clothed and discloses the truth of his nakedness, the first reaction of the emperor and his entourage was to try more than ever to maintain the deception. To quote from the story: **'"But he has got nothing on," said a little child. "Oh, listen to the innocent," said its father. And one person whispered to the other what the child had said. "He has nothing on—a child says he has nothing on!" "But he has nothing on!" at last cried all the people. The Emperor writhed, for he knew it was true. But he thought "The procession must go on now." So he held himself stiffer than ever, and the chamberlains held up the invisible train'** (*Andersen's Fairy Tales,* trs E.V. Lucas & H.B. Paull, 1963, p.243 of 311).

The reality is that any challenge to humans' carefully constructed ego-castle, their practiced way of justifying themselves, is regarded as a threat and resisted.

The scale of the change involved is also a factor. It follows from what has been said that the greater the change the greater the resistance and the greatest change of all that humans can be confronted with is a change to their world view, their framework or paradigm of thinking. It can therefore be expected that nothing will be more fiercely resisted than this 'paradigm shift'.

Even paradigm shifts vary in their degree of difficulty. A paradigm shift which involves little self-confrontation is not as difficult as a paradigm shift where significant self-confrontation is involved. For instance the shift from the agrarian age to the industrial age in early 19th century England was violently resisted, as epitomised by the Luddites who organised themselves to destroy manufacturing machinery, claiming it put them out of work. A paradigm shift that involves changing from a strategy of living in denial of the human condition to one of confronting the human condition is the most difficult paradigm shift of all to make.

The following statement by Richard Tarnas summarises the resistance that has met each major change to humans' world view: **'As if to consecrate the birth of a fundamental new cultural vision, in each case a symbolically resonant trial and martyrdom of some sort was suffered by its central prophet: thus the trial and execution of Socrates at the birth of the classical Greek mind, the trial and crucifixion of Jesus at the birth of Christianity, and the trial and condemnation of Galileo at the birth of modern science'** (*The Passion of the Western Mind,* 1991, p.395 of 544). While Socrates and Galileo introduced monumental paradigm shifts and as a result were all ruthlessly persecuted, they did not threaten humans' denial of the issue of the human condition as Christ did. Those exceptionally innocent individuals in history, such as Christ, who could **'face'** the **'fire/sun'**, confront the truth of the cooperative ideals of life and thus look into and talk openly and freely about the human condition without feeling condemned and depressed, were extremely confronting of the average human's corrupted state. As a result, the paradigm shift that Christ introduced required such a profound change that to succeed people had to completely abandon their existing strategy to life, namely their life of denial, and be 'born-again' to a state of honesty by taking up support of Christ's denial-free life.

It needs to be emphasised that the paradigm shift involved in adopting a religion, entailing a 'born-again' conversion to a life of supporting a denial-free state, is fundamentally different to the paradigm shift that occurs with the arrival of understanding of the human condition. While confrontation with the honest, denial-free state occurs in both situations, with the ability to understand the human condition the need to simply abandon living in denial—the leap to faith, the 'born-again' conversion experience—is replaced with the ability to dismantle the need for denial, dissolve with understanding the insecurity that made denial necessary. Religion is about dogma and faith while understanding of the human condition is about knowledge and the psychological amelioration that that knowledge makes possible.

The following quote by journalist Robert Howard adds to the list of world-changing ideas that have occurred with inclusions from relatively recent times. He starts where Tarnas left off, with the birth of modern science and Copernicus: **'Three major blows have dented humanity's self-esteem: Copernicus showing that the Earth was not the cen-**

tre of the universe, Darwin showing descent from animals and Freud arguing that the rational, conscious mind is not master' (*Bulletin* mag. 11 Aug. 1992).

Darwin's idea of natural selection challenged the widely accepted literal biblical interpretation of the creation and of humans' unique divine status distinct from animals—of having been **'created'** by God **'in his own image'** (Genesis 1:27)—and as a result his 1859 book, *The Origin of Species,* **'was greeted with violent and malicious criticism'** (title page, 1968 Penguin edn). Nevertheless, Darwin did not address the issue of the human condition. In fact he studiously avoided the subject of human behaviour which, at its core, is the issue of the human condition. There are no references in his book to human behaviour, apart from references to how humans have been able to alter plant and animal breeds, and a single highly pertinent acknowledgment at the very conclusion: **'In the distant future I see open fields for far more important researches. Psychology will be based on a new foundation...Light will be thrown on the origin of man and his history'** (ibid. p.458 of 476).

Freud did bring the issue of the human condition into focus, albeit much less directly than Christ. His work looked at human psychosis, humans' repression of their soul as a result of the human condition ('psychosis' literally means 'soul-illness', from the Greek and Latin roots *psyche,* meaning 'soul' and *iasis* meaning 'abnormal state or condition', and 'psychiatry' literally means 'soul-healing', from the Greek *iatrea,* meaning 'healing'). In raising the issue of the importance of humans' psyche or soul or original instinctive self—as Howard pointed out, the truth is it rivals the rational conscious mind in influence—and for daring to unearth people's psychological denials, Freud's work has been resisted and resented and he was subject to much vilification. A November 1993 cover of *Time* magazine asked **'Is Freud Dead?'** with the article itself accompanied by a picture of the psychoanalyst's couch being thrown out the window. Sir Laurens van der Post recognised that the degree of resistance to Freud's work was a measure of its honesty when he wrote, **'One could perhaps better have measured the originality of Freud's achievement by reason of the numbers of the highly intelligent, well-informed men who instantly mobilised to attack him'** (*Jung and The Story of Our Time,* 1976, p.108 of 275).

Humans have practiced many different denials and should you confront them with any one of them they naturally (as it has been essential to their lives) protect and defend that denial. It follows that with the most important denial practiced by humans, that of the denial of the issue of the human condition, the determination to protect and defend it will be the most vigorous. In fact for some people the temptation to react extremely aggressively towards those who are seen to threaten that denial can be very great.

There is one more factor to consider in assessing the difficulty people will have adapting to change, and that is the speed of the change taking place. Some changes are not as sudden as others. Humans have had some 2 million years to adjust to living in denial of the human condition however with the arrival of understanding of the human condition there is no such luxury. With such an important explanation now available to humans and the technology for rapid global communication at our disposal, the knowledge will spread with *extreme* speed once the initial resistance is overcome and so too the process of adjustment will have to be expeditious. In his aptly titled book, *Future Shock,* Alvin Toffler actually anticipated the shock that would accompany the rapid arrival of understanding of the human condition when he wrote, **'Future shock...** [is] **the shattering stress and disorientation that we induce in individuals by subjecting them to too much change in too short a time'** (1970, p.4).

The most difficult change of all for humans to have to adjust to is one that involves the rapid adoption of a completely new way of thinking—a paradigm shift of thought. Moreover, if the old way is an extremely habituated way of behaving, in fact a 2-million-year-old practice, and if what is new is totally exposing and confronting of the all-important denial that humans have been practicing, then the difficulty will be all the more intense.

These difficult aspects of change all arise when you challenge humans' major mechanism for protecting their self-worth, namely their denial of the human condition. A challenge to this particular denial will challenge the very basis of the way humans have been validating themselves, their way of avoiding the implication that they are 'bad' or unworthy, which is the implication at the heart of the issue of the human condition. It therefore has to be expected that nothing will be resisted as fiercely. Plato fully anticipated this fierce resistance when he talked about the cave prisoners having to be **'forcibly dragged'** out of denial with irrefutable logic, saying **'the process would be a painful one, to which he** [the cave prisoner] **would much object'**. He even spelt out how determinedly they might object, saying, **'if anyone tried to release them and lead them up, they would kill him if they could lay hands on him.'**

George Bernard Shaw observed that **'All great truths begin as blas-**

phemies.' Given that there is no greater truth than understanding of the human condition, exposing and dismantling humans' historic denial of the issue of the human condition can appear to be the ultimate **'blasphemy'**, **'impiety'**, irreverence, heresy, profanity, sacrilege.

There is yet another dimension to this difficulty humans have adapting to change. While virtually all adults have been living in denial of the human condition, those who are more corrupted have had to employ more denial than those who are less corrupted to keep the criticism of their corrupted state at bay. Humans have been living in various degrees of denial and as a result are variously alienated or estranged from their true self or soul. Because change is most threatening to the less secure, it is the less secure who find it most difficult accepting and adapting to change.

It follows that while most people feel the need to resist something they fear, those who are most afraid, those who, in the case of the denial of the human condition, are the more alienated in society, will put up the most determined opposition to the arrival of understanding of the human condition. As will be described in some detail in *The Demystification Of Religion* essay in this book, the more alienated are initially going to strongly resist having their disguises removed. The last great battle on Earth, referred to in the *Bible* as the Battle of Armageddon, is in fact the battle between those whose alienation is so great they are afraid to have it exposed and those who are sufficiently free of alienation, sufficiently secure in self, not to be overly afraid of the exposure and therefore able to support and defend the humanity-liberating, all-important breakthrough understanding of the human condition.

It also needs to be emphasised that alienation tends to increase with age. The longer a person lived in the corrupted, alienated world, the more encounters they inevitably had with that corruption and alienation and the more they became corrupted and alienated themselves as a result. Furthermore, the older a person, the more habituated they had become to living in denial of the issue of the human condition.

It follows from what has been said that the people most able to take up and adjust to the arrival of understanding of the human condition will be the more secure or less alienated among young people in society, while the most resistance will come from the more alienated among older people in society. It can be seen that when understanding of the human condition arrives, the worst possible 'generation gap' will appear.

Historically, for reasons already explained, young people have always been more receptive to new ideas than older people. Older people, because they are more alienated, and because they are more habituated and ego-attached to their existing ways of doing things and thinking, have always found change difficult. Christ recognised the attachment older people have to their ways of behaving and thinking when he observed, **'no-one after drinking old wine wants the new, for he says, "The old is better"'** (Luke 5:39). We have a saying for the difficulty older people have adopting change: **'you can't teach an old dog new tricks'**—and, as has been emphasised, no new concept is as revolutionary as the arrival of understanding of the human condition. Such a monumental paradigm shift will in fact so test even the adaptability of young people that initially only the soundest amongst them will be able to cope with the change. Not being as wedded or habituated as older minds to the old denial-maintaining paradigm, younger, less alienated minds are significantly more able to 'hear', consider, appreciate and then adjust to this new denial-free paradigm than minds that are more alienated or older, or indeed both.

The reality is that it is young people who have to take up new ideas. Even scientists, who are supposed to respond objectively to reasoned argument and explanation, can suffer from extreme prejudice against change as they get older. The science historian Thomas Kuhn found that **'the old scientists who became established within the dominant paradigm have to die off first: they will virtually never accept the new paradigm. Only the younger generation of scientists, who don't have the emotional attachment to the old paradigm, will be willing to change their minds'** (a reference to the work of Kuhn by Marilyn Ferguson, *New Age* mag. Aug. 1982). Physicist Max Planck succinctly described the difficulty older scientists have adopting new ideas when he said **'science progresses funeral by funeral'** (see his *Scientific Autobiography*, 1948). Charles Darwin similarly experienced first-hand this problem of older people resisting new scientific ideas, writing, **'I have got fairly sick of hostile reviews...I can pretty plainly see that, if my view is ever to be generally adopted, it will be by young people growing up and replacing the old workers'** (*Charles Darwin*, ed. Francis Darwin, 1902). Young people must take up new ideas, this is especially so with the arrival of understanding of the human condition.

There is yet another difficulty and indeed extreme danger for the more alienated, in particular some older adults, when understanding of the human condition arrives. They can wrongfully take on the role of guardians for humanity against human condition-confronting information. The following describes the nature of this danger.

Not only is understanding the human condition something new to have to adjust to, it is also a subject that the more alienated are especially 'deaf' to, and therefore cannot evaluate. A new concept is difficult enough for people to adjust to let alone one that they have learnt to live in exceptional fear and denial of and, as a result, do not want to and thus cannot 'hear'. This poses an extra problem for the more alienated—and the world, for if you cannot 'hear' and thus consider the new information then you cannot hope to understand, assess and, if it is worthwhile, appreciate it. This is problematical in itself however there is a further dimension to the problem that the more alienated pose for the arrival of human condition-confronting information, which has a compounding effect. While they suffer the most from the deaf effect these people are also the most fearful and thus intuitively aware of, sensitive to, and on their guard against any information that brings the human condition into focus. Both the deaf effect and sensitivity to any human condition-confronting information increases with alienation.

This combination of alienation or loss of 'hearing' accompanied by an acute 'radar' for any information relating to the human condition is an extremely dangerous mix when understanding of the human condition arrives. While historically there was always a need to repress and live in psychological denial of information that brought the human condition into focus but did not explain and resolve it (because such information left humans extremely depressed and unjustly condemned for their corrupt, divisive nature), when the full, dignifying, ameliorating understanding that resolves the human condition arrives, it ends that unjust condemnation. The full truth about humans dignifies humans, it lifts the 'burden of guilt' from humans through explanation and thus amelioration of human nature. Not all human condition-confronting information is dangerous. The human condition-confronting information that explains the human condition brings an end to the historic criticism of humans and is thus the *least dangerous* information possible.

The great danger is that while the more alienated among us, in particular some older adults, cannot 'hear' and thus discover that this new information is the safe and complete truth about humans, they will be the most aware and most fearful that the information is bringing the human condition into focus. They will be afraid of the information without being able to be enlightened as to its non-fearful true nature. Some of them can even go to war against the information because they see it as unbearable and dangerous for humanity, without realising its immense goodness.

The more alienated are the least able to evaluate the safety or otherwise of human condition-confronting information, and yet, because of their heightened fear of such information, they tend to put themselves forward as the guardians for humanity, the people who can assess what is good for us all. But as blind people are not the ones to tell us when the sun is coming up, the alienated are not the ones able to ascertain when the all-wonderful, human condition-liberating understanding of the human condition has arrived.

In their extreme fear and resulting hatred of human condition-confronting information, and in their delusion about their right to act as guardians for humanity against such information, some more alienated people can be tempted to take it upon themselves to eliminate the information from existence, unwittingly trying to kill the goose that lays the golden egg and destroy the information that liberates humanity from the human condition. They can make the worst mistake of any human on Earth.

. . . Apology: the dotted lines indicate text that has temporarily . . . . .
. . . . . been withdrawn because it deals with issues before the . . . . .
. . . . . courts. Once the legal restrictions end the fully restored pages . . . . .
. . . . . will be available on the FHA's website, or from the FHA. instead of living out their fear of, and anger towards human condition-confronting information they can trust in democracy. The democratic principle of freedom of expression allows new ideas to emerge and be tested by society as a whole for their value.

Democracy is the mechanism that has been developed to counter humans' natural resistance to change. It ensures that the human journey to greater understanding is not shut down by prejudice. It is the mechanism that protects humans from the insecure aspects of themselves.

The arrival of understanding of the human condition poses the ultimate threat to the status quo, the status quo being the state of living in denial of the human condition. At the very deepest level of its relevance and meaning, democracy has existed to ensure this ultimate threat to the status quo is not stifled before it can emerge. If there were zero tolerance of human condition-confronting information, then understanding of the human condition could never emerge. It was as if the founding fathers of the human race knew that one day a crisis point would be reached, a moment when all human frailties would conspire to threaten to shut down the human journey from ignorance to enlightenment, and to ward against that eventuality they implemented the practice of democracy. Democracy alone had the capability to ensure that human insecurities would not jeopardise the efforts the whole human race has contributed since consciousness emerged, and that understanding of the human condition would be able to emerge.

Where human condition-confronting information is concerned the more alienated amongst us can be especially tempted to dispense with the democratic principle of freedom of expression and to set out to destroy the information by any means available, but in fact it is where human condition-confronting information is concerned that the principles of democracy need to be the most scrupulously adhered to.

The wisest and most important legal counsel ever given was that given by a lawyer named Gamaliel during the birth of Christianity. His counsel was essentially to trust in the principles of democracy to ascertain what was of value to humanity and what was not. The apostles had been gaoled and were threatened with death for defending Christ's denial-free existence when **'a Pharisee named Gamaliel, a teacher of the law, who was honoured by all the people, stood up in the Sanhedrin [the full assembly of the elders of Israel] and ordered that the men [apostles] be put outside for a little while. Then he addressed them: "Men of Israel, consider carefully what you intend to do to these men. Some time ago Theudas appeared, claiming to be somebody, and about four hundred men rallied to him. He was killed, all his followers were dispersed, and it all came to nothing. After him, Judas the Galilean appeared in the days of the census and led a band of people in revolt. He too was killed, and all his followers were scattered. Therefore, in the present case I advise you: Leave these men alone! Let them go! For if their purpose or activity is of human origin** [ie, is a product of egocentricity, insecurity and alienation]**, it will fail. But if it is from God** [if it is sound]**, you will not be able to stop these men; you will only find yourselves fighting against God." His speech persuaded them. They...Let them go'** (Acts 5:34-38).

This was the most important counsel ever offered because it allowed the Christian movement to survive its most vulnerable, early stage when, after Christ's death, only a few people were aware of the immense importance of the movement and where the death of those few could very well have meant the end of the movement. People are sometimes tempted to think that a good idea will withstand whatever resistance it encounters, but that is not true. In John Stuart Mill's 1859 essay, *On Liberty*—a document considered a philosophical pillar of western civilisation—Mill emphasised this point when he said, **'the dictum that truth always triumphs over persecution is one of those pleasant falsehoods which men repeat after one another till they pass into commonplaces, but which all experience refutes. History teems with instances of truth put down by persecution. If not suppressed for ever, it may be thrown back for centuries'** (*American state papers; On liberty; Representative government; Utilitarianism*, 1952, p.280 of 476).

The reason I rate the survival of Christianity so highly is because it, along with the other great religions, civilised humanity and by so doing bought the necessary time for science to develop and the ameliorating understanding of the human condition to eventually be found. Living in a state of denial was such an awful thing to have to do that it was like acid infiltrating your system, and, indeed, the world that had to absorb the alienated behaviour. As mentioned, religions gave humans a means to be partially freed from their state of denial by being 'born-again' to a relatively denial-free state through their support of a denial-free prophet. While a system of total honesty and idealism required the confrontation of the issue of the human condition, religions at least allowed their supporters to acknowledge their corrupted state to an extent. Religions allowed humans to introduce some truth into their world and by so doing reduce the self-condemnation they felt from living so falsely in denial of their corrupt reality. By deferring, albeit in a limited capacity, to the truth that the denial-free thinkers or prophets of old represented, humans were introducing some truthful idealism into their life. While religions did not solve or even address the real problem on Earth of the human condition, they did offer humans some extremely precious relief; they reduced the level of self-loathing and resentment from being unfairly criticised, and thus the level of frustration, unhappiness and anger in the world. Religions made people feel better about themselves. In recent times however, humans have become so alienated that the truthfulness of prophets has become too honest to bear, too confronting, and, as a result, non-confronting forms of idealism, such as environmentalism, have developed to provide this measure of relief. More will be said about the role that has been played by religion and other non-human-condition-confronting forms of idealism, of 'pseudo-idealism', in the *Death by Dogma* essay that is to be published shortly in another book I have written titled . . . .

As Gamaliel urged 2000 years ago, humans should invest their trust in democracy to decide what is ultimately beneficial or not for society, especially when human condition-confronting information is concerned.

C2: The cave allegory was the theory, this is what happened in practice

Plato's cave allegory provides the theory of what would happen to someone who **'returns to the cave with the message'** that the human condition has been solved and through rational argument and accountable explanation makes the cave dwellers **'look directly at the light of the fire'**, directly at the issue of the human condition.

As has been emphasised, it will be the more alienated and, as it follows, older adults who will most fear and resist the change that the arrival of understanding of the human condition inevitably entails. In fact, as has been mentioned, the last great battle on Earth, the Battle of Armageddon predicted in the *Bible,* is the battle between the more alienated and the less alienated; between those whose alienation is such that they are afraid to have it exposed and as a result determinedly resist exposure, and those who are sufficiently free of alienation, sufficiently secure in self not to be overly afraid of the exposure and able to support the humanity-liberating understanding of the human condition. The more alienated, those with the greater need to live in the 'cave' of denial, will, as Plato said, maintain that **'the ascent** [out of the cave] **was not worth even attempting. And if anyone tried to release them and lead them up, they would kill him if they could lay hands on him.'**

There are people who will feel that facing the issue of the human condition and humans' resulting state of alienation is impossible, but as the following excerpt from R.D. Laing's writing emphasises, any real progress for humanity ultimately depends on confronting and dealing with the issue of alienation: **'Our alienation goes to the roots. The realization of this is the essential springboard for any serious reflection on any aspect of present inter-human life** [p.12 of 156]**...The con-**

dition of alienation, of being asleep, of being unconscious, of being out of one's mind, is the condition of the normal man [p.24]**...between *us* and It** [our soul] **there is a veil which is more like fifty feet of solid concrete. *Deus absconditus.* Or we have absconded** [p.118] **We respect the voyager, the explorer, the climber, the space man. It makes far more sense to me as a valid project—indeed, as a desperately urgently required project for our time—to explore the inner space and time of consciousness. Perhaps this is one of the few things that still make sense in our historical context. We are so out of touch with this realm** [so in denial of the issue of the human condition] **that many people can now argue seriously that it does not exist. It is very small wonder that it is perilous indeed to explore such a lost realm** [p.105]**'** *(The Politics of Experience* and *The Bird of Paradise,* 1967).

. . . .

. . . . . . Apology: As explained in Notes to the Reader, the dotted lines indicate text that has temporarily been withdrawn because it deals with issues before the courts. Once the legal restrictions end the fully restored pages will be available on the FHA's website, or from the FHA. . . . . . .

. . . .

. . . . there are two broad observations that should be made here as they bear out all that Plato had to say about what would happen to someone who **'returns to the cave with the message'** that the human condition has been solved.

. . . .

The charges made against Socrates and, later, the pioneers of Christianity, were mentioned earlier to illustrate that humans—especially the more alienated—can see any change to their way of viewing the world as a threat. Further, these people can be tempted to use any means at their disposal. . . . . to try to eliminate the perceived threat. In particular, when younger, more adaptable minds take up a valuable new idea the defenders of the status quo, the established way of viewing the world, falsely assert that the young people have been corrupted.

In the case of the early Christians, the opponents of the apostle Stephen **'produced false witnesses who testified, "This fellow never stops speaking against the holy place and against the law".'** In Socrates' case, he was charged with **'atheism'** or **'impiety'** or **'introducing strange gods'** and **'corrupting the young'**.

. . . .

The understanding of the human condition that the FHA supports does challenge the most established way humans manage their world, namely living in denial of the issue of the human condition; however, this challenge comes not from disrespect, but from enlightenment. . . . .

. . . .

Apology: As explained in Notes to the Reader, the dotted lines indicate text that has temporarily been withdrawn because it deals with issues before the courts. Once the legal restrictions end the fully restored pages will be available on the FHA's website, or from the FHA.

. . . .

. . . .
. . . .
. . . .
. . . .
. . . .
. . . .
. . . .
. . . .
. . . .

Recall that Plato said that **'they would say that his visit to the upper world had ruined his sight, and that the ascent was not worth even attempting.'** . . . .
. . . .
. . . .

. . . .
. . . .
. . . .
. . . .
. . . .
. . . .
. . . .
. . . .
. . . .
. . . .
. . . .
. . . .
. . . .
. . .

The fact is a person cannot be sound enough to confront, look into, and bring out understanding of the human condition and simultaneously be a . . . . The denial-free thinker or prophet, Jesus Christ, pointed out this obvious truth when, in responding to his persecutor's accusation of him being **'possessed by Beelzebub...the prince of demons'**, he stated, **'How can Satan drive out Satan?'** (Mark 3:22, 23). He was making the same point when discussing the differentiation between false and true prophets: **'Watch out for false prophets...by their fruit you will recognise them...a good tree cannot bear bad fruit, and a bad tree cannot bear good fruit'** (Matt. 7:15,16,18). Only true prophets can deliver insight into the human condition. Sir Laurens van der Post also made the same point when in a passage reminiscent of Plato's cave analogy, he wrote: **'He who tries to go down into the labyrinthine pit of himself, to travel the swirling, misty netherlands below sea-level through which the harsh road to heaven and wholeness runs, is doomed to fail and never see the light where night joins day unless he goes out of love in search of love'** (*The Face Beside the Fire,* 1953, p.290 of 311). Alienation cannot look into the human condition, one precludes the other. Examination of the human condition requires exceptional soundness. The truth is, unsoundness lies within the person who cannot tolerate someone looking into the human condition.

Apology: As explained in Notes to the Reader, the dotted lines indicate text that has temporarily been withdrawn because it deals with issues before the courts. Once the legal restrictions end the fully restored pages will be available on the FHA's website, or from the FHA.

. . . .
. . . .
. . . .
. . . .

. . . .
. . . .
. . . .
. . . .
. . . .
. . . .
. . . .
. . . .
. . . .
. . . .
. . . .
. . . .
. . . .
. . . .

. . . .
. . . . As was pointed out in the *Introduction,* living in denial of the human condition has been the real state of mind-controlled indoctrination. Humans have been progressively indoctrinating themselves with the denial and avoiding any form of deeper thinking for 2 million years. Effectively they have been drumming into their minds that they must not think about the issue of the human condition or any of the many other important truths that bring it into focus. What the FHA is introducing is understanding that frees—de-programs—humans from their highly mind-controlled state of denial. . . . .
. . . .

With regard to the young adults' enthusiasm and tenacious support of this new information that explains the human condition, it is not surprising that once people appreciate the importance of the information, they hold on to and defend it tenaciously. The human mind has thirsted for understanding of the human condition since the onset of rational thought. Plato foresaw this situation too, saying of those who learn to confront the **'sun'** (that is, understand the human condition): **'it won't be surprising if those who get so far are unwilling to return to mundane affairs, and if their minds long to remain**

among higher things' (p.282).

. . . .
. . . .
. . . .
. . . .
. . . . Religions catered for the insecurity created by the human condition. Therefore, by explaining and ameliorating the human condition—bringing dignifying biological understanding to humans' 'sinful', corrupted state—the need for religion is obsolete. Further, our work is concerned with bringing *understanding* to the human condition. It challenges peope to think about the human condition, question and understand themselves, a process that is the very opposite of abandoning thought about the dilemma of the human condition and, instead, deferring to some form of dogma, faith or belief.

I should mention that, like the people who attacked Socrates when they accused him of **'atheism'** . . . .
. . . .
. . . . is particularly ironic given that my work is based on the recognition of integrative meaning or God, rather than on the denial of this most fundamental and universal of truths that almost everyone else has been practicing.

In summary, older adults can find change very difficult, especially when the change involves leaving the state of denial of the issue of the human condition. The great Australian educator, Sir James Darling, gave this wise counsel about the danger of older people resisting change: **'At every time when there has been great activity and great originality, there has been opposition and tenacity from the old. Those who have grown up in another age...are terribly afraid of newness of life...they cannot adapt themselves to the new life. They are wrong, of course...** [their] **opposition is often cruel, sometimes fatal. That is what older people should remember in their criticism, for theirs is the power, usually not indeed to stop the Spring from coming, but at least to trample and to kill the first few flowers of the year...The mind of most men is not adaptable after a certain age and the onrush of a Renaissance is very rapid'** (*The Education of a Civilized Man*, 1962, p.53 of 223).

In his 1964 song, *The Times They Are A-Changin'*, Bob Dylan anticipated what would occur when the extremely confronting but wonderfully liberating understanding of the human condition arrived. The reader will notice in the fourth stanza the description of the tragic generation gap that can occur. **'Come gather round people wherever you roam/And admit that the waters around you have grown/And accept it that soon you'll be drenched to the bone** [the suffering and destruction on Earth resulting from the agony of having to live with the dilemma of the human condition are fast reaching crisis levels, we are entering end-play]/**If your time to you is worth saving/And you better start swimming or you'll sink like a stone/Oh the times they are a-changin'//Come writers and critics who prophesise with your pen/And keep your eyes wide, the chance won't come again/And don't speak too soon for the wheel's still in spin/And there's no tellin' who that it's namin'** [As has already been explained, up until now humanity has had to hide the truth of who is corrupted and alienated and who is not, because without understanding of the human condition, it would have led to unjust condemnation of those who were no longer innocent. However, with the human condition now compassionately resolved, who is alienated and who is not can be, must be, and unavoidably will be revealed.]/**For the loser now will be later to win/For the times they are a-changin'//Come senators, congressmen, please heed the call/Don't stand in the doorway, don't block up the hall/For he that gets hurt will be he who has stalled** [Once the human condition is explained there is no point trying to put the old denial back in place because the denial has become transparent to those who have grasped the explanation of the human condition.]/**The battle outside ragin'/Will soon shake your windows and rattle your walls** [the liberating, but also exposing, truth about humans is on its way]/**For the times they are a-changin'//Come mothers and fathers throughout the land/And don't criticise what you can't understand/Your sons and your daughters are beyond your command/Your old road is rapidly agein'/Please get out of the new one if you can't lend your hand/Oh the times they are a-changin'//The line it is drawn, the curse it is cast** [the exposing truth about human nature is out]/**The slow one now will later be fast/As the present now will later be past/The order is rapidly fadin'/And the first one now will later be last/For the times they are a-changin'** [as has already been explained, having, been repressed up until now, innocence now comes to the fore to lead humanity home].**'**

A quote from John Stuart Mill, included earlier, emphasised that there is no guarantee that a worthwhile new idea will be able to survive persecution, that persecution can be, as Darling said, **'fatal'**. The following quote from the findings of the science historian Thomas Kuhn makes the same important point: **'In science** [according to Kuhn] **ideas do not change simply because new facts win out over outmoded**

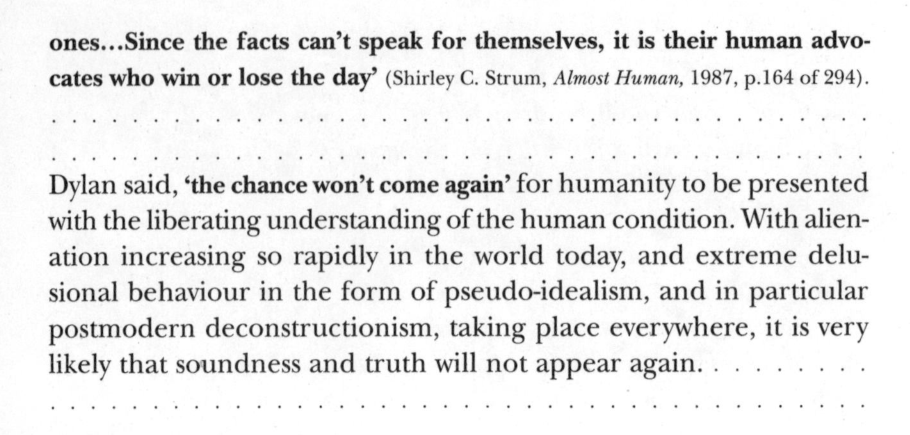

ones...Since the facts can't speak for themselves, it is their human advocates who win or lose the day' (Shirley C. Strum, *Almost Human,* 1987, p.164 of 294).

. . . .

. . . .

Dylan said, **'the chance won't come again'** for humanity to be presented with the liberating understanding of the human condition. With alienation increasing so rapidly in the world today, and extreme delusional behaviour in the form of pseudo-idealism, and in particular postmodern deconstructionism, taking place everywhere, it is very likely that soundness and truth will not appear again. . . . .
. . . .
. . .

The second observation that bears out all that Plato had to say about what would happen to someone who **'returns to the cave with the message'** that the human condition has been solved concerns the treatment recently given to the late Sir Laurens van der Post.

I mentioned in the *Introduction* that Sir Laurens van der Post has been of such importance in my journey that I dedicated *Beyond* to him. He is the person I have quoted most often in my books. I also mentioned in the *Introduction* that in his London *Times'* full-page obituary he was acknowledged as **'a prophet'**, and that given the truly exceptional profundity and integrity of his thinking—which has been amply exhibited in quotes I have already included from his writing—he was clearly an exceptional prophet, an exceptionally unevasive, denial-free, penetrating thinker. In 2001, the publishers John Murray released a posthumous biography of Sir Laurens by journalist J.D.F. Jones, titled *Storyteller: the many lives of Laurens van der Post.* A review of that biography written by author and journalist Christopher Booker, a long-time acquaintance of Sir Laurens, appeared in the London *Spectator* on 20 October 2001. It begins: **'The key to what makes this biography of the writer Laurens van der Post so unusual is betrayed in a tiny footnote on p.357. In a flash of vanity, J.D.F Jones cannot resist claiming credit for an obscure review he had written of one of Sir Laurens's autobiographical books back in 1983, in which the "present author" (himself) had gone out "on what was then a slender limb and declared that he did not believe a word of this 'flabby and embarrassing stuff'." What makes this so odd is that, 16 years later, after Sir Laurens's death in 1996, Mr Jones should have pushed himself forward to become his "authorised biographer", to be given access to most, but not all, of van der Post's private**

papers. Without revealing his belief that van der Post was a complete fraud, he won the confidence of the author's daughter Lucia, who had worked with him at the *Financial Times*. He received an advance of £50,000 from Sir Laurens's publishers. He then set out single-mindedly to strip away every last shred of the reputation of the man whom the headline to his 1983 review had called "van der Posture". It must be the only occasion in history when someone has managed to hijack the position of "authorised biographer" to produce what is nothing but an utterly ruthless hatchet job.'

The review goes on to say, **'Alarm bells began to ring when Mr Jones appeared to be making only the most perfunctory efforts to interview all those who had been closest to Sir Laurens'**. The review also refers to a number of claims made by Jones, such as that Sir Laurens van der Post was **'descended from a 17th-century Hottentot princess-turned-prostitute'**, and that **'he romanticised his famous "discovery" of the Kalahari Bushmen in the 1950s.'** The review refers to the **'lucrative serialisation** [of Jones' book] **in that same Sunday tabloid, along the lines of "we expose the secret sex life of Prince Charles's guru"'**, and says that **'when Jones is eager to recycle any salacious anecdote about Laurens's love-life, however distorted or improbable, it is almost comical how much less demanding he is of the evidence for his own stories than he is for those told by the man he obsessively tries to blacken as a fantasist and a liar.'** The review continues, **'So relentless is this denigration that it reminds one just how easy it is to turn anyone's life into a negative caricature if one sets out to do so...he** [Jones] **is so determined not to see anything positive in his subject...he is wholly incapable of understanding those qualities in Laurens which evoked such extraordinary response from millions of readers, such love from his wide circle of friends, and which gave him his unique position in the inner life of our age.'** Booker concludes, **'Certainly there are those who relish the idea of a hatchet job being done on the "self-styled mystic"...** [but] **ultimately this book exposes the limitations not of van der Post, only of its author.'**

The injustice inflicted upon Sir Laurens van der Post further illustrates Plato's assertion that if anyone tried to **'lead them up** [out of the cave of denial]**, they would kill him if they could lay hands on him.'** In this instance, an utterly vicious attack was made on his character. The attack had to be in the form of character assassination because it is not possible to attack his work, the marvellous insights into the human condition that he is so famous for. . . . .
. . . .

. . . .

Apology: As explained in Notes to the Reader, the dotted lines indicate text that has temporarily been withdrawn because it deals with issues before the courts. Once the legal restrictions end the fully restored pages will be available on the FHA's website, or from the FHA.

. . . .

. . . .

. . . .

. . . .

. . . .

. . . .

. . . .

Sir Laurens van der Post has been, as it were, fed to the dogs. In Australia, the national newspaper, *The Australian,* printed a review of Jones' book under the headline **'Charming Charlatan'**. The reviewer, journalist Luke Slattery, was so seduced by the book that he suggested that readers **'might be inclined to offer a wheelbarrow of his** [Sir Laurens van der Post's] **books to the nearest second-hand shop, or to junk them all'** (19–20 Jan. 2002). In the review, Slattery even asserts that **'Jones is never unfair to his subject'**. While academic Peter Alexander, in his review of Jones' book in Australia's other leading newspaper, the *Sydney Morning Herald,* was sufficiently seduced by the book for his review to be titled **'A life's work of lies'**, he at least acknowledges Jones' inability to appreciate the essence of Sir Laurens van der Post (namely his phenomenal ability to tell the truth about humans), pointing out that Jones **'responds to Van der Post's verbal aerobatics by asking, in evident exasperation: "What does that mean?"'** (16–17 Feb. 2002). As Booker noted, Jones was **'wholly incapable of understanding those qualities in Laurens which evoked such extraordinary response from millions of readers'**. Being unable to deal with the essence of your subject's life's work in a biography is a fundamental flaw and any review of Jones' book should be prefaced on that point. Further, as Booker pointed out, it is not merely that Jones does not understand Sir Laurens' life's work, he despises it, describing it as **'flabby and embarrassing stuff'**. The fact that Jones has no understanding of Sir Laurens' work and yet is contemptuous of it, together with the evidence that Jones lobbied to be Sir Laurens' authorised biographer, should lead any reviewer to recognise, as Booker did, that Jones' book is a hatchet job and certainly not a biography.

. . . .

. . . .

. . . I think it is appalling that Sir Laurens van der Post's detractors waited until after his death. I can only imagine what a devastatingly effective response he would have made to their cowardly attack had he been alive to defend himself.

Similar attempts were made to assassinate the characters of Arthur Koestler and Carl Jung following their deaths. Koestler, like Sir Laurens van der Post, was another exceptionally sound, denial-free thinker or prophet, and was quoted earlier courageously acknowledging the truth of integrative meaning. The reader of *Beyond* will recognise many of his extraordinarily truthful insights, while much will be said about the exceptional integrity of Jung's mind in the next essay.

The haters of the truth placed the head of Christ's mentor, John the Baptist, on a platter and forced Plato's mentor, Socrates, to drink poison. The world of prophets has been one of brutally enforced estrangement, but that brutality was only a measure of how much their truth exposed, condemned and hurt people. Sir Laurens van der Post let so much truth out he left the world of lies a wasteland. No wonder they hated him. He destroyed their psychological home. He dragged them out of the safety of their cave of denial.

. . . .
. . . .
. . . .
. . . .
. . . .
. . . .
. . . .
. . . .
. . . . . .

It is worthwhile repeating Christ's summation of the way some people respond to those who bring light to the dark, cavernous state of denial humans have lived in to protect themselves from exposure to the issue of the human condition: **'the light shines in the darkness but...everyone who does evil hates the light, and will not come into the light for fear that his deeds will be exposed.'** Indeed an awareness must have come down through the centuries as to how to go about denigrating prophets because it appears Christ was also subjected to the technique. To warn, as he did, of **'ferocious wolves' 'in sheep's clothing'** (Matt. 7:15), and of being **'like lambs among wolves'** (Luke 10:3 & Matt. 10:16)

. . . .
. . . .
. . . . In fact the *Bible* records that **'looking for evidence against Jesus so that they could put him to death...Many testified falsely against him'** (Mark 14: 55,56).

It was mentioned that the ferocity of the attacks on truthsayers could be so intense that people who were not aware of the motivation for the attack could easily be persuaded simply by its ferocity that the subject of the attack *must* be evil. In fact truthsayers could be attacked with such ferocity that even those who did know the motivation for it could be disconcerted. . . . .
. . . .
. . . .
. . . .
. . . .
. . . .
. . . .
. . . .
. . . .
. . . .
. . . .
. . . .
. . . .
. . . .
. . . . My brother Simon, Tim Macartney-Snape and I all went to Geelong Grammar School, the ethos of which was precisely what Plato advocated for education: **'Plato's ideal education system is primarily structured so as to produce philosopher-kings** [denial-defiant thinkers]**.'** The emphasis at Geelong Grammar School during our time was on protecting and cultivating the innocence of students, rather than concentrating on their intelligence and academic achievement. Geelong Grammar cultivated soul over intellect. To quote from an article in the school's magazine titled *What we profess and practice:* **'Primal innocence, like primal Eden, is destroyed: yet both can be restored; the Divine Image lives on, the burden and the glory of mankind, and true education consists in its recognition and its restoration'** (M.D. de B. Collins Persse, *The Corian*, Apr. 1982). One academic year was spent living in huts in the mountains, bushwalking every weekend after studies. Competition in sport and in study was played down, and many non-academic activities were fostered. Above all service to society was emphasised over self-interest and personal gain. No doubt this background made us potentially vulnerable to the more mean and ruthless ways of the world. In fact Sir James Darling, whose 30-year tenure as headmaster of Geelong Grammar established the school's ethos, was well aware of this problem of vulnerability. In his

1962 book, *The Education of a Civilized Man* he wrote that: **'The objective** [of education] **is a development of the whole man, sensitive all round the circumference...** [A] **criticism** [of developing such sensitivity]**...is that the sensitive man cannot survive in the hard modern world. In a sense this is true...he may at first sight appear less well equipped to deal with life than his more callous or superficial fellow...But the future, someone has said, lies not with the predatory** [selfish] **and the immune** [alienated] **but with the sensitive** [innocent] **who live dangerously** [defy the denial practiced by the cave world]. **There is a threefold choice for the free man...He may grasp for himself what he can get and trample the needs and feelings of others beneath his feet: or he may try to withdraw from the world to a monastery...: or he may "take arms against a sea of troubles, and by opposing end them"...** [And so] **There remains the sensitive, on one proviso: he must be sensitive *and* tough. He must combine tenderness and awareness with fortitude, perseverance, and courage'** (pp.32-34 of 223). I might mention that while this ethos of cultivating soul over intellect was highly unusual in education and tended to strand students in a lonely state of idealism, Geelong Grammar School became one of the most highly regarded schools in the world. The future king of England, Prince Charles, was sent there for part of his education.

Christ was aware of the dangers of innocence being too trusting and abused as a result, for he warned of the need to be **'shrewd'**, something that is very difficult when you are sheltered, innocent and thus naive about the capacities of the toughened, angry, ruthless, alienated world. He said, **'I am sending you out like sheep among wolves. Therefore be as shrewd as snakes and as innocent as doves'** (Matt. 10:16).

Plato also foresaw that naivety would be a problem. He said, **'Nor will you think it strange that anyone who descends from contemplation of the divine to the imperfections of human life should blunder and make a fool of himself, if, while still blinded and unaccustomed to the surrounding darkness, he's forcibly put on trial in the law-courts or elsewhere about the images of justice or their shadows, and made to dispute about the conceptions of justice held by men who have never seen absolute justice'** (p.282).

Astonishingly enough, as can be seen from the above quote, Plato predicted that we would be **'forcibly put on trial in the law-courts'**.

. . . .

. . . .

. . . .

. . . .

. . . .

. . . .

The traditional practice in court of swearing on the *Bible* **'to tell the whole truth and nothing but the truth'** is a mockery because all resigned humans are dedicated to lying. Humanity has been living in denial of all the truths that bring the issue of the human condition into focus; it has been lying on a massive scale. What we are introducing is information that does not lie, that does not comply with this denial, and now we are asking the legal system to defend us. Yet the legal system has been upholding the lying world, the world of denial. In the same way as mechanistic science, our system of justice has been complying with that historic need to deny any truth that brings the issue of the human condition into focus. . . . .
. . . .

It is true that while the dignifying understanding of the human condition had still to be found there was some need to lie, to block out and live in denial of and, in the case of truth-acknowledging prophets, to fabricate a way of repressing their overly confronting truth. As Booker's review of the Sir Laurens van der Post biography said, **'certainly there are those who relish the idea of** [ie there are those who need] **a hatchet job being done'** on someone as profound in his thinking as Sir Laurens van der Post.

There are two points to be made about this justified need to repress and deny truths that brought the human condition into focus.

Firstly, there always had to be a balance between denying confronting truth in order to protect humans from suicidal depression and allowing some confronting truth to survive. This balance was necessary because there had to be some confronting truth, some honesty in the world, or the world would become unbearably dark and sinister. It was also necessary if the full dignifying truth about humans was to be found. For example, if Sir Laurens van der Post's books had not been allowed to be published, I doubt I would have had the strength to stand so alone against the world of denial and eventually find the greater dignifying understanding of the human condition. I am deeply indebted to Sir Laurens van der Post, which is why *Beyond* is dedicated to him. With regard to the world becoming unbearably dark and sinister, if you study the so-called 'great' literary works, what is great about them is that they manage to defy, to some degree, the denial that almost everyone is practising. They punch holes, as it were, through the heavy layer of block-out and alienation that is blanketing the world. While humans needed to live in denial, they also recognised the need for some honesty.

How much human condition-confronting information a society could tolerate depended on the average level of alienation in that society. The more alienated and insecure a society the less human condition-confronting information it could tolerate. Obviously the responsibility of society was always to try to tolerate as much **'sun'**—human condition-confronting information—as it possibly could.

Secondly, as has already been emphasised, it was always necessary to allow for the possibility of the full dignifying understanding of the human condition to emerge. To completely repress human condition-confronting information would mean destroying any chance humanity had of achieving its freedom from the human condition. Not to allow for that possibility would be akin to killing the goose that lays the golden egg.

As has been stressed, the democratic principle of freedom of expression was put in place to ensure that some human condition-confronting truth would always be allowed in society (assuming not everyone became completely alienated), and that ultimately understanding of the human condition would be able to emerge.

While it is acknowledged that some fabrication or unfair argument or misrepresentation has been necessary when opposing human condition-confronting information (since the repression could not be achieved using truthful evidence, as it was the truth that was being repressed), there has always also been the need and responsibility to allow for some freedom of expression of human condition-confronting information in society. . . . .
. . . .

It was explained earlier that where human condition-confronting information is involved there is a great danger that the more alienated, in their extreme fear and resulting hatred of such information, and in their extreme 'deafness' and resulting inability to assess and appreciate the possible importance of such information, may take it upon themselves to act as guardians for humanity against the human condition-confronting information. Overriding the principle of democracy, they may set out to destroy the information by any means

possible. . . . .
. . . .

As has been carefully explained, there is no more confronting and difficult material for humans to have to adjust to than the arrival of understanding of the human condition. Again, as George Bernard Shaw observed, **'all great truths begin as blasphemies'**, and there is no 'greater truth' than understanding of the human condition. Exposing and dismantling humans' historic denial of the issue of the human condition can appear to be the ultimate **'blasphemy'**, **'impiety'**, irreverence, heresy, profanity, sacrilege. Feeling it to be such a threat, people can set out to destroy the threat by any means at their disposal, no matter how undemocratic their actions. However, if in their fury, people allow themselves and/or are allowed by society, to abandon the democratic principle of freedom of expression, then the worst possible crime can be committed.

Our democratic principles were arduously formulated, fought and died for, and enshrined in law and constitution, to ensure that when humanity embarked on the final crucial stage of our journey to self-understanding, prejudice would not prevent that all-important understanding emerging. The principles were not developed to be dispensed with the moment self-confrontation loomed; quite the contrary, it is precisely at that point that these principles should be scrupulously adhered to.

. . . .

. . . .

Apology: As explained in Notes to the Reader, the dotted lines indicate text that has temporarily been withdrawn because it deals with issues before the courts. Once the legal restrictions end the fully restored pages will be available on the FHA's website, or from the FHA.

. . . .

. . . .

. . . .

. . . .

. . . .

. . . .

If those of us who are supporting these ideas can stand up to and defeat this resistance then humanity will be liberated from the human condition, because these ideas do resolve the human condition. As has been stressed, the evidence for that is that the subject is being so openly discussed here and that people are already able to confront the human condition using these understandings.

In fact, what is required now is not repression of these ideas but support and development of them. Now that the fortress walls of denial have been breached, humanity needs as much truth as possible. Indeed humanity needs a veritable banquet of understanding about the human condition, because these understandings, while dignifying, are initially extremely confronting and destabilising, so significant bridging understanding and associated psychological counselling is required to help people manage the change. This banquet of understanding is also required because the sooner we can put an end to the argument and resistance to the truth about humans, the sooner we can resolve the dangerous situation that the world has arrived at where alienation and its effects are destroying humanity and the planet. The great battle between those who want to stay in the darkest corner of the pit of their cave **'prison'** of denial, with all its dangerous distortions and angry hurt from having to live in such a dark state, and those who want to enter the denial-free sunlit, true world—the long-anticipated battle of Armageddon—needs to be a short battle, and in fact it can be.

C3: Humanity's departure from the cave to life in the sun

At the beginning of the previous section a quote from Plato's *Republic* was included that described the difficulty humans would have leaving 'the cave' of denial. The quote ended with the comment that **'the process would be a painful one, to which he** [the prisoner in the cave] **would much object, and when he emerged into the light his eyes would be so overwhelmed by the brightness of it that he wouldn't be able to see a single one of the things he was now told were real.'** Plato followed this comment with the following description of how humans can gradually adjust to a life 'in the sun' free of denial outside the 'cave': **'Certainly not at first, because he would need to grow accustomed to the light before he could see things in the world outside the cave. First he would find it easiest to look at shadows, next at the reflections of men and other objects in water, and later on at the objects themselves. After that he would find it easier to observe the heavenly bodies and the sky at night than by day, and to look at the light of the moon and stars, rather than at the sun**

and its light. The thing he would be able to do last would be to look directly at the sun [the truth of integrative meaning or God], **and observe its nature without using reflections in water or any other medium, but just as it is. Later on he would come to the conclusion that it is the sun that produces the changing seasons and years and controls everything in the visible world, and is in a sense responsible for everything that he and his fellow-prisoners used to see. And when he thought of his first home and what passed for wisdom there and of his fellow-prisoners, don't you think he would congratulate himself on his good fortune and be sorry for them?'** (p.280).

The final sentence emphasises just how wonderful, brilliant and above all meaningful the world really is once you can see it free of denial. What passed for wise insight in the 'cave' of denial was in truth superficial shallowness. There is a comment in the *Bible* that makes the same point: **'The Lord knows that the thoughts of the wise are futile'** (I Cor. 3: 20 & Psalm 94:11).

The *Encarta* entry states that Plato anticipated humanity's **'escape into the sun-filled setting outside the cave…the transition to the real world, the world of full and perfect being.'** Given that the understanding of the human condition that is now available survives . . . .

. . . .

. . . . and given that humans can develop the patience and perseverance needed to overcome the historic denial—the patience **'needed to grow accustomed to the light'**—then humans *will* discover an incredible freedom, **'the world of full and perfect being.'**

With understanding of the human condition found, humans can be shown that it is now safe to confront the truth about themselves and by so doing end their alienated state of living in denial of the issue of the human condition. Plato talked about the need to **'put sight into blind eyes'**, and proceeded to say, **'this capacity** [of **a mind…to see clearly**] **is innate in each man's mind'**, but once lost **'the mind as a whole must be turned away from the world of change** [away from the world that denies the absolute, unchanging truths or realities] **until it can bear to look straight at reality, and at the brightest of all realities which is what we call the Good…this business of turning the mind round might be made a subject of professional skill, which would effect the conversion as easily and effectively as possible. It would not be concerned to implant sight, but to ensure that someone who had it already was turned in the right direction and looking the right way'** (p.283). With these perceptive words Plato was anticipating the practice of psychiatry, which will become one of the most important activities in the world now that understanding of the human condition is found. Humans can and must turn their minds from the practice of denial to the new truthful way of thinking. When Plato referred to **'someone who had it** [the capacity to see clearly] **already'** he had already commented that this capacity **'is innate in each man's mind'**, so the 'someone' is in fact everyone. All humans were born innocent, with a perfect instinctive orientation to cooperative meaning.

Scottish psychiatrist R.D. Laing referred to this need to **'turn the mind round'** from its alienated state when he wrote: **'Our capacity to think, except in the service of what we are dangerously deluded in supposing is our self-interest, and in conformity with common sense, is pitifully limited: our capacity even to see, hear, touch, taste and smell is so shrouded in veils of mystification that an intensive discipline of un-learning is necessary of *anyone* before one can begin to experience the world afresh, with innocence, truth and love**[p.23 of 156]**...True sanity entails in one way or another the dissolution of the normal ego, that false self competently adjusted to our alienated social reality: the emergence of the "inner" archetypal mediators of divine power, and through this death a rebirth, and the eventual re-establishment of a new kind of ego-functioning, the ego now being the servant of the divine, no longer its betrayer** [p.119]**'** *(The Politics of Experience* and *The Bird of Paradise,* 1967).

With regard to being honest about humans' corrupted state, to living with the truth rather than in denial of it, to acknowledging people's alienation, to differentiating people according to their level of soundness, which was referred to earlier in this essay when talking about the 'meek inheriting the Earth' and the 'first being last and the last first'—it has to be emphasised that the more innocent or less corrupted or 'meek' have no interest in power and glory. It is difficult for people living in denial of the human condition and living off egocentric, competitive and aggressive forms of reinforcement not to project their way of viewing the world onto everyone else, but the truth is there are people sufficiently innocent and thus secure in self, people in whom our species' original instinctive self or soul is sufficiently uncorrupted, for them not to be ego-centric, not to have their mind centred around ego and in need of the reinforcement to be gained from fame, fortune, power and glory. In fact all humans will have a selfless rather than a selfish mindset when they finally leave the cave. As Plato said, **'Will our released prisoner hanker after these prizes or envy this power or honour? Won't he be more likely to feel, as Homer says, that he would far rather be "a serf in the house of some**

landless man", or indeed anything else in the world, than live and think as they do?' (p.281). Christ was also aware of this when he said **'If anyone wants to be first, he must be the very last, and the servant of all'** (Mark 9:35).

In the *Introduction* it was emphasised that people need to adopt an attitude of patience and perseverance to overcome their state of denial. Plato made the same point—the cave dweller **'would need to grow accustomed to the light before he could see things in the world outside the cave.'** The fact is, with such a massive paradigm shift as is involved in changing from living in denial to living free of denial, there is a great deal of psychological adjustment that has to take place, which takes time. The existence of people living with and in support of these understandings in the FHA demonstrates that these understandings of the human condition do enable people to at last safely begin confronting the issue of the human condition and begin living in a denial-free, truthful world.

This psychological adjustment takes time, in fact the full adjustment will take a number of generations, but this does not mean that people who are in various stages of freeing themselves from living in denial cannot participate fully in the new, exciting, humanity-liberating paradigm. People can live fully in support of these critically important understandings while slowly confronting the truth that the understandings contain. The more alienated the individual, the slower the process of confronting these understandings has to be. To support the understandings you only have to have established in your mind that they are indeed the liberating understandings that humanity has been searching for. Once you acknowledge that then you know that these understandings are worthy of your support and you can offer that support while avoiding overly confronting the truth the understandings contain. The last chapter in *Beyond* describes the various activities of the human condition-resolved new world. It shows everyone participating in different ways, each according to their particular level of insecurity with the truth. Individuals can now live an extraordinarily meaningful existence despite their yet-to-be-ameliorated insecurities and by so doing they can avoid impeding humanity's liberation from the human condition. Once people recognise this is possible, their relief and enthusiasm for this truthful, all-meaningful and liberating way of living will carry all before it. Soon from one end of the horizon to the other an army will appear in its millions to do battle with human suffering and its weapon will be understanding.

It should be emphasised that the new way of living that is now possible is fundamentally different to, in fact the very opposite of that in a religion or cult, where people deferred to or put their faith and trust in a deity or figure of reverence, or even in a belief system. The freedom that understanding of the human condition brings is not dependent on dogma or belief or faith, in fact it is such a fundamental liberation that it takes humans beyond the situation where they need dogmatic forms of reinforcement or religious faith and belief. Humans no longer have to deny, escape and transcend the issue of their corrupted self, deferring to others for the management of their lives. They can *understand* and begin confronting their corrupted condition, ultimately end their insecure state, and begin managing their lives through their ability to understand existence.

What is being introduced means the end of faith and belief and the beginning of knowing. Knowledge is the opposite of faith, dogma and belief. Humans can *know* the truth now. In fact their faith, trust, hope and belief that one day understanding of ourselves would be found has been fulfilled. As Bronowski predicted, **'Knowledge is our destiny. Self-knowledge, at last bringing together the experience of the arts and the explanations of science, waits ahead of us.'** The convergence of the humanities and science—ultimately of theology and biology—eventually had to occur. Religions themselves looked forward to their fulfilment, and, paradoxically, obsolescence. For example, in 'Genesis' in the *Bible* it says that one day **'you will be like God, knowing'** (3:5), and in Chapter 9 of the *Lotus Sutra,* Buddha (Siddartha Gautama 560–480 BC) says, **'In the future they will every one be Buddhas. And will reach Perfect Enlightenment'** (*The Lotus of the Wonderful Law,* tr. W.E. Soothill, 1987, p.148 of 275). 'Revelation', the last book in the *Bible,* refers to the arrival of a new world free of the horror of human suffering under the duress of the human condition, stating **'There will be no more death or mourning or crying or pain, for the old order of things has passed away'** (Rev. 21:4).

The finding of understanding of the human condition takes humanity beyond the insecure state of the human condition, the state of insecurity that religions catered for. It takes humanity as a whole from the insecure state of adolescence (where humans search for their identity, for understanding of who they are, specifically for understanding of their divisive nature, the human condition) to the secure state of adulthood—hence the name for the organisation that has been established to promote and develop the understanding of the human condition that is now available, the Foundation for

Humanity's Adulthood.

During the Middle Ages the Italian saint, Thomas Aquinas, produced his monumental work *Summa Theologica* in which he **'sought to close the gap between reason and faith, holding that reason can prove the existence of God, and that nothing in Christian teaching is contrary to reason'** (*The Last Two Million Years,* 1974 edn, p.291 of 488). Of course Aquinas has been proved right—reason has proved the existence of God, and nothing in Christian teaching turns out to be contrary to reason. I say 'of course' because faith and reason, religion and science, are clearly two different perspectives of the one reality and so it is obvious that they had to be ultimately reconcilable. The concern with Aquinas' view was that while the human condition had still to be explained there remained a need to maintain the denial of it that protected humans from suicidal depression. Humans needed to face away from the **'fire/sun'**. 'God' needed to be either totally denied or viewed as a mystical, abstract concept that was not too confronting. The condemning truth of the cooperative, integrative meaning of life needed to be evaded and denied. To a large degree humans had to be protected from reason making their reality transparent—that is, until understanding of the human condition was found. For many people it was a case of the whole truth or no truth. People needed an intellectual escape from the truth of their circumstances, a way to mentally evade the unconfrontable. As a result a whole series of intellectual evasions of Aquinas' obvious truth developed, such as:

'Rationalism', as promoted by René Descartes, with his dictum that **'doubt itself cannot be doubted'** (*The Last Two Million Years,* 1974 edn, p.292 of 488), which *denies* that anyone can be sure of what is true, in which case no one can be sure that there is any such thing as an ideal state to become corrupted and alienated from;

'Empiricism', as promoted by John Locke and David Hume, which held **'that nothing exists but sensations—there is no God to underlie them'** (ibid. p.293), which *denies* there is any profound truth, no ultimate morality, and therefore nothing to condemn humans;

'Socialism', as promoted by Karl Marx with his belief that **'The philosophers have only interpreted the world in various ways; the point is to change it'** (*Theses on Feuerbach,* 1845), which *denies* the truth that the object of human life is to try to confront the issue of the human condition and ultimately find the dignifying, reconciling, ameliorating understanding of it, and instead maintain that the object is simply to stop humans being non-ideal, deny them their reality, by dogmatically imposing ideality upon them;

'Existentialism', as promoted by Jean-Paul Sartre, which **'stresses man's freedom to determine his own future'** *(The Last Two Million Years,* 1974 edn, p.293 of 488), that is be free from, live in *denial* of, the depressing issue of your corrupt reality and instead simply enjoy being whoever you are.

The 'New Age', 'Self-Improvement', 'Peace', 'Green' and 'Feminist' movements were called **'social revolutions'** and posed as **'alternative cultures'** but they failed to revolutionise anything, they failed to introduce any alternative to the artificial, superficial, destructive, unequal, egocentric, dishonest world humans have been living in. The truth is that while they moderated some excesses, these movements brought no fundamental change, no 'revolution', no 'alternative' to the horror of our species' plight. These 'movements', led by false prophets, seduced people with the belief that they could bring about an equitable, sensitive, caring, corruption-free, healed **'New Age'** without having to confront and solve the human condition. They sought to dogmatically impose idealism rather than attempt to understand why humans have not been ideally behaved. In fact their proponents opposed trying to confront, understand and by so doing reconcile the issue of the human condition. Instead of leading humans to peace they were leading humanity away from it. These pseudo-idealistic movements were merchants of delusion, they cultivated rather than clarified a *denial* of reality. . . . .

. . . .

. . . .

. . . .

Most recently, 'Postmodern Deconstructionism', promoted by Jacques Derrida and Michel Foucault and others, has taken lying to its most sophisticated level, asserting that life **'has no absolute truth or meaning'** *(Macquarie Dict.* 3rd edn, 1998), that is there is no truth or meaning so there is nothing from which to be corrupted and alienated. It is the ultimate *denial* because by eliminating meaning and truth the concepts of innocence, corruption, alienation and the human condition become meaningless.

Thankfully, with the human condition explained and humans at last dignified, all intellectual escapism is unnecessary. All the lying stops. Because humanity's predicament was largely one of wanting 'all the truth or no truth', our world has largely been one of no truth, all lies, a totally dark cave existence, an almost completely empty, meaningless, virtually dead realm. This quote about intellectualism from Sir Laurens van der Post's writing was included in the *Introduction:* **'Once asked then which people he** [psychoanalyst Carl Jung] **had found most difficult to heal, he had answered instantly, "Habitual liars and intellectuals"...Jung maintained that the intellectualist was also, by constant deeds of omission, a kind of habitual liar'** (*Jung and the Story of Our Time,* 1976, p.133 of 275). Intellectuals were **'hard to heal'** because, being necessarily the 'hardest bitten' by the dilemma of the human condition, they took it upon themselves to be the custodians of denial, the keepers of the lie.

Although humanity had to evade the issue of the human condition while it was unsolved, now that it is solved *everything* changes. Humanity suddenly moves from a situation where there is virtually no truth to a situation where all the truth is revealed. Templeton prize-winning Australian biologist Charles Birch recently said that the **'meeting of science and religion is as yet no bigger than a cloud on the horizon'** (*Sydney Morning Herald,* 27 Feb. 1998). Since it is with religions that the truths about humans have been kept, albeit in safely abstract forms of description, such as 'God', 'soul' and 'sin', and in science that those truths have been investigated, albeit without them being acknowledged, then the 'meeting' between science and religion that Birch is talking of is the time when the reconciling, non-abstract and non-evasive scientific understanding of the human situation is finally synthesised.

Birch's metaphor of this reconciliation of science and religion being like a cloud on the horizon is apt because the cloud is the precursor of an immense storm of lightning and thunder. In fact there are many accounts describing the arrival of the dignifying and thus liberating, but, at the same time, all-exposing and thus immensely confronting truth about humans using precisely this imagery. For example, it was mentioned in the *Introduction* that Sir Laurens van der Post titled his book that anticipates the arrival of the naked truth about humans as **'The Voice of the Thunder'**, while the *Bible* describes the arrival of the truth about the human condition as being like a great storm with **'lightning, which flashes and lights up the sky from one end to the other'** (Luke 17:24, see also Matt. 24:27). Bob Dylan's 1964 song *When the Ship Comes In* anticipates what it will be like when the truth about humans finally arrives. He says it will be like a **'hurricane'** arriving: **'Oh the time will come up when the winds let up and the breeze will cease to be breathing/Like the stillness in the wind before the hurricane begins/**

The hour that the ship comes in/And the sea will split and the ships will hit/And the sands on the shoreline will be shaking/And the tide will sound and the waves will pound/And the morning will be a-breaking.'

What is being introduced brings the *real* 'culture shock', 'future shock', 'brave new world', 'tectonic paradigm shift', 'gestalt switch', 'turning point', 'renaissance', 'revolution', or 'sea change' humanity has long anticipated. The most exciting and the most challenging adjustment humans have ever had to make lies directly ahead of humanity now, over the next half century. Difficult as it will be, given how afraid of the naked truth some people are, adjusting to the truth about ourselves can be done. Humans would never have had the strength to pursue the immense journey to find knowledge, ultimately self-knowledge, if we had not always believed that when we did finally find that truth we would be able to cope with it. What makes truth-day, honesty-day, exposure-day, come-clean-day, self-confrontation-day, the day of reckoning, the day when people's alienations will be exposed—'judgment day' in fact—bearable is that it is actually a 'day' of great compassion, a state free of any condemnation or judgment. To quote an anonymous Turkish poet, judgment day is **'Not the day of judgment but the day of understanding'** (*National Geographic,* Nov. 1987).

Being able to understand why humans have been so competitive, aggressive and selfish removes the underlying insecurity in human life. The source psychosis of all human psychoses is repaired, allowing the corrupt aggressive, egocentric and selfish behaviour of humans to abate and eventually disappear forever.

Real reconciliation of the poles or duality of human life of good and evil in all their manifestations—such as of instinct and intellect, soul and mind, conscience and conscious, subjectivity and objectivity, ignorance and knowledge, dogma and logic, mysticism and rationalism, religion and science, faith and reason, holism and mechanism, idealism and realism, Yin and Yang, left wing and right wing, socialism and capitalism, women and men, young and old, black and white, innocence and corruption, soundness and alienation, happiness and unhappiness, frivolity and discipline, fragility and toughness, naturalness and artificiality, play and work, spiritualism and materialism, poverty and wealth, country and city, Abel and Cain, honesty and falseness, unevasiveness and evasiveness, instinctualism and intellectualism, altruism and egotism, sensitivity and insensitivity, the non-sexual and the sexual, peace and war, love and hate, selflessness and selfishness, cooperation and competition, the integrative and the divisive, etc, etc —has always depended on truth, which ultimately is understanding.

The proverbs assert that 'understanding is compassion', **'the truth will set you free'** (*Bible,* John 8:32), 'honesty is therapy' and 'in repentance lies salvation', but humans have never been able to 'understand' themselves, know 'the truth' about themselves, be 'honest' about their condition, explain why they have been divisively rather than cooperatively behaved and in so doing end their insecurity, ameliorate their lives and thus be able to 'repent' and change their ways.

Humans' divisive nature is not an unchangeable or immutable state as many have come to regard it, rather it has been the result of the human condition, the inability to understand themselves, and therefore it disappears when that understanding is found—which thank goodness it now is. The denial can now end, all the lying can stop. In Plato's imagery, we can now, as the song *Sunshine* from that immensely optimistic 1960s rock musical *Hair,* says, **'Let the sunshine/ Let the sunshine in/The sunshine in'** (lyrics by James Rado & Gerome Ragni).

The journey of human thought has been illustrated with the works of many great thinkers, yet the imagery and concepts expressed by those thinkers consistently echoes that of Plato's. Whitehead was right when he said the history of philosophy has been **'a series of footnotes to Plato'**.

Resignation

The subject of the human condition, together with humans' denial of it and the processes involved in its resolution, were summarised in the *Introduction* and elaborated upon in the first essay, *Deciphering Plato's Cave Allegory;* however, many unanswered questions remain. Many of these questions are answered in this essay, in the course of explaining what has been the most important psychological event in human life. This all-important event has dictated the whole nature of existence for adult humans from time immemorial and yet it has been an event that has not even been acknowledged prior to now. This event was an act of resignation, albeit reluctant, and therefore has been termed *Resignation.*

Some of the key outstanding questions about our species' historic denial of the human condition and the effects of that denial are:

- Were we born with this state of denial of the subject of the human condition?
- If we were not born with it, when did we adopt it?
- What precisely was so frightening and depressing about the subject?
- How exactly did the mind go about blocking it out?
- How have the Members of the Foundation for Humanity's Adulthood (the FHA is the organisation that has been established to promote and develop the understanding of the human condition that is now available) been able to overcome their denial and survive confronting the human condition?
- How much truth and beauty have humans buried in order to live in denial—just how superficial has humans' alienated state been?

- Now that we can understand the human condition, how much can humans who have grown up in denial hope to dismantle their denial and free themselves of all the blindnesses, insensitivities and superficialities that accompanied it?
- How different will humans of the near future—who will grow up free of denial of the human condition—be to humans of today?
- What will the world free of the human condition be like—how different will it be from today's world?

In addressing the first two questions, humans were not born with this denial. As will be explained in this essay, at about 12 years of age humans began to try to understand the dilemma of the human condition. However with humanity unable, until now, to explain this deepest of issues, by about the age of 15 they finally realised that they had no choice but to *resign* themselves to a life of living in denial of the depressing subject. While humanity was unable to acknowledge either the issue of the human condition or the resulting need to resign to a life of denial of it, it was not possible to describe and explain the unresigned life of children and young adolescents. Now that we can explain the human condition their world can at last be safely examined.

The elephant in the living room that only young people could see

While most people are effectively 'deaf' to discussion of the human condition, there is a particular group of people that can hear this information. This group is young adolescents. A study of their circumstances makes it starkly clear why they are not only able to hear, but why, when and how the denial of the issue of the human condition arises.

Lisa Tassone's wonderfully honest letter to the FHA, written when she was 16, has already been quoted in the *Introduction* but is worthy of another inclusion because it demonstrates just how easily young adolescents are able to access the information in my books. **'Before stumbling upon *Free: The End Of The Human Condition* that was discreetly shoved in the back of the philosophy section, I was at the end of my road. I had experienced a year of complete and utter pain, confusion, anger and**

frustration. When I finally took the plunge to seek medical help (as I was suicidal), I was diagnosed with severe depression and put on medication. After reading your book (which I stayed up till 2am reading, I just couldn't put it down), I have been one of the fastest recovering depressants around. No wonder why. If everyone knew your insights, so much would be resolved. The purpose of this letter is to thank you for your courage in publishing your sure-to-be controversial work, and for basically recovering and saving this 16 year old. Not only is your work the absolute truth and has restored my faith in humanity, it has given me inspiration to help others. I may seem young to know what I'm talking about but, well, I do. I have tested all your work and others and yours always held up' (Brisbane, 4 Oct. 1999).

While well-educated adults typically find my books difficult to read and understand, this 16-year-old clearly had no such difficulty. To explain how this is possible, we need to consider the situation young people encounter as they enter into adolescence, and face the world.

Considering humans have had to live in denial of the issue of the human condition but were not born with this denial already 'wired' into their mind (as will become clear in subsequent material), it follows that each new generation encountered and somehow adjusted to the world of denial. Adults who were already living in a fraudulent state of denial could not admit that they were, so it is to be expected that each new generation was greatly perplexed by what was occurring in the adult world around them.

This essay describes and explains this perplexing situation and its outcome.

The question of why humans are not ideally behaved, of how and why humans became corrupted, is the issue of the human condition. It is this issue that my first two books, *Free: The End Of The Human Condition* and *Beyond The Human Condition,* are primarily concerned with explaining. This explanation was summarised in the previous essay, *Deciphering Plato's Cave Allegory.*

In the *Introduction* it was explained that historically there has been no explanation for the riddle of the human condition; for why humans have been competitive, aggressive and selfish when the ideals are to be cooperative, loving and selfless. While humans could not explain the riddle of the human condition they had no choice other than to block the whole issue from their minds, because not to was to become dangerously—even suicidally—depressed. Eventually, humans learnt that the only way to achieve that block-out was simply to accept humans' divisive reality as normal and deny the whole concept and truth of cooperative ideality. If you make no acknowledgment of the cooperative, integrative meaning of life—if there is no state of ideality—then there is no conflict of human divisiveness with universal integrativeness, and therefore no issue, no dilemma of the human condition to become depressed about.

For humans to block out and deny the concept of cooperative ideality they actually had to block out and deny two concepts: firstly, an instinctive expectation within themselves of encountering a cooperative, loving, selfless world; and secondly, a conscious awareness within themselves from observing the world around them that life's meaning or purpose is to be cooperative or integrative. (Note, the biological explanation for how humans acquired an instinctive self or 'soul' that is orientated to behaving cooperatively and lovingly was also summarised early in the previous essay, *Deciphering Plato's Cave Allegory*, and is explained in detail in *Beyond* in the chapter 'How We Acquired Our Conscience'. Similarly, the physics that explain that the meaning of existence is to be cooperative or integrative was summarised at the beginning of the *Plato* essay, and is explained in detail in *Beyond* in the chapter, 'Science and Religion'.) In order, therefore, to save themselves from suicidally depressing thoughts about why they were not ideally behaved, it was necessary for humans to live life in denial of both the existence within themselves of instinctive expectations of a cooperative, loving world and the existence of a cooperative or integrative purpose to life itself.

That being the case, the question is, what happened to each new generation of humans when they arrived in a world of people practising such denial? Having not yet adopted denial, young people were still aware of both their instinctive self or soul's expectations of encountering a cooperative, loving world, and of the cooperative or integrative purpose to existence. It is easy to see that they would find their way of thinking and viewing the world completely at odds with the view and behaviour of those around them. The world around them was no longer cooperative, loving and selfless but instead extremely competitive, aggressive and selfish, and it was a world where almost all adults were denying there was anything fundamentally wrong with this non-ideal state of affairs.

This conflict of views placed the new generation in an extremely perplexing predicament. Young people were faced with having to try to reconcile their point of view with the view being presented by the adults around them and their associated world. While they were affected by the lack of cooperative, loving ideality in the world they encountered from birth onwards, it wasn't until their early teenage years that young people tried to *understand* why the world wasn't ideal. Their struggle with this dilemma is what historically has preoccupied and distressed the minds of young adolescents.

Young adolescents struggled mightily with the dilemma—essentially the question of the human condition, the question of why the world was not ideally behaved when they instinctively expected it to be. This struggle continued for a number of years until eventually, at around the age of 15, they realised there was no answer to the question and they had no choice but simply to deny the problem existed. They resisted this desertion from the truth for as long as possible, but finally accepted that the only sensible way of coping was to resign themselves to a life lived in denial of the depressing issue of the human condition—and of all the associated states and truths that reminded them of it. They accepted that the only way to cope with the problem was to block out those expectations of ideality, namely their loving, idealistic soul and the condemning truth of integrative meaning, and take up an attitude of denial like countless generations had done before them.

Each generation valiantly resisted resigning to a life of denial of soul and of the truth of integrative meaning because doing so meant living an immensely fraudulent and penalised life thereafter. In resigning, a person was evading the real issue before us as a species, which was the issue of the human condition; evading the most fundamental of truths about life, namely the truth of integrative meaning; and evading also the 'magical', all-sensitive, cooperative, selfless, loving—ideal—world of our instinctive self or soul. The price of resignation could hardly have been higher, it meant becoming an evasive, blocked-out, alienated, superficial and artificial being.

When you resigned you were effectively accepting the death of both your soul and mind. By blocking out the soul's central expectation of a loving world you were blocking out the whole sensitivity base of your being, while any thinking that tried to progress from denial of that most fundamental of truths, integrative meaning, was obviously going to be flawed and incapable of reaching any profound truth.

Dying in soul and mind so you could stay alive in the non-ideal upset world of reality became an inevitable and inescapable horror for humans and while resigning yourself to a life of denial was the only solution for almost all humans, accepting it as a way of life was not at all easy. The great (relatively unevasive) Scottish psychiatrist R.D. Laing clearly described just how fraudulent the resigned world of adults has been when he wrote: '[In the world today] **There is little conjunction of truth and social "reality". Around us are pseudo-events, to which we adjust with a false consciousness adapted to see these events as true and real, and even as beautiful. In the society of men the truth resides now less in what things are than in what they are not. Our social realities are so ugly if seen in the light of exiled truth, and beauty is almost no longer possible if it is not a lie'** (*The Politics of Experience* and *The Bird of Paradise,* 1967, p.11 of 156).

Young unresigned adolescents were able to see how shallow and fraudulent the resigned world was, and determinedly resisted resigning. Eventually, however, the horrific depression that came from trying to face down ideality without understanding of the human condition forced them to accept resignation.

In fact, they were to learn that the issue of the human condition was so depressing that they would have to dedicate almost their whole existence to escaping and distracting themselves from thinking about it.

The human mind has *had* to be dedicated to the task of denial, because the evidence for the dilemma of the human condition is not easy to avoid. Firstly, the theme of existence—integrative meaning—is on display all around us, every object we look at is a hierarchy of ordered matter, an example of the development of order or integration of matter. Secondly, all of life is deeply connected with our instinctive heritage, 'a friend of our soul'—an association that reminds our mind of all the truths associated with our soul's true world. Thirdly of course, human's competitive, selfish and aggressive—divisive—behaviour, which is in such conflict with integrative meaning, is also on display before us every day of our lives. The human race has prided itself on its capacity to think but its real mental skill has been in avoiding thinking. Despite humans' bravado about being intellectual beings, the truth is humans have spent most of their time avoiding thought.

This dedication, necessary as it was, has been yet another penalty to living in denial because it has limited humans' ability to be operational in the world and to savour the world. The distinguished British child psychiatrist and psychoanalyst, D.W. Winnicott, has described how the mind can repress subjects, force them out of conscious awareness into the subconscious, and how preoccupying and thus limiting it is maintaining that repression: **'The word "unconscious"...has been used for a very long time to describe unawareness...there are depths to our natures which we cannot easily plumb...a special variety of unconscious, which he** [Freud] **named *the repressed unconscious*...**[there is] **the fact that what is unconscious cannot be remembered because of its being associated with painful feeling or some other intolerable emotion. Energy has to be all the time employed in maintaining the repression, and it can easily be seen that if there is a great deal of an individual's personality that is repressed, there is relatively little energy left for a direct participation in life'** (*Thinking about Children,* 1996, p.9 of 343).

What made resignation especially difficult was that each new generation of adolescents had to accept it without the benefit of having older people who were already resigned, in particular their parents, acknowledge and talk to them about the psychological crisis they were going through. Unable to explain their divisive state, humans could not afford even to admit there was a problem associated with it. Not being able to confront and acknowledge what was taking place in a young adolescent's life, all that parents or adult friends could do was offer sympathy. Mothers could stroke their son or daughter's brow, but they could not acknowledge or in any way talk about the crisis their teenager was going through. There was 'an elephant in our living room', an all-important issue in human life that nobody was talking about, and young adolescents just had to discover for themselves why it was so necessary to ignore 'the elephant', why adults found the human condition such an unconfrontable, off-limits subject, and why each new generation of humans had to resign to a life of blocking out the subject of the human condition.

Given the truly awful world and state they were having to resign themselves to, and that they had to go through the agony of resigning without being able to talk to anybody about it, life leading up to resignation was a hellish existence for young adolescents.

Two of the most popular books amongst FHA Members are Sir Laurens van der Post's *A Story Like the Wind* (1972) and its sequel, *A Far Off Place* (1974). The books are a fictional account of two young people negotiating resignation. Although the books do not talk in terms of negotiating resignation, they do contain a powerful selection of symbolic messages associated with resignation. For example, the two central characters lose their parents (they learn that they are on their own negotiating resignation), they have to cross an immense desert (face entering the wasteland of the resigned world of adulthood), and their friends include a Kalahari Bushman and a faithful dog (they derive soul-strength from the relative innocence of the Bushman and the dog). In *A Far Off Place,* Sir Laurens describes the unreality of the world—the 'far off place'—that humans have resigned themselves to: **'He felt acutely that he was not going out into the real world at all, but entering not even a fortress so much as a new kind of menagerie, a prison in which partial forms of life were being preserved in a condition of unreality'** (p.406 of 413).

So horrific is the act of resignation that it often resulted in glandular fever. This illness has been termed the 'kissing disease' because it most commonly occurs at about the same time as, and has thus evasively been blamed on, the occurrence of puberty. This is an 'evasive' explanation because for glandular fever to occur a person's immune system has to be extremely rundown, yet at puberty the body is physically at its healthiest. Therefore for glandular fever to break out, adolescents must be under extraordinary psychological stress, much greater than the stresses that could possibly be associated with the physical adjustments to puberty. The stresses that cause glandular fever in young adolescents are those associated with having to resign.

The destructive effect of the silence from the world of resigned adults

While the issue of the human condition, the issue of the imperfection of the world around them, was devastatingly depressing for nearly all young adolescents, the loneliness of what they were going through was almost as difficult.

Adult humans live virtually in total denial of the issue of the human condition. In fact humanity is *a species in denial,* effectively a whole race affected by a type of amnesia; were there intelligence in outer space we would be known as the alienated species, so much is alienation our dominant characteristic. 'Conspiracy theorists', people who believe there is a great conspiracy on Earth, are often accused of paranoia, yet in fact there *is* a great conspiracy on Earth, but it is not some secret plan hatched by an elite few to dominate the world, as these theorists often maintain, rather it is this great conspiracy of silence about the human condition.

Young adolescents in each new generation tried to understand the paradox and hypocrisy of human life that was visible to them everywhere they looked. For a while they asked their parents why the world wasn't ideal. They asked, 'Mum, why do you and Dad shout at each other?' and 'Why are we going to a lavish party when that family down the road is poor?' and 'Why is everyone so unhappy and preoccupied?' and 'Why are people so artificial and false?' and 'Why do men kill each other?' Basically they were asking the fundamental question 'Why isn't the world ideal?' Parents, unable to explain the riddle of the human condition and fully resigned to a life dedicated to denying and evading the whole depressing subject, have not, until now, been able to answer these questions. In fact such questions have always made parents feel so awkward that young people soon learnt to stop asking. Nevertheless these questions are the truly important questions about human life, as George Wald pointed out in the *Introduction,* **'The great questions are those an intelligent child asks and, getting no answers, stops asking.'** An FHA Member employed as a nanny recalled these questions from a 13-year-old girl in her charge, **'I was always crushed as a kid because Mum and Dad couldn't answer all my questions, weren't you?'** and, **'Why are adults so silent. Why can't they remember what it was like as a kid?'** (personal communication, Aug. 2000). In the soundtrack to the classic 1973 film, *American Graffiti,* there is an achingly beautiful snippet of dialogue between the legendary 1970s disc jockey Wolfman Jack and a 13-year-old boy who has rung Wolfman's radio program. It appears at the beginning of the song *To The Aisle:* 'Wolfman: **Hellooo** Boy: **Yeahhh** Wolfman: **How old are you?** Boy: **I'm thirteeeen, how old are youuu?** Wolfman: **I'm only 14** Boy: **Oh boy, I love you Wolfman.'** While it is hard to convey the nuances in the voices, you sense a feeling of overwhelming appreciation by the boy of Wolfman's efforts to relate to his world.

The silence practiced by the resigned adult world produced an extremely confusing state of affairs for young adolescents to cope with, adjust to and try to understand. In fact the silence, the denial, was so great they almost invariably could not defy it. Eventually, the coercive effect of the silence combined with the pain of the depression that arose from trying to confront the human condition forced them to resign. Once resigned, the answer to the question of why the falseness existed became self-evident and they became part of the lie and problem encountered by the next generation. As the late

Jim Morrison of the rock band, The Doors, said, **'Nobody gets out of this world alive'** (by 'alive' he meant alive psychologically, not physically). The situation is described in the *Bible* metaphorically in the words, 'the sins of the father carry on from generation to generation' (see Exodus 20:5, Deut. 5:9). (It should be emphasised here that, from the greater perspective and tragic as it was, adults' resigned dishonest way of living was unavoidable and thus not a 'sin' in the sense of being something bad, evil or wrong. Understanding the human condition allows humans to understand themselves; it brings healing compassion into their lives; it lifts the historic burden of guilt from the human situation.)

Once resigned to living evasively you could no longer see the evasion, but in their unresigned state children and young adolescents were able to see the full horror of the human condition. The 19th-century novelist, George Eliot (the pen-name of Marian Evans), acknowledged the situation when she wrote that, **'Childhood is only the beautiful and happy time in contemplation and retrospect; to the child, it is full of deep sorrows, the meaning of which is unknown** [to adults living in denial of the human condition].'

A *Newsweek* article that discussed childhood stress was bordering on the truth when it said, **'Parents are frequently wrong about the sources of stress in their children's lives, according to surveys by Georgia Witkin of Mount Sinai Medical School; they think children worry most about friendships and popularity, but they're actually fretting about the grown-ups'** (May 1999).

Antoine de Saint-Exupéry articulated the child's point of view beautifully in his celebrated 1945 book *The Little Prince,* when he had the Little Prince say, **'grown-ups are certainly very, very odd'** (p.41 of 91).

Resigned adults have not been able to deal with childhood trauma because they have not been able to deal with the dilemma posed by human nature. Being resigned, they were committed to blocking out the truth of what it was that was tormenting children, namely the false world of adults. Resigned and unresigned minds have lived in two completely different worlds. Children and adults were like different species, each almost invisible to the other. Robert Louis Stevenson described the situation thus, **'And so it happens that although the paths of children cross with those of adults in one hundred places every day, they never go in the same direction; nor do they even rest on the same foundations'** (quoted in the introduction to the 1997 Spanish film, *The Colour of the Clouds*).

The 'silence' of resigned adults had a devastating effect on young adolescents. Psychologists coined the word 'codependent' to describe someone who is **'reliant on another to the extent that independent action is no longer possible'** *(Macquarie Dict.* 3rd edn, 1998). By far the main reason for codependent situations amongst humans—in fact the word 'codependent' would almost certainly not have been created had this reason not existed—was that adult humans lived in denial of and were silent about their resigned, alienated state. Of course this denial, lie, silence was not something that people could avoid. In fact, for the most part people have not been aware of their alienation. If the alienated realised they were alienated they would not be alienated—they would not have succeeded in blocking out and denying their corrupted condition. To the extent that people had some awareness of their alienation and its dishonest, artificial, superficial and extremely evasive (of so many fundamental truths) behaviour they tended to believe that the reason for their behaviour was self-evident to others. Being largely unaware of the existence of alienation and that it varied in degree from person to person, everyone tended to believe that everyone else was like themselves, and therefore through experiencing life the way they were experiencing it, others would empathise with their behaviour. The truth of the extreme differences between people in their degree of alienation has been one of the main truths that humans have lived in denial of.

The fact is, people have differed greatly in their degree of alienation, which, incidentally, is a situation that will continue until understanding of the human condition heals alienation. This difference in alienation was most pronounced between those who had not yet resigned to a life of denial of the human condition and those who had. The difficulty for those who had not yet resigned and were not alienated, and also for those who had resigned but were less alienated than others, was they had no way of either understanding or appreciating the distorted behaviour of people who were more alienated than themselves. If the more alienated would not be honest about their dishonest, artificial, superficial and evasive behaviour—which, as explained, they were not able to be until now—then the more innocent and thus less alienated were at a loss to appreciate their behaviour. This placed the more innocent and less alienated in a potentially codependent position. If the alienated would not admit that their behaviour was false or non-ideal and carried on as if there were nothing unnatural or distorted about it, then the more innocent had to decide which of the two of them was right. Was the more alienated person's insinuation that their behaviour was not false and non-ideal right, or was the more innocent person's view that it was false and non-ideal right? How did the more innocent resolve this question?

As has been mentioned, part of the resigned adults' world of denial was denial of the truth that humans once lived instinctively in a coöperative state. This past cooperative period means that humans now carry within them an instinctive expectation of encountering a cooperative, loving, trusting world. Someone with an uncorrupted, instinctive self or soul, that is, someone who is innocent, therefore expects others to be cooperative, loving and trustworthy. In encounters between the innocent and the alienated where the alienated said, in effect, that they were not at fault then, in their instinctive generosity and trust, the more innocent were left believing they must have been at fault, they must have been in the wrong. In their naivety, generosity and trust, they would question their own view, not the more alienated view. The more innocent did not know people were lying. Their more trusting nature made them susceptible to believing the more alienated were right, rather than accepting their own view of the situation.

Another element that put the more alienated person in a much stronger position than the more innocent one was the fact that, while innocence has not known about the state of alienation, because it has never experienced it, alienation has known about the state of innocence, because it came from there. While those in the alienated state could not acknowledge the state of innocence, because it was too confronting and exposing of their alienation, they intuitively knew what was going on, knew that the innocent person was struggling with their lies. While the world of denial and its alienation was a devastating mystery to innocents, it was no mystery to the alienated and as such they were not going to be affected by the innocents' point of view in the struggle between what the innocents were saying was right and what the alienated were saying was right.

Since the emergence of consciousness and with it the human condition, the more alienated have, sometimes unwittingly and sometimes deliberately, bluffed, seduced, and 'sucked in' the more innocent by their more corrupted behaviour; they have made the innocent doubt their thoughts, they have even made them believe and suffer from feeling that they must have been the cause of the alienated person's behaviour, such was the generosity and trust of innocence. These experiences of guilt, pain and hurt had the potential to destroy innocence, thereby adding to the ranks of the corrupted and alienated. This has been the essential pattern of behaviour in so-called 'dysfunctional' families. Also, as is explained in *Beyond* in 'Stage 2' of the 'Illustrated Summary of the Development and Resolution of Upset', and as will be explained in the first section of the next essay, 'Bringing peace to the war between the sexes', men have had a more egocentric role in the battle to overcome the human condition than women, which left women in a more innocent position in the battle and thus often codependent to men.

Various codependent situations occurred in relationships as a result of these differing levels of innocence and alienation in humans, but the most pronounced was the codependency between unresigned young adolescents and resigned adults because it was in this situation that the difference in alienation was the most extreme. This already extreme codependency was further compounded when the alienated adults were the adolescent's parents, because parents were the people in whom innocents placed their greatest trust. The greater the difference in the levels of alienation and the greater the degree of trust in the relationship, the more extreme and devastating the codependency would be. Even very young children suffered from being codependent to their parents; in fact, if resigned adults realised the degree to which children blamed themselves for their parents' behaviour and situation, they may have been too afraid to have had children. You can sense, in Lisa Tassone's letter, the relief from at last being told what is going on in the resigned adult world; at last someone had 'spilt the beans', let the truth out, broken the silence.

The Simon and Garfunkel's 1964 song *The Sound of Silence* contains the line, **'Fool, said I, you do not know—silence like a cancer grows.'** The silence of the resigned adult world has been murderous to young people. One reason children so enjoy Roald Dahl's books and Saint-Exupéry's *Little Prince* (which has sold more than 50 million copies and was the third most-read book of the 20th century, after the *Bible* and the *Koran)*, is because they acknowledge that the world of adults has been a monstrously weird and awful place. Dahl and Saint-Exupéry took the child's view of the adult world and tried to alleviate children's suffering by exposing the truth about the horribly false, vicious and distorted adult world. Completely unaware of their own terrifying, corrupt and dishonest state, many parents feared the horror in children's fairy tales would terrify and corrupt their children, but in fact children found some relief in the honesty of the stories. Adults use humour in a similar way, indeed humour only exists because of human alienation. For the most part, adult humans maintain a carefully constructed facade of denial but every now and then they make a mistake and the truth of their real situation is revealed, providing the basis for humour. When someone falls over, for instance, it's humorous because suddenly their carefully constructed image of togetherness disintegrates.

Now that the human condition has finally been explained, it is at last safe to completely and permanently break the terrible silence of the adult world of denial; it is safe to tell children why it was necessary for the adult world to be so corrupted, dishonest and awful, and in the process admit all the denied truth about that world. Parents can at last compassionately tell children the truth about themselves—and the result will be that children will no longer have to die inside themselves in a sea of silence and lies, nor will they have to resign. This will enable a new kind of human, free of soul-hurt and psychological damage, to grow up and populate the world.

Trying to confront the human condition within

Significantly, in addition to struggling with the unacknowledged and unexplained question of why the world around them was divisive when they expected it to be integrative, young adolescents faced another difficulty—that of understanding their own non-ideal, divisive state.

Before looking at this difficulty, it is necessary first to explain how young adolescents became upset—hurt, corrupted and divisively behaved—in the first place.

While humans were living in denial of the human condition and all the truths that brought the human condition into focus (especially the truth that our species once lived in a cooperative, loving, all-sensitive state), they were unable to see how devastating it would be for infants encountering their immensely corrupted, alienated world from a perspective free of that denial. It was pointed out earlier (in the description of the horrific effect the silence of the resigned adult world had on unresigned new generations) that humans have almost no instinctive expectation of encountering a corrupted, alienated world, in fact, quite the contrary. As briefly explained in the previous essay, and fully explained in *Beyond,* the corrupted state of humans only began to emerge some 2 million years ago, and only became well established roughly a million years ago, which means that the great majority of our species' past—from approximately 8 million years ago to a million years ago—was spent living in a predominantly cooperative, loving state. This means our species' instinctive heritage, and thus instinctive expectation, is essentially one of encountering a cooperative, loving world, a world that is behaviourally completely at odds with the world humans are now born into.

The main assault on an infant's expectation of a cooperative, gentle, all-sensitive and loving world comes from the inability to treat them in an exceptionally unconditionally loving way. As is explained in *Beyond* in the chapter 'How We Acquired Our Conscience', and as was summarised early in the preceding essay, the prime mover or main influence in the development of our species' past cooperative, utterly integrated state was the nurturing of infants. It was the unconditional love given to infants, by their mothers in particular but also by their fathers and their society, that gave rise to adults who were trained to behave unconditionally selflessly. This training in selflessness eventually became instinctive and our species became cooperatively behaved and integrated. Tragically, since the *necessary* battle with the human condition began to develop some 2 million years ago, it has not been possible for any child to be given the amount of unconditional love or nurturing it instinctively expects.

So, not only have infants had the shock of encountering an almost completely corrupted, alienated world, they have also had the shock of not receiving anything remotely approaching the amount of unconditional love their instincts expected. If we add to this equation the extreme sensitivity of the innocent, uncorrupted human, we can start to appreciate just how devastating an experience it is for infants and young children encountering humans' immensely corrupted, alienated, preoccupied and comparatively loveless world.

Certainly, since the human condition first emerged there must have been some natural selection for individuals who could somehow cope with this shock, so that infants coming into the world today have some instinctive readiness for the horror of the world they are to encounter. However, despite this instinctive 'toughening', the shock must still be immense and the resulting emotional and psychological devastation comparably immense. This emotional and psychological 'damage' has taken many forms, including block-out or denial of the perceived 'wrongs', anger and resentment towards the 'ill-treatment', withdrawal, depression, even schizophrenia and autism, whereby children mentally dissociate from the pain. And this has been the lot of all infants and young children since the struggle with the human condition emerged in the human species. We *are* an immensely heroic species to have endured such angst. This horrific situation that awaited infants was another reason that parents, had they not been living in denial of the horror of their world, would likely have had difficulty mustering up the courage to have children. Indeed, the rapidly escalating rate of adolescent suicide and the burgeoning increase in childless adults are symptoms of a society experiencing an extreme problem with parenting.

The truth is, whatever corruption, 'heartache', suffering, hurt and resulting soul-damage that befell us in later life, it was insignificant compared to the soul-damage we all incurred to varying degrees in our first few extremely vulnerable and, as it is acknowledged, 'impressionable' and 'formative' years. In fact, this major adjustment to the corrupt, alienated world produced what we have referred to as character or personality, each person's unique version of the soul-corrupted, human condition-distorted and afflicted state. The die of our character is cast during our infancy and early childhood.

The extraordinarily honest South African writer Olive Schreiner beautifully articulated the vulnerability and impressionability of infants and young children in *The Story of an African Farm:* **'The souls of little children are marvellously delicate and tender things, and keep forever the shadow that first falls on them...The first six years of our life make us; all that is added later is veneer'** (1883, p.193 of 300). A similar quote, derived from a Jesuit saying, is read out at the beginning of each of the *Seven Up* documentaries: **'Give me a child until he's seven and I'll give you the man'**. William Wordsworth mirrored this sentiment with, **'The child is the father of the man'**, while the Roman poet Virgil said, **'As the twig is bent so the tree inclines.'** In recent years a community advertisement on Australian television reminded parents that **'a happy childhood lasts a lifetime'**.

It was during infancy and early childhood—the period when humans first encountered the corrupt world and somehow adjusted to the shocks their instinctive expectations received—that most of the corruptions that have existed in the world were passed on to subsequent generations.

With this appreciation of the hurt and damage the psyche or soul suffered in early childhood, it is now possible to look at the effect this damage had on young adolescents approaching resignation.

As was mentioned earlier, young adolescents not only discovered the human condition without (the inconsistency of the corrupt external world with the cooperative, loving ideal world), they also discovered it within (their own lack of ideality, their selfishness, their upset anger and alienation arising from their hurt and damaged early childhood). It was the depression that resulted from trying to make sense of this corruption within themselves that finally forced young adolescents to resign.

When they were about 15 years old, humans came to fully experience and realise just how confronting cooperative ideality was of their own upset, corrupted, non-ideal state. Trying to understand a corrupt external world was difficult and depressing enough, trying to reconcile cooperative ideality with their own selfish divisiveness, with the upset anger, egocentricity and alienation within themselves, brought on overwhelming self-criticism and consequent depression and was thus the final straw in trying to resist resignation. Unable to explain their corrupted state—namely the human condition—humans could not disprove the utterly condemning implication that they were bad, evil, worthless beings.

In fact, were the depression induced by this implicit criticism to increase any further it would have resulted in suicide. It was at this crisis point that young adolescents recognised they had no responsible choice other than to concede defeat and take up the tactic of denying, evading and blocking out the cooperative ideal state and truths—and the whole depressing issue of the human condition which that state and those truths were giving rise to—despite how extremely penalised, fraudulent and mind, soul and truth-destroying a tactic it was.

To have resigned and be living in denial of the cooperative ideals of life is to have repressed and thus forgotten just how confronting and condemning innocence and its idealistic world was.

To help resurrect in the resigned adult reader's mind an awareness of just how exposing and condemning ideality was for young adolescents who were in the midst of resignation, we can look at some of the reactions humans of all ages have had to the world of ideality that innocence represents. These reactions reveal just how fragile the human situation has been, just how sensitive humans have been to the criticism that innocence and its idealism has represented.

Bullying in the playground is one such reaction as it is often a case of children with a more hurt background attacking more nurtured, innocent children because of the unbearable criticism the more innocent children represented of their less innocent state. Wearing sunglasses, while practical, has also been a way of attacking and holding at bay the innocence of the wholesome, natural sunlit world. Remember Plato's analogy of humans living in a cave away from the sunlight?

Surrounding themselves with people, objects and environments that did not confront humans with their own corrupted condition also helped in keeping the depressing issue of the human condition at bay. Even the recent popularity of palm trees in landscaping is to a large degree because they are spiky and punk-looking, objects as alienated as humans currently are.

So important has blocking out, evading and denying the subject of the human condition been for humans' sanity, so delicate has their situation been, so insecure have humans been that if they were confronted by innocence they would attack and even try to destroy it. An example of this extreme reaction is explained in detail in *Beyond,* in the chapter titled 'The Story of Homo', which explains that hunting animals was an activity largely concerned with the satisfaction gained from attacking the criticising innocence of animals. Certainly most of the food in hunter-gatherer societies was supplied by women's gathering, which prompts the question of why did men need to go hunting? The answer is to get even with innocence for its unjust criticism of their lack thereof. I remember seeing a cartoon depicting two cement truck drivers gleefully dumping their load of concrete on a tiny road-side daisy. This is merely an adult version of children burning ants and tormenting pets.

The extent of the satisfaction corrupted humans could derive from retaliating against the unjust condemnation that innocence represented is revealed in this comment by W.D.M. Bell, an African big-game hunter of the early 1900s: **'There is nothing more satisfactory than the complete flop of a running elephant shot in the brain'** (*African Safari,* P. Jay Fetner, 1987, p.113 of 678). Another sport hunter made his feelings of satisfaction from being able to 'get even' with unjustly condemning innocence perfectly clear when he said: **'Next thing I knew, a large male chimpanzee had hoisted himself up out of the underbrush and was hang-**

ing out sideways from the tree trunk, which he was clutching with his left hand and left foot. Looking down my barrel at ten yards was man's closest relative, an ape, which, when mature, has the intelligence of a three-year-old child. Wouldn't I feel like a murderer if I shot him? I had some misgivings as my globular front sight rested on the ape's chest and my finger on the trigger. But then, gradually, insidiously, my thinking took a different turn. I thought of the gorge-lifting sentimentality—most of it commercially inspired—that has come to surround chimpanzees. I thought of the long list of ridiculous anthropomorphic books about the "personalities" of these apes. I thought of that chimp who fingerpainted on TV and sold his "works" for so much money he wound up having to pay income tax. I thought of one ape who was recommended for a knighthood, the ape who was left his master's yacht, the ape who was elected to parliament in some banana republic; and various other apes who were made astronauts and honorary colonels. Gathering like storm clouds in my mind, these thoughts roused me to such a pitch of indignation that there appeared to be only one honorable course of action. I blasted that ape with downright enthusiasm and have felt clean inside ever since' (ibid. p.117–118). In Peter Beard's astonishing book, *The End of the Game* (1963), he reproduced a page from the journal of the famous African white hunter, J.A. Hunter, where Hunter records having dispatched **'996 Rhinos'** from **'August 29th 1944 to October 31st 1946'** (p.137 of 280).

The main preoccupation of humans has been with finding ways to cope with the criticism the world of idealism and purity made of their corrupted, impure, non-ideal state. Evasion, denial and retaliation against innocence saved humans from the suicidal depression that could result from attempting to confront the issue of the human condition. Humans have lived within a tiny comfort zone outside of which there has been mental terror, fearful depression. The horrible reality is that each new generation of humans had to learn this awful truth. They had to accept they needed to resign themselves to a life of denying the whole issue of the human condition, despite the high price involved of dying in soul and mind.

It can be seen from what has been described so far in this essay that, while the innocence of children, like all forms of true beauty, has been one of life's greatest inspirations and joys, it has also exposed and thus criticised and condemned the no-longer-innocent adult world that could not explain why it had become corrupted. This criticism that children's innocence presented to adults was yet another contributing factor to the schism that has existed between adults and children that has made parenting so difficult.

In the final stanzas of his poem *Children,* H.W. Longfellow (1807–1882) acknowledges the comparative innocence of children and the inspiration that they offer the corrupted world of adults: **'For what are all our contrivings/And the wisdom of our books/When compared with your caresses/And the gladness of your looks?//Ye are better than all the ballads/That ever were sung or said/For ye are living poems/And all the rest are dead.'**

The comic actor W.C. Fields (c.1879–1946) illustrated the confronting effect of children's innocence, when asked whether he liked children he responded, **'Ah yes.... boiled or fried'**. On another occasion he reported, **'I never met a kid I liked'**.

Living amongst young children is akin to living with the truth. In adults' resigned state, children's innocence dangerously exposed the mechanism they were depending on to cope with their non-innocent, corrupted state, which was their denial of the truth that there is an ideal, innocent state. Referring to children as 'kids' was really a derogatory, retaliatory 'put down', a way of holding their confronting innocence at bay.

These examples show how confronting the innocent ideal state has been to the corrupted state and it was precisely this confrontation—between their own corrupted state and the innocent ideal state that they could still access—that young adolescents found unbearably depressing.

The depression

The depression that resulted from trying to confront the issue of the human condition ultimately made resignation unavoidable for practically all young adolescents. So deep and dark was the depression that individuals, after resigning, made a silent agreement never to again allow themselves to recall the experience or the depressing states, truths and issues that caused it.

Depression was the main feature of resignation. Indeed, while humans do not normally acknowledge it, depression, or the blues, melancholia, the sickness unto death or the noonday demon, as it has variously been termed, has been one of the main ailments of human life for the past 2 million years.

Depression is the darkest of human states and as such it is extremely rare to find such an honest description of it as occurs in Gerard Manley Hopkins' poem, *No Worst There is None.* The poem was written in the late 1800s in what is now archaic English, but there is no doubting what Hopkins is talking about: **'No worst, there is none. Pitched past pitch of grief/More pangs will, schooled at forepangs, wilder wring/Comforter, where, where is your comforting?/Mary, mother of us, where is your relief?/My cries heave, herds-long; huddle in a main, a chief/Woe, wórld-sorrow; on an áge-old anvil wince and sing—/Then lull, then leave off. Fury had shrieked "No ling-/ering! Let me be fell: force I must be brief"/O the mind, mind has mountains; cliffs of fall/Frightful, sheer, no-man-fathomed. Hold them cheap/May** [any] **who ne'er** [have never] **hung there. Nor does long our small/Durance deal with that steep or deep. Here! creep/Wretch, under a comfort serves in a whirlwind: all/Life death does end and each day dies with sleep.'** The word 'hung' is the perfect word for depression, for the state that there is 'no worse' than.

In *The Moral Intelligence of Children,* Pulitzer prize-winning author Robert Coles provides a remarkably honest description of the agony of resignation and the difficulty adults have had in helping adolescents during this period. He wrote: **'I tell of the loneliness many young people feel, even if they have a good number of friends...It's a loneliness that has to do with a self-imposed judgment of sorts: I am pushed and pulled by an array of urges, yearnings, worries, fears, that I can't share with anyone, really...This sense of utter difference...makes for a certain moodiness well known among adolescents, who are, after all, constantly trying to figure out exactly how they ought to and might live...I remember...a young man of fifteen who engaged in light banter, only to shut down, shake his head, refuse to talk at all when his own life and troubles became the subject at hand. He had stopped going to school, begun using large amounts of pot; he sat in his room for hours listening to rock music, the door closed. To myself I called him a host of psychiatric names: withdrawn, depressed, possibly psychotic; finally I asked him about his head-shaking behavior: I wondered whom he was thereby addressing. He replied: "No one." I hesitated, gulped a bit as I took a chance: "Not yourself?" He looked right at me now in a sustained stare, for the first time. "Why do you say that?"...I decided not to answer the question in the manner that I was trained to reply...an account of what I had surmised about him, what I thought was happening inside him...Instead, with some unease...I heard myself saying this: "I've been there; I remember being there—remember when I felt I couldn't say a word to anyone"...I can still remember those words, still**

remember feeling that I ought not have spoken them: it was a breach in "technique." The young man kept staring at me, didn't speak, at least with his mouth. When he took out his handkerchief and wiped his eyes, I realized they had begun to fill' (1996, pp.143–144 of 218).

When Coles says that **'I heard myself saying this: I've been there; I remember being there'**, and that that acknowledgment was **'a breach in technique'**, he is acknowledging that resignation has been something so dark humans have had to forget it, and indeed it has been something they have had a responsibility to forget if they were to effectively delude themselves that there was not another condemning ideal world.

The phrase **'I've been there'** is also used by the Australian poet Henry Lawson in his 1897 poem, *The Voice from Over Yonder*, which is about the depression that results from trying to think about why human life is at odds with the Godly, cooperative ideals of life: **'Say it! think it, if you dare!/Have you ever thought or wondered/Why the Man and God were sundered?/Do you think the Maker blundered?'/And the voice in mocking accents, answered only: "I've been there."'** The unsaid words in this final phrase are, 'and I'm not going *there* again'; the 'there' and the 'over yonder' of the title being the state of depression.

Resignation poetry

The poet Theodore Roethke was referring to resignation and the unhappiness of having to leave the magic world of our soul when he wrote, **'So much of adolescence is an ill-defined dying/An intolerable waiting/A longing for another place and time/Another condition.'**

The following examples of 'resignation poetry'—written in the midst of resignation—express the torturous process adolescents go through in accepting the death of their soul's true world and adopting the false, all-but-dead, deluded, blocked-out, resigned, alienated world. In the adult reader's resigned state of denial of another true world, it might be tempting to think these poems were somehow influenced by the FHA, however, they were written without any knowledge of the FHA or its literature, in fact before the FHA was established and long before my first books were published. Of course, parents of adolescent children only have to enter their own young teenager's room (bunker!) and ask if they may read some of their offspring's personal writings; if they are allowed, they will very likely be shown material of a similar nature. Alternatively, they can offer this essay and my other writings and see for themselves whether they respond as Lisa Tassone did.

The following astonishingly honest poem was sent to the FHA in February 2000 by 27-year-old Fiona Miller after she had just read *Beyond* and become aware of the FHA.

With the poem Fiona attached the comment, **'I dug out this poem I wrote in my diary when I was about 13 or 14 years old...It has always sounded very depressing to me whenever I have read it and so I have not shown anyone since leaving school...Maybe this was the "transition point"** [a term I had used about resignation in writings I had given Fiona] **for me when instead of trying to fight forever I just integrated very nicely!!??'**

This is the poem: **'You will never have a home again/You'll forget the bonds of family and family will become just family/Smiles will never bloom from your heart again, but be fake and you will speak fake words to fake people from your fake soul/What you do today you will do tomorrow and what you do tomorrow you will do for the rest of your life/From now on pressure, stress, pain and the past can never be forgotten/You have no heart or soul and there are no good memories/Your mind and thoughts rule your body that will hold all things inside it; bottled up, now impossible to be released/You are fake, you will be fake, you will be a supreme actor of happiness but never be happy/Time, joy and freedom will hardly come your way and never last as you well know/Others' lives and the dreams of things that you can never have or be part of, will keep you alive/You will become like the rest of the world—a divine actor, trying to hide and suppress your fate, pretending it doesn't exist/There is only one way to escape society and the world you help build, but that is impossible, for no one can ever become a baby again/Instead you spend the rest of life trying to find the meaning of life and confused in its maze.'**

Fiona's comment, that her poem **'always sounded very depressing to me whenever I have read it'**, indicates that prior to reading *Beyond* she could not remember the cause of the depression. In fact when I spoke with her after receiving this exceptionally honest poem she said she had always thought it was a result of homesickness when she first went away to boarding school. This lack of memory is an example of the phenomenon explained earlier, that humans retained little memory of having resigned after they had done so because resignation was such a dark and traumatic time in their lives and remembering what happened defeated the purpose of the denial that they have committed themselves to.

The second point arises from Fiona's description of the resigned, false state as being **'divine'**. This is exceptionally perceptive because while it is such a fake, soul-denying state, resignation was nevertheless the only responsible option available and as such was, in the greater sense, something so beautifully courageous that it was in fact divine. The greater truth is humans had to be incredibly brave to suffer becoming false in order that the human species, and thus the human journey to enlightenment, could continue. The suffering was endured in the hope that one day, in some future generation, we would discover the greater dignifying, and thus liberating, understanding of humans' divisive condition—a hope which has finally been fulfilled.

This great paradox of life under the duress of the human condition, where humans had to be prepared to 'lose themselves' (suffer becoming resigned) in order that one day our species might 'find itself' (find understanding of the human condition) is marvellously expressed in Joe Darian's 1965 song, *The Impossible Dream*, from the play *The Man of La Mancha*. In the words of the song, humans had **'to be willing to march into hell for a heavenly cause'**, suffer a life of fraudulent denial in order that humanity might one day find the dignifying understanding of their divisive nature and by so doing achieve the seemingly **'impossible dream'** of liberating themselves from the deeply depressing conclusion that their divisive nature means they are bad, evil, worthless beings.

Darian's song is marvellously descriptive of the agonising paradoxes of life under the duress of the human condition. It is one of humanity's great pieces of expression and now that we can understand the human condition, the lyrics can be clearly appreciated: **'To dream the impossible dream, to fight the unbeatable foe/To bear the unbearable sorrow, to run where the brave dare not go/To right the unrightable wrong, to love pure and chaste from afar/To try when your arms are too weary, to reach the unreachable star/This is my quest, to follow that star/No matter how hopeless, no matter how far/To fight for the right without question or pause/To be willing to march into hell for a heavenly cause/And I know if I will only be true, to this glorious quest/That my heart will lie peaceful and calm, when I'm laid to my rest/And the world will be better for this, that one man scorned and covered with scars/Still strove with his last ounce of courage, to reach the unreachable star.'**

Incidentally, the honesty of Fiona Miller's poem is very similar to

R.D. Laing's honest description of the post-resigned adult human state: **'The relevance of Freud to our time is largely his insight and, to a very considerable extent, his *demonstration* that the *ordinary* person is a shrivelled, desiccated fragment of what a person can be. As adults, we have forgotten most of our childhood, not only its contents but its flavour; as men of the world, we hardly know of the existence of the inner world'** (*The Politics of Experience* and *The Bird of Paradise,* 1967, p.22 of 156).

The second resignation poem was written by FHA Member Eric Crooke in 1983 when he was 12 years old, many years before he became aware of the FHA: **'Growing Up: There is a little hillside/Where I used to sit and think/I thought of being a fireman/And of thoughts, I thought important/Then they were beyond me/Way above my head/But now they are forgotten/Trivial and dead.'**

'Life' leading up to resignation

While full resignation occurred at about the age of 15, from the moment of birth onwards mini-resignations repeatedly took place as block-outs/denials/evasions were progressively implemented to cope with the hurt experienced from the many traumatic encounters with the non-ideal, 'imperfect' world associated with the human condition. In *House of Cards,* a 1993 film based on a screenplay by Michael Lessic, one of the characters makes this intuitive comment about how sensitive and vulnerable innocent children have been to the horror of the alienated state of adults: **'I used to watch Michael** [a character in the film] **about two hours after he was born and I thought that at that moment he knew all of the secrets of the universe and every second that was passing he was forgetting them'**. As mentioned, humans are born instinctively aware of the truth of cooperative or integrative meaning, and of a world compliant with it, but then the reality of the non-ideal world strikes and they have to begin repressing those truths to cope with that new reality.

As has been explained, most of the adjustments to the reality of the imperfection of life under the duress of the human condition took place in our first few, 'formative' years when we had very little ability to cope and when our mind was most 'impressionable'. Struggling with the imperfections of life at a very young age, young children were aware of the dilemma of the human condition. While resigned adults learnt to deny the existence of this dilemma, the truth is that an unresigned mind needs only a small sample of life to see the truth of it.

To illustrate, the following two conversations took place between an FHA Member who works as a nanny and her young charge, a boy of almost four years. First conversation: 'Child: **I'm Captain Hook and I'm going to kill you with my sword.** Nanny: **Why?** Child: **Because you're a good girl.** Nanny: **No, I'm Peter Pan and I will get the lost boys and Tinkerbell to save me from you.** Child: **No, you're not Peter Pan. You're Wendy and you're a good girl.** Nanny: **Why do you want to kill a good girl?** Child: **Because I don't like good girls.** Nanny: **Why don't you like them?** Child: **Because they are good.** Nanny: **But why don't you like good girls.** Child: **I like bad girls. Good girls make me feel bad.**' Second conversation: 'Child: **Do you have a heart?** Nanny: **Yes, I have a really big heart, really big so I can love lots.** Child: **Bet it's not as big as mine!** Nanny: **Probably not. Children have *really really* super huge hearts.** Child: **Well I don't, and I'm going to cut yours out.** Nanny: **Ow, why?** Child: **Because I don't have a heart any more, so I'm going to cut you in half and rip your heart out.** Nanny: **Ow, I think that would hurt me a lot.** Child: **So!**' (personal communication, Dec. 1999).

The existence of child prodigies in the realm of classical music also confirms that children have been aware of the dilemma of the human condition, and of the greater truth that, contrary to appearances, humans were not fundamentally evil beings. Classical music appealed to the shared awareness in humans of the greater truth of humans' true divinity, that stood above the terrible suffering the human condition inflicted on humans. Indeed if humans really had believed they were evil and not a part of 'God's work' they would have disintegrated with guilt—gone insane or suicided—long ago. While humans could not relieve their condition they could relate to the greater truth that humans are not evil, rather that they are involved in an immensely heroic journey to find the dignifying understanding of their divisive condition. Classical music was, as Charles Hart wrote in Andrew Lloyd Webber's *The Phantom of the Opera* (1986), **'the music of the night'**, the music humans had to consol themselves while having to live in Plato's dark cave of denial. Children could not empathise with, and express awareness of the subtleties about the human condition if they did not know of it.

Most significantly, while humans were *aware* of this dilemma from a very young age, by about the age of 11 they had sufficient understanding of the world to try to make sense of the fundamental dilemma of the human condition; to use philosophical thought to try to *understand* why there was imperfection in the world, and indeed in themselves.

School teachers acknowledge that at around 10 or 11 years of age students change from being boisterous—the so-called 'noisy nines'—to being sobered and introspective, and until they are about 15 years old they became, as one teacher described it, **'harder and harder to reach'**.

At this latter age young people finally surrendered to the silence of the world and accepted that it was impossible, indeed dangerously depressing to continue trying to confront and understand the problem of the human condition. They finally resigned to a life of evading the issue of the human condition.

Of course after resignation there was an immense change in personality. Adolescents suddenly became outward-looking, self-distracting, escapist, seekers-of-reinforcement-through-sexual-attention and success, extroverted and artificially buoyant. They became superficial young adults, carbon-copies of virtually every other grown-up; they **'just integrated very nicely'**, to quote Fiona Miller. Sadly, under the duress of the human condition, 'growing up' has really been all about 'dying down' in soul and mind.

As part of my research for a book I am writing that explains and describes all the stages of maturation a human experiences under the duress of the human condition, I asked a school teacher to describe what she and other colleagues consider the stages of childhood and early adolescence. This is her response: **'Six and seven-year-olds are considered to be very compliant but by eight children are starting to test the waters and challenge the world a little.'** She said **'the eight-year-olds can be annoying and a little naughty'**, while **'nine and ten-year-olds can be hard to handle as they seem to hit a phase of recklessness'** and **'they are considered naughty'**. She said **'Teachers love teaching 11 and 12-year-olds because it is during this stage that children become civilised.'** She commented that **'Teachers consider years nine and ten, when students are 14, 15 and 16 years old, the most difficult to teach. The adolescents seem to be at complete odds with what is expected of them. Most teachers are terrified of these completely uncooperative ages'** (personal communication, 1997). In light of these comments it is interesting to note that the Australian playwright, Richard Tulloch, wrote a popular play titled *Year 9 Are Animals* (1987).

With the issue of the human condition acknowledged it is not difficult to interpret these stages of childhood and adolescence. By the age of eight, children had begun to wrestle with the agony and frustration of not being able to know why they and the world were not pure and innocent. Within a year or two, at the ages of nine and ten, they had become so frustrated with the problem that they began to physically lash out at the world. This reactionary, 'noisy nines', hitting-out stage did not last long because they soon learnt the futility of such a response. By the ages of 11 and 12 they had realised that the only satisfactory way to cope with the problem was to find understanding of why they and the world were not ideal. Behaviourally they changed direction entirely. From being extroverted and reacting against the world, they became sobered, introspective and deeply thoughtful searchers. They tried to understand the human condition. This search for answers about human life lasted only a couple of years before they realised that neither they nor humanity as a whole had any answers to their questions. To make matters worse, they learnt that trying to confront the issue of the human condition led to deep depression. By about the age of 15 they reached the crisis point where any greater depression would make them suicidal, and they finally accepted that they had no responsible alternative but to resign themselves to a life of denial of the human condition, as fraudulent and soul-and-mind-destroying as that was.

The agony of resignation is evident in this quote from a newspaper article titled *Who Took Sarah's Self Away*: **'She turned 13, half woman half child, part of the confusion was that one day everything was fine and then the next day you're miserable. But you can't put your finger on what has changed to make everything suddenly so wrong. I went down to my grandparents' farm with my parents one year and ran around and had a wow of a time. I thought they were all the greatest. Next year I went there again and hated everything and everyone. I looked at them and thought "god they're so old and folksy I can't stand it". It's almost like doing a complete 360 degrees in a matter of a few years. It was a bizarre time. Why do so many fearless out going girls turn into parent-hating, unhappy insecure ghosts of their former selves during their early teens?'** (*Sydney Morning Herald*, 25 Apr. 1996).

The agony is also apparent in the explanation Ella Hooper offered for naming her popular Australian band, *Killing Heidi:* **'Heidi...being a happy young girl...and killing the Heidi within means growing up'** (interview, Sydney radio 2Day FM, 29 Jan. 2001).

Having to resign to a life of denial and thus death of the world of your soul was a desperately sad decision to make. As will be explained in the later section, 'The moment of resignation', when girls resigned they were resigning to a life of having to be to some degree a 'sex object', involving destruction of their innocence, their soul. Young girls are renown for wanting to have a horse and those who are lucky enough to acquire one seem to love it dearly. It was as if horses were young girls last soul-friend before they died; died in soul. For their part, at resignation young boys had to accept the prospect of becoming angry, egocentric and alienated beasts. When young boys sat on the edge of the pond with their little toy sailing boats heading out into the pond dipping through the waves it was as if they were being themselves transported away from the terrible shores that they were having to live on, escaping to another world free of the horrors that this world held for them.

Eventually, all the beauty of the soul's world just had to be let go. The 1970 Lennon/McCartney Beatles' song *Let It Be,* voted one of the most popular songs of the 20th century, is an anthem to humans' historic need to resign themselves to **'letting it** [the subject of the human condition] **be'** until the time—which has now arrived—when **'there will be an answer'**. It is an anthem to humans' historic need to resign to a life of leaving the subject of the human condition alone and instead living in hope and faith that at some time in the future understanding of the human condition would be found. Here are the lyrics: **'When I find myself in times of trouble/Mother Mary comes to me/ Speaking words of wisdom/Let it be/And in my hour of darkness/She is standing right in front of me/Speaking words of wisdom/Let it be**//[chorus]//**And when the broken hearted people/Living in the world agree/ There will be an answer/Let it be/For though they may be parted there is/ Still a chance that they will see/There will be an answer/Let it be**//[extended chorus]//**And when the night is cloudy/There is still a light that shines on me/Shine until tomorrow/Let it be/I wake up to the sound of music/Mother Mary comes to me/Speaking words of wisdom/Let it be**// [extended chorus]**.'**

The symbolism of this song is similar to that of Mark Seymour's *The Holy Grail,* Bono's *I Still Haven't Found What I'm Looking For* and John Lennon's *Imagine* (as mentioned in the *Introduction),* in particular the common hope of **'waking up to the sound of** [the] **music'** of humanity's freedom from the human condition. Seymour's words were, **'Woke up this morning from the strangest dream/I was in the biggest army the world had ever seen/We were marching as one'**, Bono sang of **'the kingdom come/Then all the colours will bleed into one'**, while Lennon referred to a time when **'the world will be as one'**. The words from *Let It Be,* **'And when the night is cloudy/There is still a light that shines on me/Shine until tomorrow'**, express the hope (**'there is still a light'**) that, despite the horror of humanity's plight, understanding of the human condition will emerge (**'shine until tomorrow'**). The same hope appeared in Seymour's song, where it ends with the words, **'I'm still here, I'm still a fool for the Holy Grail'**, as in Bono's song, **'I'm still running** [hoping]**'** and **'You know I believe it** [believe that the time would come when **all the colours will bleed into one**]**'**. Lennon also challenged the listener to **'imagine'** [humans' freedom from the human condition].

While school teachers have almost invariably been resigned adults living in denial of both the human condition and the process of resignation that established the denial, they were aware that students were **'harder and harder'** to reach from the years 12 to 15 years old. An article written by an education reporter describes this tendency: **'It is known as the "turn-off" syndrome, and it is the sort of problem most teachers and many parents know only too well. Bright and promising students who seem to have the world at their feet, turn 13 or 14 and stop dead in their tracks. They lose interest in schoolwork and start to fail examinations. Many cannot wait until they reach 15 so they can drop out'** (*Sydney Morning Herald,* 18 May 1985).

Pink Floyd's 1979 song, *Another Brick in the Wall,* defiantly expresses the pre-resigned students' point of view, that they do not want to have to accept the world of lies that adults reside in: **'We don't need no education/We don't need no thought control/No dark sarcasm in the classroom/Teachers leave the kids alone/Hey teacher leave us kids alone/All in all it's just another brick in the wall/All in all you're just another brick in the wall.'** Paul Simon's 1973 song, *Kodachrome,* expressed similar sentiments with the line, **'When I think back on all the crap I learnt in high school it's a wonder I can think at all'**. In truth schools have been death camps for children, places where, tragically, adults have inducted each new generation into the adult world of evasion and denial that was necessary for almost all people if they were to cope with the unresolved dilemma of the human condition. Schools provided the platform for adults to pass on 'The Great or Noble [necessary] Lie'.

Psychiatrist R.D. Laing once wrote that, **'we are driving our children mad more effectively than we are genuinely educating them'** (*The Politics of Experience* and *The Bird of Paradise,* 1967, p.87 of 156). This comment recognises that children have been taught alienation, which in truth is a state of madness, a state of extreme disconnection from what is true.

The school system did offer one great kindness however in that it allowed young people to be with others like themselves, and not solely with adults.

While humanity was unable to deal with the subject of the human condition, the adult world was unable to explain what happened in children's minds as they grew up. However, with the human condition safely explained and humans' divisive state understood, we can at last honestly explain the very distinct 'stages with ages' that a child growing up experiences. In fact we can explain all the stages of maturation humans go through in their life—and indeed all the stages humanity as a whole has been and continue to go through. These stages have been broadly acknowledged as 'infancy', 'childhood', 'adolescence' and 'adulthood', but even these broad descriptions have never been clearly articulated. In *Beyond,* where the biology of the human condition is fully explained, these stages are at last made clear. Essentially, infancy is when consciousness appears and humans discover self. In childhood they play with the power of free will that consciousness brings. In adolescence, the stage humanity has been in for 2 million years, they go in search of their identity, an understanding of who they are, specifically an understanding of why they and humanity have not been ideally behaved. To end humanity's insecure adolescent state and enter secure adulthood, the intellect had to find understanding of the human condition. With that understanding found, humanity can leave the insecure upset adolescent stage and enter the peaceful maturity of adulthood. Infancy is 'I am', childhood is 'I can', adolescence is 'But who am I?' and adulthood is 'I know who I am'.

Being a member of a species that had not reached the maturity of adulthood meant that *individual* humans reaching their *own* adulthood were psychologically stranded in adolescence; they were adults who were still ignorant to their true identity. In the development of systems, one can progress to subsequent stages without having completed an earlier stage, but the subsequent stages are then severely compromised by the incompleteness of the earlier stage. The actress Mae West articulated this phenomenon of arrested and subsequent corrupted or damaged development as it affected the egocentric male gender, when she said **'If you want to understand men just remember that they are still little boys searching for approval'**.

The lyrics of *The Logical Song,* from Supertramp's 1979 album *Break-*

fast in America, bravely acknowledge that humans start their journey happy, all-sensitive and innocent only to become corrupted—as the song describes, as '**vegetable**[s]'. More significantly, in concluding with the repetition of the plea '**please tell me who I am**', the song emphasises humans' frustration at not being able to understand themselves. The lyrics are: '**When I was young, it seemed that life was so wonderful/a miracle, oh it was beautiful, magical/And all the birds in the trees, well they'd be singing so happily/so joyfully, so playfully watching me/But then they sent me away to teach me how to be sensible/logical, responsible, practical/And they showed me a world where I could be so dependable/clinical, intellectual, cynical//There are times when all the world's asleep/the questions run too deep/for such a simple man/Won't you please, please tell me what we've learned/I know it sounds absurd/but please tell me who I am//Now watch what you say or they'll be calling you a radical/a liberal, fanatical, criminal/Won't you sign up your name, we'd like to feel you're/acceptable, respectable, presentable, a vegetable!//At night, when all the world's asleep/the questions run so deep/for such a simple man/Won't you please, please tell me what we've learned/I know it sounds absurd/but please tell me who I am/Oh, won't you help me and tell me who I am/Who I am, who I am, who I am.**'

At last we can now explain what human life is really all about. What emerges is the story of humans, the most fascinating of all stories—because it is our own story. The 'Great Lie' can end, which, from an innocent's point of view, means all the 'bullshit', the dishonesty of the world of denial, can finally disappear. Of course from the resigned adults' perspective, the reason they *had to* practice such dishonesty can at last also be compassionately explained to all the world. Honesty can come to the world of lies and it will be like the coming of the summer rains to a parched and barren land.

Ships at sea

It has been explained that young adolescents can read my books with ease because they are unresigned and do not suffer from the deaf effect. Yet there is another smaller category of people who also can 'hear' description and analysis of the human condition. These people are what we in the FHA have come to term 'ships at sea'.

Occasionally an individual refused to resign when they should have—they refused to 'pull into port when the storms out at sea were raging'. Such people, perhaps one in a hundred, chose to hold onto an awareness of the idealistic true world rather than adopt the relieving but false option of blocking it out and living in denial. Although this allowed the person to retain their capacity to recognise and acknowledge the world of ideality and all the truths that emanate from that honest position, it also meant their corrupt self was constantly facing condemnation from idealism—they were constantly being 'tossed about at sea'. People who refused to resign when they should have held the moral high ground in terms of not having given in, but the price was often madness; 'ships at sea' chose honest madness over alienated stability.

Every year the Foundation receives a few letters from such 'ships at sea'. Invariably the writers express immense relief at finally having the truthful world that they have held onto acknowledged and explained. The following is a recent example. While the author of this letter has given the FHA permission to publish her letter, we have withheld her name to protect her privacy. The underlinings and emphasis are her own. She wrote:

'Dear Sirs, I'm taking the time to write to thank you for your book "Beyond The Human Condition". I can not begin to put into words what this explanation has done for me.

Many years ago my Dad looked at me in his deep strange way and said—that rainbow you're chasing isn't there and you will be very unhappy when you discover that. Words to that effect. I loved the ground my Dad walked on and everything he said was to me "the truth of God" [note the innocent's instinctive total trust in the world, which was explained earlier]**. He was right about the rainbow. For the last several years it just seemed easier and it seemed that others were happier with me when I was selling out to the "Big Lie" ("The Big Lie" is what I call it). Material success came and I was friends with the type of people who say things like this, "all our friends live in big houses". I remember that statement the best because I felt at that very moment "the wheels fall off". Misery set in—I believed I was genetically predisposed to addictive chemicals (until now). I'm not exaggerating when I say <u>I had</u> to use drugs to cope with the whole set up (society), "The Big Lie", the <u>upset</u>. I depended on the heroin haze to get thro' each day. I've always known my soul is too sensitive to cope with the state of the world. A few weeks ago my hubby found me bawling in front of the TV. An elephant had been shot on "Foreign Correspondent". There are many incidents like this. (I had been "clean", that means no drugs) for 3 months. My**

hubby and I went to the Gold Coast for our holiday—big mistake that one except that's where I bought your book—don't go there. After speaking to the maintenance guy of our hotel following noticing there was no birds—yes you read right we saw 2 ibises on the whole of Surfers Paradise. The hotels/motels/units put some poison on their roofs—kills the birds. Each day I'm dying a little more. When I hear that stuff I feel like something gets torn out of my heart.

Anyway your book is brilliant. I finished it about one week ago and coincidentally I found my purpose and a way out of the ugliness and hate. I'm going into business—kind of non-profit eco thing. I'm happy and I feel good—after reading "Beyond The Human Condition" I can feel myself evolving and I can see in others an undescribable "knowing". I think we (as humanity) are waking from the nightmare. This gives me hope, courage and purpose.

Keep up the great work guys. I'm still slaving for the government but not for much longer—at the moment the world is geared up to "suck your energy". I recommend you listen to Dire Straits "Industrial Disease" and "Telegraph Road". Truly wonderful lyrics and it always gives me confidence that I'm not alone in my inability to tolerate the "human condition". There are others out there that can see the destruction.

I've raved on. Really I just wanted to say thanks—thanks for contributing this excellent literature to humanity—thank you for telling the truth and thanks for gifting us with such an insightful Foundation. If you have any I would love a "Adam Stork" sticker!!

I may write to you again soon, my new direction is still in the early research stage, any ideas/contribution (non-financial) would be appreciated.

Thanks again...29 years old' (31 May 2000).

We can see how her father tried to warn her about the difficulty of holding out for 'rainbows'—idealism—and the immense struggle that doing so caused her, but we can also see the exceptional sensitivity and truthful awareness she has retained as a result of resisting resignation.

In *Consider Me Gone,* the singer and songwriter Sting composed these words about the danger of trying to confront the dilemma of the human condition: **'I've spent too many years at war with myself/The doctor has told me it's no good for my health/To search for perfection is all very well/But to look for heaven is to live here in hell'** (from the 1985 album *The Dream of the Blue Turtles*).

In *Giovanni's Room* (1956) the award-winning American author James Baldwin wrote: **'Perhaps everybody has a Garden of Eden, I don't**

know; but they have scarcely seen their garden before they see the flaming sword. Then, perhaps, life only offers the choice of remembering the garden or forgetting it. Either, or: it takes strength to remember, it takes another kind of strength to forget, it takes a hero to do both. People who remember court madness through pain, the pain of the perpetually recurring death of their innocence; people who forget court another kind of madness, the madness of the denial of pain and the hatred of innocence; and the world is mostly divided between madmen who remember and madmen who forget. Heroes are rare.'

The 'ships at sea' are those referred to by Baldwin as the **'people who remember' 'the garden' 'of their innocence'**, and as a result **'court madness through** [the] **pain'** of it. As was mentioned in the *Introduction,* the biblical story of **'the Garden of Eden'** is a metaphysical reference to our species' time in cooperative innocence, the time before the corrupt state of the human condition emerged. It was also explained that the **'flaming sword'** wielded by the angel Gabriel to prevent the banished Adam and Eve from returning to **'the Garden of Eden',** was a metaphor for the depressing dilemma of the human condition that has kept humans out of the Garden of Eden—kept humans from returning to alienation-free soundness and sanity.

In this quote Baldwin courageously admitted how, once people are resigned, they have a **'hatred of innocence'**. As was explained earlier, this was because of the unjust condemnation innocence presented to those more alienated. Someone once said to me, **'as far as I'm concerned the word "innocence" doesn't exist'**! That is how confronting and thus loathsome a concept it has been for some people.

Innocence was attacked, denied and repressed by the resigned world because of the condemnation innocence represented to the corrupted world of adults, just as it was within each human when they resigned. The trees, the wind, all of nature is 'a friend' of our original, innocent instinctive self or soul and by association has condemned humans for their lack of innocence; in response, humans have retaliated mightily against nature. The environmental or green movement, like almost everything in the resigned world of denial, proved to be empty rhetoric because it did not address the real issue involved in humans' destruction of nature, namely the human condition, and in failing to do so was not a real attempt to end humans' destruction of the natural world. Only by confronting and solving the human condition could the criticism the natural world represented be resolved, and the need to attack it be removed. In truth the green movement was just more 'bullshit', another 'brick in the wall' of the denial and delusion humans have lived off in the resigned world.

Baldwin's belief that some people—the **'heroes'** as he calls them—could both **'remember'** and **'forget'** is recognition that some people were less arrogant and deluded about their resigned, false, alienated state than others, and, as a result, were capable of at least alluding to the existence of another true world. That *was* a heroic thing to do because once you were resigned, breaking the silence, admitting to your alienation, being a little bit honest, 'ratting on' your condition, undermining or betraying yourself, *did* require considerable courage. The great literature of the resigned world, such as the examples given from the writings of Patrick White and Alan Paton and others in the *Introduction,* is 'great' precisely because of its honesty in breaching the fortress of denial in the face of the overwhelming necessity to maintain that fortress.

The courage required to disown your resignation was different to the courage of 'ships at sea' who avoided resigning when their corrupted condition dictated that they should. Similarly the **'pain'** associated with avoiding resignation was different to the **'pain'** from the loss of clarity and integrity that came with having resigned and become alienated.

There has been another heroic response to the human condition, in this case one that involved despising and mocking the resigned structure, despite being part of it—being as we say, 'larger than life'. The character Zorba from Nikos Kazantzakis' classic 1947 book (and subsequent film) *Zorba the Greek* was someone who was larger than life. He was often irresponsible, and even destructive, but in his crazy way defied, and even to some degree escaped, the confines of the human condition. It was actually the degree of honesty in two individuals, one a 'ship at sea' and the other a 'larger than life' character, who helped save me from total bewilderment and estrangement when I was defying the resigned world of denial as a young adult. I will talk a little about my journey in the essay titled *The Demystification Of Religion.*

Olive Schreiner's extraordinarily honest recollection of the world of the pre-resigned mind

Olive Schreiner, whose work is mentioned throughout this book, is a renowned South African writer who lived from 1855 to 1920. She is one of the three most denial-free, unevasive, honest female thinkers I have encountered, the other two being author Simone Weil and anthropologist Dian Fossey. On her death bed, Olive Schreiner wrote an amazingly honest description of the world of the pre-resigned mind. It is extremely rare to find such an articulate description of the state of mind that precedes resignation because, as described, once resignation has occurred it is nearly impossible to revisit the issues that made it necessary.

The following extract, marvellously titled *Somewhere, Some Time, Some Place,* is from a 1987 collection of Schreiner's writings titled, *An Olive Schreiner Reader: Writings on Women and South Africa,* edited by Carol Barash.

'When a child, not yet nine years old, I walked out one morning along the mountain tops on which my home stood. The sun had not yet risen, and the mountain grass was heavy with dew; as I looked back I could see the marks my feet had made on the long, grassy slope behind me. I walked till I came to a place where a little stream ran, which farther on passed over the precipices into the deep valley below. Here it passed between soft, earthy banks; at one place a large slice of earth had fallen away from the bank on the other side, and it had made a little island a few feet wide with water flowing all round it. It was covered with wild mint and a weed with yellow flowers and long waving grasses. I sat down on the bank at the foot of a dwarfed olive tree, the only tree near. All the plants on the island were dark with the heavy night's dew, and the sun had not yet risen.

I had got up so early because I had been awake much in the night and could not sleep longer. My heart was heavy; my physical heart seemed to have a pain in it, as if small, sharp crystals were cutting into it. All the world seemed wrong to me. It was not only that sense of the small misunderstandings and tiny injustices of daily life, which perhaps all sensitive children feel at some time pressing down on them; but the whole Universe seemed to be weighing on me.

I had grown up in a land where wars were common. From my earliest

years I had heard of bloodshed and battles and hairbreadth escapes; I had heard them told of by those who had seen and taken part in them. In my native country dark men were killed and their lands taken from them by white men armed with superior weapons; even near to me such things had happened. I knew also how white men fought white men; the stronger even hanging the weaker on gallows when they did not submit; and I had seen how white men used the dark as beasts of labour, often without any thought for their good or happiness. Three times I had seen an ox striving to pull a heavily loaded wagon up a hill, the blood and foam streaming from its mouth and nostrils as it struggled, and I had seen it fall dead, under the lash. In the bush in the kloof below I had seen bush-bucks and little long-tailed monkeys that I loved so shot dead, not from any necessity but for the pleasure of killing, and the cock-o-veets and the honey-suckers and the wood-doves that made the bush so beautiful to me. And sometimes I had seen bands of convicts going past to work on the roads, and had heard the chains clanking which went round their waists and passed between their legs to the irons on their feet; I had seen the terrible look in their eyes of a wild creature, when every man's hand is against it, and no one loves it, and it only hates and fears. I had got up early in the morning to drop small bits of tobacco at the roadside, hoping they would find them and pick them up. I had wanted to say to them, "Someone loves you"; but the man with the gun was always there. Once I had seen a pack of dogs set on by men to attack a strange dog, which had come among them and had done no harm to anyone. I had watched it torn to pieces, though I had done all I could to save it. Why did everyone press on everyone and try to make them do what they wanted? Why did the strong always crush the weak? Why did we hate and kill and torture? Why was it all as it was? Why had the world ever been made? Why, oh why, had I ever been born?

The little sharp crystals seemed to cut deeper into my heart.

And then, as I sat looking at that little, damp, dark island, the sun began to rise. It shot its lights across the long, grassy slopes of the mountains and struck the little mound of earth in the water. All the leaves and flowers and grasses on it turned bright gold, and the dewdrops hanging from them were like diamonds; and the water in the stream glinted as it ran. And, as I looked at that almost intolerable beauty, a curious feeling came over me. It was not what I *thought* put into exact words, but I seemed to *see* a world in which creatures no more hated and crushed, in which the strong helped the weak, and men understood each other, and forgave each other, and did not try to crush others, but to help. I did not think of it, as something to be in a distant picture; it was there, about me, and I was in it, and a part of it. And there came to me, as I sat there, a joy such as never besides have I experienced, except perhaps once, a joy without limit.

And then, as I sat on there, the sun rose higher and higher, and shone hot on my back, and the morning light was everywhere. And slowly and slowly the vision vanished, and I began to think and question myself.

How could that glory ever really be? In a world where creature preys on creature, and man, the strongest of all, preys more than all, how could this be? And my mind went back to the dark thoughts I had in the night. In a world where the little ant-lion digs his hole in the sand and lies hidden at the bottom for the small ant to fall in and be eaten, and the leopard's eyes gleam yellow through bushes as it watches the little bush-buck coming down to the fountain to drink, and millions and millions of human beings use all they know, and their wonderful hands, to kill and press down others, what hope could there ever be? The world was as it was! And what was I? A tiny, miserable worm, a speck within a speck, an imperceptible atom, a less than a nothing! What did it matter what *I* did, how *I* lifted my hands, and how *I* cried out? The great world would roll on, and on, just as it had! What if nowhere, at no time, in no place, was there anything else?

The band about my heart seemed to grow tighter and tighter. A helpless, tiny, miserable worm! Could I prevent one man from torturing an animal that was in his power; stop one armed man from going out to kill? In my own heart, was there not bitterness, the anger against those who injured me or others, till my heart was like a burning coal? If the world had been made so, so it was! But, why, oh why, had I ever been born? Why did the Universe exist?"

And then, as I sat on there, another thought came to me; and in some form or other it has remained with me ever since, all my life. It was like this: You cannot by willing it alter the vast world outside of you; you cannot, perhaps, cut the lash from one whip; you cannot stop the march of even one armed man going out to kill; you cannot, perhaps, strike the handcuff from one chained hand; you cannot even remake your own soul so that there shall be no tendency to evil in it; the great world rolls on, and *you* cannot reshape it; but this one thing only you can do—in that one, small, minute, almost infinitesimal spot in the Universe, where your will rules, there where alone you are as God, *strive* to make that you hunger for real! No man can prevent you there. In your own heart strive to kill out all hate, all desire to see evil come even to those who have injured you or another; what is weaker than yourself try to help; whatever is in pain or unjustly treated and cries out, say, "I am here! I, little, weak, feeble, but I will do what I can for you." This is all you can do; but do it; it is not nothing! And then this feeling came to me, a feeling it is not easy to put into words, but it was like this: You also are a part of the great Universe; what you strive for something strives for; *and nothing in the Universe is quite alone;* you are moving on towards something.

And as I walked back that morning over the grass slopes, I was not sorry I was going back to the old life. I did not wish I was dead and that the Universe had never existed. I, also, had something to live for—and even if I failed to reach it utterly—somewhere, some time, some place, it was! I was not alone.

More than a generation has passed since that day, but it remains to me the most important and unforgettable of my life. In the darkest hour its light has never quite died out.

In the long years which have passed, the adult has seen much of which the young child knew nothing.

In my native land I have seen the horror of a great war. Smoke has risen from burning homesteads; women and children by thousands have been thrown into great camps to perish there; men whom I have known have been tied in chairs and executed for fighting against strangers in the land of their own birth. In the world's great cities I have seen how everywhere the upper stone grinds hard on the nether, and men and women feed upon the toil of their fellow men without any increase of spiritual beauty or joy for themselves, only a heavy congestion; while those who are fed upon grow bitter and narrow from the loss of the life that is sucked from them. Within my own soul I have perceived elements militating against all I hungered for, of which the young child knew nothing; I have watched closely the great, terrible world of public life, of politics, diplomacy, and international relations, where, as under a terrible magnifying glass, the greed, the ambition, the cruelty and falsehood of the individual soul are seen, in so hideously enlarged and wholly unrestrained a form that it might be forgiven to one who cried out to the powers that lie behind life: "Is it not possible to put out a sponge and wipe up humanity from the earth? It is stain!" I have realised that the struggle against the primitive, self-seeking instincts in human nature, whether in the individual or in the larger social organism, is a life-and-death struggle, to be renewed by the individual till death, by the race through the ages. I have tried to wear no blinkers. I have not held a veil before my eyes, that I might profess that cruelty, injustice, and mental and physical anguish were not. I have tried to look nakedly in the face those facts which make most against all hope—and yet, in the darkest hour, the consciousness which I carried back with me that morning has never wholly deserted me; even as a man who clings with one hand to a rock, though the

waves pass over his head, yet knows what his hand touches.

But, in the course of the long years which have passed, something else has happened. That which was for the young child only a vision, a flash of almost blinding light, which it could hardly even to itself translate, has, in the course of a long life's experience, become a hope, which I think the cool reason can find grounds to justify, and which a growing knowledge of human nature and human life does endorse.

Somewhere, some time, some place—even on earth!' (pp.216–220 of 261)

To focus on a technicality after such an inspirational discourse is a shame but I do want to address Schreiner's comment that she was **'not yet nine years old'** when she recalls she was able to gain a philosophical appreciation of the dilemma of the human condition. According to my thinking and research this seems premature. I mentioned earlier that nine-year-olds are normally still in the hitting-out-in-frustration stage, yet to enter the deeply-thoughtful stage, let alone plumb it to the extent of being able to reach some appreciation of the meaningfulness of our brutal world. Schreiner was recalling an event 50-odd years after it occurred so possibly her memory was not accurate and she mistook the age she was when it occurred; however, her memory of all the details of what took place seem so clear that her claim of being **'not yet nine'** deserves to be trusted. It is possible that someone who retained exceptional innocence and sensitivity and was also exceptionally intelligent could develop such an early appreciation of the dilemma of the human condition.

Although being resigned is a bit like being pregnant, in that you either are or are not resigned and cannot deny and admit the truth at the same time, it seems that in Schreiner's case resignation had been partially resisted. The subtlety involved is that for people who should resign, the degree to which they can understand the human condition is the degree to which they can avoid resignation. Most people can find no reconciling understanding of the human condition but Schreiner describes having been able to arrive at some awareness of a meaningfulness to human life, which, while not actual understanding of the human condition, is nevertheless a form of reconciling knowledge, and it seems that it was this that allowed her to be so exceptionally honest in her writings. This is not to deny that she was also a person of exceptional moral courage.

Acknowledging the fundamental questions about human life: **'All the world seemed wrong to me'**; **'Why did everyone press on everyone and try to make them do what they wanted? Why did the strong always crush the**

weak? Why did we hate and kill and torture? Why was it all as it was? Why had the world ever been made? Why, oh why, had I ever been born?', Schreiner says she could not accept that **'The world was as it was!'**, without **'hope'**. She **'began to think and question myself'**, and discovered the human condition without (**'so hideously enlarged and wholly unrestrained a form that it might be forgiven to one who cried out to the powers that lie behind life: "Is it not possible to put out a sponge and wipe up humanity from the earth? It is stain!"'**) and within (**'Within my own soul I have perceived elements militating against all I hungered for'** and **'you cannot even remake your own soul so that there shall be no tendency to evil in it'**), and also discovered the immense depression those truths lead to (**'the darkest hour'** where **'the whole Universe seemed to be weighing on me'** and where **'The band about my heart seemed to grow tighter and tighter'**, asking **'why, oh why, had I ever been born? Why did the Universe exist?'**).

Schreiner tried to resist resignation, she **'tried to look nakedly in the face those facts which make most against all hope'**, saying **'I have tried to wear no blinkers. I have not held a veil before my eyes'**.

To hold back resignation Schreiner held onto **'the consciousness which I carried back with me that morning'** where, in **'a flash of almost blinding light'**, she saw that **'*nothing in the Universe is quite alone*'**; we are **'a part of the great Universe'** that **'strives for' 'something'** that we can **'hope'** for which is **'that glory'** of an integrative or cooperative destiny where, through **'a growing knowledge of human nature'**, we will produce **'a world' 'Somewhere, some time, some place'** in which **'creatures** [will] **no more** [be] **hated and crushed, in which the strong help the weak, and men understand each other, and forgive each other, and do not try to crush others, but to help'**—a reconciliation, which has now occurred, that leads to a lifting of the human condition and **'a joy without limit'**.

As one prepares to die, evasion finally becomes useless and the deepest truths can sometimes emerge. It seems that impending death transferred Schreiner back behind the walls of denial to the real questions and thoughts about human life. All the relatively evasive middle years were discarded and her original clear view of the real truth and dilemma about life came to the fore.

A history of analysis of resignation

Resignation is the most important psychological event to occur in human life and yet it is very rarely acknowledged and almost never discussed and analysed.

In the extremely insecure, evasive, alienated times we currently live in, resignation is almost totally denied. Only recently on an Oprah Winfrey program titled, *Is Your Child Depressed?*, Dr Siegler, author of the latest evasive, mechanistic book on adolescence, *The Essential Guide to the New Adolescence: how to raise an emotionally healthy teenager*, arrogantly proclaimed to depressed adolescents sitting before her that their depression was nothing more than their **'puberty hormones...overwhelming them'** (15 May 2000). Similarly, a recent newspaper article was published under the heading 'Lost Generation—Adolescence is a vulnerable stage of life. But in Australia, it is potentially fatal. Youth suicide is on the increase'. It gave the following explanation for youth suicide: **'"They** [adolescents] **are having to come to terms with a huge amount of change", says Bronwyn Donaghy, author of *Leaving Early*, a book on youth suicide. "From the changes going on with their bodies; the transition into sexual beings; changing relationships with parents; exams and looking for employment"'** (*Sydney Morning Herald*, 24 Aug. 2000). There is no acknowledgment in these words of the real issue involved of resignation. What has been said is all evasive 'bullshit' or 'cave speak', to use Plato's allegory.

The reason depression and youth suicide are increasing worldwide at such a rapid rate is because of the equally rapid increase in alienation around the world. The level of dishonesty that humans are now practicing is so great that new generations arriving in the world find it almost unbearable. Resigned alienated people are blind to their level of falseness but it is visible to the innocent and to young people who have not yet resigned.

A 1999 book by clinical psychologist Dr Michael Yapko, titled *Hand-Me-Down Blues,* records that **'someone born since 1945 is likely to be up to 3 times more depressed than their parents and 10 times more than their grandparents'**. A recent book about depression states that **'Depression is the leading cause of disability in the U.S. for people over the age of five'** (Andrew Solomon, *The Noonday Demon,* 2001).

Jeff Kennett, former Premier of Victoria, recently established a

National Institute for Depression in Australia. It is obvious that without any acknowledgment or awareness of the real issue of the human condition this institute can hope to achieve little beyond offering comfort and support. In fact to young, pre-resigned adolescents it will be just another fraudulent adult enterprise that they will have to endure, another brick in the wall, another load of 'bullshit'.

Humanity is approaching end-game, the situation where the mind cannot adapt to any more dishonesty without going completely mad. Despite the necessary brave front humans have put on, the truth is that alienation/depression/loneliness is now an epidemic on Earth. As Nobel laureate Albert Camus observed, **'This world is poisoned by its misery, and seems to wallow in it. It has utterly surrendered to that evil which Nietzsche called the spirit of heaviness** [depression]**'** (*The Almond Trees*, an essay, 1940; first pub. in *Summer*, 1954, p.34 of 87).

There has been scarcely any truthful acknowledgment and analysis of resignation in our society. The following is the full collection of analysis of the phenomenon of resignation that I have found since 1975 when I first started to actively write about the human condition.

In his 1974 book, *He: Understanding Masculine Psychology*, Robert A. Johnson, a highly regarded practitioner of Jungian principles, describes firstly the agony of resignation, then life in the soul-destroyed, schizoid, alienated world of reality, and finally the hope of eventual reconciliation and return to an upset-free, healed state of unity of self. **'It is painful to watch a young man become aware that the world is not just joy and happiness, to watch the disintegration of his childlike beauty, faith, and optimism. This is regrettable but necessary. If we are not cast out of the Garden of Eden, there can be no heavenly Jerusalem...According to tradition, there are potentially three stages of psychological development for a man. The archetypal pattern is that one goes from the unconscious perfection of childhood, to the conscious imperfection of middle life, to conscious perfection of old age. One moves from an innocent wholeness, in which the inner world and the outer world are united, to a separation and differentiation between the inner and outer worlds with an accompanying sense of life's duality, and then, hopefully, at last to satori or enlightenment, a conscious reconciliation of the inner and outer once again in harmonious wholeness...we have to get out of the Garden of Eden before we can even start for the heavenly Jerusalem, even though they are the same place. The man's first step out of Eden into the pain of duality gives him his Fisher King wound...Alienation is the current term for it'** (pp.10,11 of 97). (The 'Fisher King' referred to is a character in the great European legend of King Arthur and his knights of the round table.)

In his 1949 book, *The Origins and History of Consciousness,* Erich Neumann, an analytical psychologist who has been described as Carl Jung's most gifted student, fully recognised the battle of the human condition that took place when consciousness first emerged in our human ancestors; namely the battle between their already established non-understanding, **'unconscious'**, instinctual self and their newly emerging **'conscious'**, intellectual self. Since then this battle has been re-fought in each human's life at the time of resignation, and as such the following account can also be applied to what occurs at resignation. Neumann wrote: **'Whereas, originally, the opposites could function side by side without undue strain and without excluding one another, now, with the development and elaboration of the opposition between conscious and unconscious, they fly apart. That is to say, it is no longer possible for an object to be loved and hated at the same time. Ego and consciousness identify themselves in principle with one side of the opposition and leave the other in the unconscious, either preventing it from coming up at all, i.e., consciously suppressing it, or else repressing it, i.e., eliminating it from consciousness without being aware of doing so. Only deep psychological analysis can then discover the unconscious counterposition'** (p.117 of 493).

While Neumann understood the essential conflict involved in the human condition he did not explain the reason for the conflict. That explanation is given in the preceding essay in the section, 'How our soul became corrupted'. Essentially, while unconscious instincts can *orientate* animals' behaviour so that they survive, the conscious intellect needs to *understand* cause and effect for it to be able to know how to behave. When the conscious intellect sets out to find understanding of existence, the already established instinctive orientations unwittingly, as it were, challenge that search, try to stop it, and a battle between the instinct and intellect, the **'unconscious'** and **'conscious'**, arises. For any who are interested, I have written an essay about Neumann's book, *The Origins and History of Consciousness,* which the FHA is happy to make available.

For Robert A. Johnson and Erich Neumann to have been so penetrating of the issue of the human condition they must be denial-free thinkers, contemporary scientist-prophets.

The moment of resignation: how humans change from believing in a cooperative, selfless, loving world, to believing in a competitive, selfish, aggressive, egocentric, must-win, power-fame-fortune-and-glory-obsessed world

It is now necessary to take analysis of what happens at resignation to a deeper level.

It has been explained that essentially what happens at resignation is that the attempt to confront ideality without the ability to explain humans' lack of it meant that corrupted humans had no choice but to live in denial of the soul's ideal world and the associated issue of the human condition.

The important question remains of how exactly did the transition from belief in cooperative, selfless, loving ideality to belief in a competitive, selfish, aggressive, egocentric way of living take place? How did humans manage to change from believing in selfless cooperation to believing in selfish competition? It is an immense transition to make given they are such opposing positions—and the transition is made in such a short time. In fact, resignation takes place in one particular moment, and the major adjustments involved occur over only a few days.

To recap on the lead up to the moment of resignation. At about 11 or 12 years of age young adolescents become sufficiently able to understand the world to begin to try to explain and thus understand the key question about human life of why humans are not ideally behaved. The first thing they learn in this quest is that the adult world has no answer to this question, indeed, for some reason, adults do not even want to consider it. As a result of this lack of response from the outside world, young adolescents retreat inwards with their thoughts. Finally, at about the age of 14 or 15, their thinking about the human condition begins to focus on the question of their own lack of ideality. They discover the dilemma of the human condition exists within, as well as without. It is this encounter with the question of their own lack of ideality that brings young adolescents into contact with the fearful depression that trying to think about the human condition causes any human who is not entirely sound.

Without the explanation for why corrupted humans are not evil, worthless, meaningless beings, adolescents can only conclude that their non-ideal, corrupted state *is* an evil, worthless state; that they *are* a blight on the face of the Earth. As the depression from this thinking deepens, they realise that if they become any more depressed they will be suicidal. At this point they desperately search for a way of saving themselves from their predicament. It is at this point that they realise the only thing they can do is adopt an attitude of denial of the whole issue, find a way to block it from their mind.

Of course adolescents try to resist living a life of denial of this all-important and fundamental question in human life of the human condition because it is such a fraudulent, dishonest, escapist, artificial, deluded position to take up. To re-quote Fiona Miller's resignation poem, adopting the denial means **'You will never have a home again...Smiles will never bloom from your heart again, but be fake and you will speak fake words to fake people from your fake soul...you will be fake, you will be a supreme actor of happiness but never be happy/Time, joy and freedom will hardly come your way and never last...You will become like the rest of the world—a divine actor, trying to hide and suppress your fate, pretending it doesn't exist.'**

While the adolescent tries to resist, the fear of depression from thinking they are worthless and meaningless eventually becomes greater than their fear of living an utterly dishonest existence, and the adolescent resigns to living a life of denial of the issue of the human condition.

The question is how exactly is this denial achieved? Essentially, to deny the issue of the human condition involves two actions. Firstly, as has already been stressed in this essay, you have to deny the existence of ideality. If there is no ideal state then there is no dilemma, no conflict, no issue with not being ideal. However, it is not sufficient to merely deny the existence of cooperative ideality, it is also necessary to believe that competitiveness is a valid, meaningful way of behaving. You have to deny cooperative ideality—in particular the integrative meaning of life and the existence of an instinctive expectation within humans of living cooperatively and lovingly—*and* you have to believe that competitiveness is valid and meaningful. Of course, in order to believe that competitiveness is meaningful, an explanation or justification has eventually to be found for why it is meaningful. Without the true explanation of the human condition, the explanation for why humans are divisively behaved, contrived, false excuses have to be devised and believed.

Essentially, to deny the issue of the human condition the mind has to make an amazing switch from believing in cooperative ideality to believing in false excuses for humans' competitive, aggressive and selfish behaviour. Competitiveness has to be embraced and cooperative integrativeness denied. The question is how does the mind make this amazing switch? We need to look at what happens at the actual moment of resignation, and in the hours and days immediately following.

Thinking increasingly deeply about the pertinent issue of their own lack of ideality, the adolescent eventually reaches a moment of perfect clarity on the matter. They reach a moment when, on one hand, they can see perfectly clearly the truth of cooperative ideality, and, on the other hand, just how corrupted or non-ideal they are as individuals. At this moment they are thinking entirely truthfully and profoundly, delving right to the bottom of the dilemma of their condition and seeing the full implication of their apparent worthlessness. At this point, depression reaches its peak; in fact the depression at this moment must be incredibly intense, as though their whole body is going to dissolve, disintegrate with agony and pain.

When walking through bushland near a school a doctor friend and I came across a lone student sitting huddled on a track with his head resting on his arms. He looked to be about 13 or 14 years old. He wasn't crying, or looking to be emotionally distressed by some particular event at school, it was something entirely different. When I asked him if he was alright he raised his head slightly, enough for us to see an expression of overwhelming, unreachable despair. I remember both my friend and I felt very strongly that we were intruding, that no one could reach where he was, and that he wanted to be left alone. It was a depression from another realm that he was wrestling with.

It is at that unprotected moment of clarity about the extent of their lack of ideality and resulting extreme depression that the mind of the adolescent finally becomes receptive to the option of adopting denial of the issue of the human condition, incredibly false as they know it to be. Despite knowing full well at that stage that it is an outrageous lie to believe that there is no such truth as cooperative ideality, or to believe in contrived excuses for competitiveness, their need to escape the depression is so incredibly great that their mind actually welcomes the opportunity to believe in the lies. In fact it does not just welcome the lies, it embraces them as if embracing their mother. Such is their fear of revisiting the depression they have just experienced. The negatives of becoming a false person have no currency at that moment when the need for relief is so desperate.

Once the mind resigns itself to blocking out the truth of cooperative meaning and becomes determined to believe that competitiveness is meaningful, it doesn't take long to find contrived excuses for competitiveness. The mind sees that, 'Virtually everyone else is behaving selfishly and competitively, so such behaviour must simply be human nature, an entirely permissible, natural way to behave', and, 'Humans are only animals and animals are always competing with, fighting and killing each other, so that's why we are.' The adult world of resigned humans, who are past masters at living in denial, readily offer the newly resigned adolescent refinements to these excuses. As was described in the *Plato* essay, in the section 'The contrived excuses for humans' divisive behaviour', the 'animals are competitive and that's why we are' excuse became greatly refined. Darwin's idea of natural selection was misrepresented as evolution being concerned with the 'survival of the fittest'. This misrepresentation is termed Social Darwinism, and it was further developed by Edward O. Wilson with his theory of Sociobiology, which argued that human selfishness is due to their need to perpetuate their genes. The current expression of this excuse that genetics is a selfish process is 'Evolutionary Psychology', which maintains that even acts of selflessness in human behaviour—human morality—can be explained in terms of genetic reciprocity, of 'you scratch my back and I'll scratch yours'. Robert Wright expounded this theory of Evolutionary Psychology in his 1994 book, *The Moral Animal: Why We Are The Way We Are.*

Once the adolescent has finally given in, resigned and allowed themselves to embrace the lies, they do experience some relief. They learn that delusion does work for them. As tentative as the situation is to begin with—because at that stage they are still capable of remembering that they are lies that they are adopting—the new thought process has gained a foothold. It is then only a matter of reinforcing the thought process and the blocking out or denial of the truth of cooperative meaning and belief in a selfish, competitive, survival-of-the-fittest world become inscribed habituated pathways in their brain.

From that moment onwards, maintaining the denial becomes a minute by minute growing preoccupation, until, after only a few days, recalling the truth of cooperative ideality and the question it raises of their corrupted condition becomes a forgotten issue. The resigned adolescent determines that the terrible 'dark night of the soul', as resignation has been aptly described, that awful moment when the issue of the human condition was seen starkly, will never occur again. Once the escape route is accepted, going back to confrontation with the human condition becomes an anathema—which is why it is so hard for people who are resigned to take in or 'hear' any of the analysis of the human condition in my books. It is a realm that resigned minds are absolutely determined never to allow themselves to go near again.

After only a few days it is almost impossible to get people who have become resigned to consider there is such a thing as a cooperative purpose to existence, or to consider humans were once instinctively orientated to living cooperatively. Resigned minds believe with a passion that humans were once brutish and aggressive 'like other animals'. They *will not* allow themselves to believe that humans have an instinctive self or soul orientated to cooperative behaviour. And they believe with all their being that the meaning of life is to be competitive, and that succeeding in competition with other humans *is* the way to achieve a secure sense of self-worth. They actually believe that 'winning *is* everything', that success in the form of power, fame, fortune and glory *is* meaningful. And it does sustain them, not because competition and winning is meaningful, but because it is keeping their mind away from the few steps of logic it would take to bring them back into contact with the depressing issue of the human condition. Their mind in effect says 'I *am going to* believe in, live off, and enjoy this new way of viewing the world; I simply do not care if it is false.'

Once their brain has replaced the truth with the lie of a selfish, competitive meaning to existence then winning in that claimed selfish, competitive battle *is* reinforcing. Success becomes measured not by how integrative, how cooperative, loving and selfless you are, but by how well you are able to succeed in an alleged competitive, survival-of-the-fittest world. A complete transition occurs; from believing in a cooperative, selfless, loving world, resigned men especially became believers in a competitive, selfish, aggressive, must-win, power-fame-fortune-glory obsessed existence.

In saying that men especially become obsessed with an egocentric, must-win strategy, it needs to be explained that the form of winning or success in a claimed battle of the survival-of-the-fittest has been different for resigned men and resigned women. To understand this difference the roles men and women played in humanity's recent, 2-million-year heroic journey to find understanding of the human condition first needs to be explained.

The different roles men and women have played is explained in detail in the first section of the next essay, 'Bringing peace to the war between the sexes'. What follows is a very brief summary of that explanation. Nurturing was the priority activity for humanity during its infancy and childhood stage, which lasted from 10 million years ago, when, as was explained in the *Plato* essay, our ancestors began to develop love-indoctrination, to 2 million years ago when consciousness emerged. With the emergence of consciousness the priority activity changed from nurturing to having to defy and ultimately overthrow the ignorance of our instinctive self or soul. What our instinctive self or soul was ignorant of was our intellect's need to search for knowledge. Since this threat of ignorance was a challenge to the group, namely humanity, and since men were the group protectors, it was men who had to take on the task of championing the intellect and defeating the ignorance of our soul. With this change in priorities humanity changed from being a matriarchal society to a patriarchal one.

Tragically, the only methods available to men to defy the soul's implication that the intellect was wrong to search for knowledge was to attack the soul, block out its implied criticism, and try and prove the implied criticism wrong. In doing their job of battling ignorance men became angry, alienated and egocentric. They lost their innocence and became corrupted. Unable to explain their loss of innocence, men began to resent and attack innocence in all its forms because of innocence's implied criticism of men's lack of innocence. Since it was not women's role to overthrow ignorance, and since it was important that women stay relatively free of the battle in order to retain as much innocence as they could to nurture a subsequent generation, women did not directly and actively participate in the fight against ignorance. This meant that women were largely unaware of the cause of the upset in men and therefore tended to be unsympathetic towards it. Being unjustly condemned by the relative innocence and naivety of women, men then retaliated and attacked women. Since women reproduced the species men could not destroy them and instead violated women's innocence or 'honour' by rape; they invented 'sex', as in 'fucking' or destroying, as distinct from the act of procreation. What was being fucked or destroyed was women's innocence. In this way women's innocence was repressed and they came to share men's upset. While sex was originally for procreation it became 'perverted', used to attack the innocence of women. In time the image of innocence in women, the physical beauty that 'attracted' sex, also became a means of inspiring the journey to self-understanding. This inspirational aspect meant that while at base sex was rape, on a nobler level, it became an act of love.

It needs to be explained how it was that the image of innocence, women's physical beauty, became 'attractive'. As will be summarised in the first section of the next essay, 'Bringing peace to the war between the sexes', and as was explained in the *Plato* essay, during the nurturing phase of humanity's development, youthfulness was a highly sought after trait with which to mate with because the cooperative training of the love-indoctrination process wore off with age. Youthfulness became associated with cooperativeness and, since cute, neotenous, childlike features of a domed forehead, snub nose and large eyes were the attributes of youthfulness, they became sought after features. Our ape ancestors self-selected for cooperativeness, however as explained when the human condition emerged with consciousness some 2 million years ago men began to resent and attack women, because women's innocence and naivety about what men were doing was perceived as criticism. Suddenly, instead of the neotenous image of innocence in women being cultivated because it was a sign of cooperativeness, it became a target for sexual destruction.

What this means is that throughout the battle to find understanding of the human condition, women were being forced to suffer the destruction of their soul, their innocence, while at the same time their image of innocence was being cultivated. We evasively described neotenous features as 'attractive' to avoid saying that what was being attracted was destruction, through sex, of women's innocence. It can be seen that since all other forms of innocence were being destroyed, this image of innocence—'the beauty of woman'—was the only form of innocence to be actively cultivated during humanity's adolescence. Women's beauty became men's *only* equivalent for, and measure of, the beauty of their lost pure world. Consequently, women are now highly adapted to their role of supporting men in their battle against ignorance and inspiring them with their beauty. Stark evidence of this adaption is the countless number of women's magazines, almost entirely dedicated to making women 'more attractive'; they are sex-object manuals. It was little wonder men fell in love with women. The 'mystery of women' was that it was only the physical image or object of innocence that men were falling in love with; the illusion was that women were psychologically as well as physically innocent.

Understanding the different roles men and women played in humanity's journey to enlightenment, and the effects of those roles, it is possible to explain the different effects resignation had on young men and women. The form of winning or success in the claimed competitive battle of the survival-of-the-fittest that resigned humans participated in was different for men and women.

Since it is men who took up the task of championing the ego, defying and defeating the ignorance of the instinctive world of our soul, when men resigned they became *ego-centric;* they became pre-occupied with winning affirmation through power, fame, fortune and glory. Young pre-resigned men are not egocentric. They have a conscious thinking self or ego that needs reinforcement but it is not centred or focussed or preoccupied with gaining adulation and admiration for self. It is not self-centred and selfish; quite the opposite, it is focused on the issue of why humans are not concerned for others and selflessly behaved.

In comparison, women have been responsible for helping men and inspiring them with their beauty. It can now be explained that it is only resigned women who live off the reinforcement from men's attention to their sex-object self. Young pre-resigned women are not preoccupied with being a successful sex-object; quite the opposite, they are preoccupied with such questions as why humans mis-use one another sexually. I remember talking to a group of 16-year-old school girls, almost all of whom were dressed in a way that showed they were preoccupied with gaining boys' attention. There were two or three however that were dressed in simple, functional clothes and were not wearing make-up; girls who were not yet preoccupied with their sex-object self. These girls were still interested in questions about the extreme imperfections of the human situation (such as the glaring inequalities between humans) and, in general, still deeply interested in discussing the human condition. In fact the other girls tried to discourage these girls from their line of thinking and questioning, making comments such as, 'Don't you know there is no answer to those questions', 'Why don't you get real and enjoy yourself; there is a spunk in Year 12 and he's coming to the party on Saturday night.' Once girls resign they are so intent on deriving reinforcement from men's attention to their sex-object self that many who become anorexic and/or bulimic do so in an effort to make themselves more attractive. Women's reinforcement from men has been their escape from the agony of the human condition.

It can be seen that resignation brought about a complete reorientation of a person's life. The pre-resigned mind and the resigned mind are a world apart. Thankfully resignation and with it male egocentricity and female sex-object preoccupation can now subside and eventually end. With understanding of the human condition resignation is no longer necessary.

Resigned men and women adopted different ways of winning or succeeding in the claimed competitive battle of the survival-of-the-fittest that resigned humans decided they had to be part of. Of course it helped that nearly every other human around them was believing in, and living off the lies. The universal conspiracy of denial came to their aid; *they were initiated into the cult of denial, brainwashed by themselves and almost everybody around them to believe in the great lie.*

This then is how resigned people came to quickly accept humans' divisive reality as normal and dismiss the truthful world of their soul. In so doing they lost the ability that unresigned minds have of thinking truthfully.

It should also be explained that after living a false, seemingly meaningless resigned existence for a number of years resigned adults could become so disenchanted with their selfish, aggressive, ugly, and in many ways destructive life that they could decide to abandon that way of living and take up support of some form of idealism. Having resigned and abandoned the world of the soul, they could become, as the revealing term says, 'born-again' supporters of the idealistic world of the soul again. They could become 'born-again' to religion, to supporting the left wing in politics, to being dedicated environmentalists, feminists, activists for the rights of indigenous people, or animal liberationists. These are pseudo forms of idealism because real idealism depended on defying and ultimately defeating the unjustly condemning idealism of the world of the soul, not on caving in to it. It is true that the battle to defy and defeat ignorance was corrupting and when people became overly corrupt they had to give up fighting ignorance and try and bring some soul and its world of soundness back into their lives. For those who had become overly corrupted, excessively angry and destructive, the adoption of a born-again strategy was a responsible reaction. The problem is, unable to explain and thus confront and admit their extremely corrupted and alienated state, they were using the born-again-to-'idealism' lifestyle to delude themselves that what they were doing was actually ideal. They deluded themselves that they held the 'moral high ground' when the truth is the opposite. They even used their born-again lifestyle to delude themselves that they were uncorrupted people. There was an extremely deluded, selfish aspect to their behaviour; a desire to, as their critics say, 'feel-good' about themselves. The truth is the born-again state is the most dishonest and alienated state humans could adopt. More will be said about the immense danger of the delusion of pseudo-idealism in the essay *Death by Dogma* to be published in my next book.

It should be emphasised that the great value of religions, compared to other forms of pseudo-idealism, is that in religion there is a high degree of honesty, a significant acknowledgment of the alienated state. This honesty is contained in the prophet around whom the religion is founded. By acknowledging the prophet and his denial-free thoughts a person's own lack of honesty and soundness is indirectly acknowledged. Yet the problem with religions, and why they have in recent times become unpopular, is precisely this honesty. The more alienated people became, the less confronting honesty they could bear. Born-againers wanted more guilt-free forms of idealism to support, as this quote acknowledges: **'The environment became the last best cause, the ultimate guilt-free issue'** (*Time* mag. 31 Dec. 1990).

While people who are resigned and not born-again to 'idealism' are living a false existence, they are still participating in the battle to defy the soul's ignorance as to the true goodness or worthiness of humans. Those who are resigned and not born-again *are* 'bullshitting', living dishonestly; however those who have effectively quit the all-important battle and are pretending to be ideal are 'double bullshitters'.

With understanding of the human condition it is not hard to understand what has been referred to as 'prejudice' against 'idealism'. At a certain point the lies became suffocating, unbearable; especially the lie that humans' lack of ideality means they are evil, inferior and worthless, and most especially the lie that people practicing born-again 'idealism' are themselves ideal.

A pre-resigned adolescent

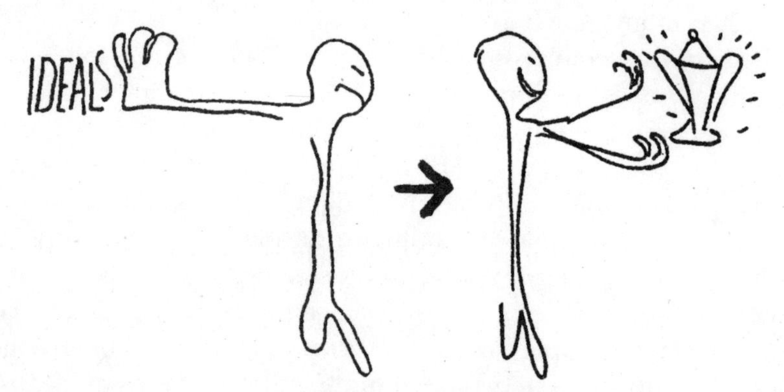

The moment of resignation

A bullshitter: a selfish, power-fame-fortune-and-glory seeking resigned adult

DRAWINGS1996
JEREMY GRIFFITH

A double bullshitter: a born-again resigned adult

The extent of humans' fear of the issue of the human condition

In the earlier section, 'A history of analysis of resignation', Erich Neumann's concluding sentence, that **'only deep psychological analysis can then discover the unconscious counterposition'**, emphasises just how committed the resigned mind becomes to denial. This extreme commitment is a direct measure of the extent of the mind's fear of the depressing issue of the human condition.

Ronald Hayman's 2000 book, *Life of Jung,* documents Carl Jung's journey into this **'unconscious counterposition'**—the truthful world that humans repressed when they became resigned—and described how those depressing depths nearly destroyed Jung yet also allowed him to think truthfully and thus effectively; most of his important concepts being found while he was in that truthful state and domain. Hayman wrote: **'He** [Jung] **claimed to have acquired the knack of catching unconscious material "in flagrante", and his** [1963] **book *Memories, Dreams, Reflections* suggests his behaviour was heroic—that he was making a dangerous expedition into the unconscious for the sake of scientific discovery. Several dreams involved subterranean staircases and caverns, which suggested that his fantasies were located somewhere underground. In December 1913, he says, he decided to drop downwards. "I let myself fall, It was as if the floor literally gave way underneath me and I plummeted into**

dark depths"...It took about three years to recover from the breakdown...It was during Jung's breakdown that he arrived at some of his most important concepts...Had it not been for his breakdown, Jung might never have developed the technique he called active imagination, based on conversations with his anima [the soulful, more female side of himself] **and with fantasy figures. He told patients to draw or paint characters from dreams or fantasies, and to interrogate them. This was like praying to an internal god, "for there are answers inside you if you are not afraid of them". It was a matter of "letting the unconscious come up".'**

In his 1976 book, *Jung and the Story of Our Time,* Sir Laurens van der Post offered this account of Jung's journey into the unconscious: **'Jung no longer looked for the answer** [to what would **make** humans **whole**] **vicariously through the neuroses of others. More and more he looked into his own deeply wounded self...no physician has taken the task of healing more seriously than he did. He was, in all this, quite alone...He was denounced and abandoned by most of his former colleagues. He had to face, alone, the unknown in this unconscious universe to which he had been brought. He was bombarded by symbols and images demanding that he should return with them from whatever fathomless depths they had come...He found himself turning to the child in himself as if instinct, too, was exhorting him to become like the child which the New Testament exhortation makes imperative. In this way he hoped to emerge from darkness into the light of which the Kingdom of Heaven is the supreme image...Although he recognised in the dreams and fantasies psychological material and patterns that he had encountered only in the most schizoid and psychotic of his patients, he felt he had to accept them also as part of himself...No-one could possibly know better than he the dangers of succumbing to such dark forces...So, on the afternoon of December 12 of that year, 1913, sitting in his chair at his desk, he made one of the bravest decisions, I believe, ever recorded in the history of the human spirit. He committed himself absolutely to this equinoxial urge from within and in doing so apparently subordinating reason to unreason, and risking the sacrifice even of sanity to insanity. But he had always wanted to know how the human spirit would behave if deprived of all preconditioning and left entirely to itself. He had an intuition that no real beginning would be possible unless he had some experience of what mind and imagination did if allowed to act naturally and freely on their own. And he was about to find out in a way which a world which does not recognise the reality of "these mountains of the mind and their cliffs of fall, frightful, sheer, no-man-fathomed" of which Manley Hopkins had spoken, cannot measure. His whole spirit must have**

reeled with an inverted vertigo and horror of what he was about to do. He put it to me once, without hint of laughter. "I said to myself 'Well Jung, here you go.' And it was as if the ground literally gave way under me and I let myself drop." That was the greatest of his many moments of truth, and so far did he fall and so unfamiliar and frightening was the material that he found as a result, that there were many moments when indeed it looked as if insanity might have overcome sanity' (pp.153-156 of 275).

Sir Laurens van der Post described how Jung had to return to his pre-resigned childhood state in order to reconnect with the truthful world. He also acknowledges that to allow his **'mind and imagination'** to think **'naturally and freely'**, Jung had to abandon the **'preconditioning'** of the resigned world of evasion and denial.

It is now possible to understand this observation Sir Laurens made in *Jung and the Story of Our Time:* **'Few of us to this day recognise the imperative of courage in the life of the imagination and how it alone can make us free from fear and open to the fullness of reality. Its "cliffs of fall, frightful, sheer, no-man-fathomed" demand a heart as brave as that of any soldier going into battle or any mountaineer pioneering a new way up Everest.'** To look into the human condition *was* a frighteningly dangerous and therefore almost impossibly difficult task for most people; however, there were rare individuals for whom this was not the case. As is explained more fully in other essays in this book, those rare individuals who could confront the issue of the human condition with relative safety have historically been referred to as 'prophets'. Prophets were people who were sufficiently loved or nurtured in their upbringing, and also sufficiently sheltered from corrupt reality during that upbringing that, as a result, their **'child within'** or soul or instinctive self escaped being hurt/damaged/corrupted. They remained sufficiently 'innocent' that when they entered adolescence they did not feel overly criticised and depressed by the cooperative ideals of life and thus did not have to resign to a life of living in denial of those ideals; they did not have to live in denial of the issue of the human condition. (Note, I said above that Jung had to 'return into his pre-resigned state to reconnect with the truthful world'. The implication is that Jung was a resigned individual and yet in the *Plato* essay he was named as a prophet. This apparent anomaly will be clarified shortly when resigned and unresigned prophets are explained.)

Denial-free, unevasive thinkers or prophets could look into and think truthfully about human life. What was needed to look into the human condition was not courage, but innocence, with all its soundness and security of self. Where the unresigned, denial-free thinkers or prophets needed courage was to defy the almost universal practice of denial and endure the persecution that the resigned world subjected them to for exposing its resigned, corrupted, false state.

The reason for including these quotes about Jung's terrifying journey into the **'unconscious counterposition'** is to illustrate just how justifiably fearful most adults have been of that **'counterposition'**—how fearful they have been of the pre-resigned, truthful world of our integrative-meaning-orientated original instinctive self or soul. It is a world that historically has been associated with terrifying self-confrontation, condemnation and thus depression. Considering almost all adults learnt to live in total denial of the truthful world that surrounds them it is little wonder that young, pre-resigned adolescents were abandoned by the adult world and left alone with their truthful thoughts.

A measure of the extent of this block-out or denial is the way in which resigned adults swagger about confidently imposing their superficiality on everybody and everything; the more deluded amongst them spouting esoteric, intellectual truthlessness everywhere they went, as if they were not living in a state of fraudulent dishonesty at all. The deluded arrogance or hubris of resigned adults has often been immense. The denial-free, unevasive thinker or prophet, Jesus Christ, described such extreme delusion thus: **'They like to walk round in flowing robes and love to be greeted in the market-place and have the most important seats in the synagogues and the places of honour at banquets. They devour widows' houses and for a show make lengthy prayers. Such men will be punished most severely'** (Luke 20:46). Delusion/alienation/denial will not be **'punished'** with the arrival of the understanding of the human condition because at base there was a valid and understandable reason for it, but thankfully denial and its manifestations now become unnecessary in human life.

Until now, resigned adults have been justifiably terrified of looking inwards. Just how afraid many people have been is revealed in the extreme negativity of this response to the possibility of self-confrontation: **'If you spend too much time with your head up your bum in search of existential angst, all you'll find is dark and dirty'** (Dr Don Edgar, *Sydney Morning Herald,* 5 May 2000). When it comes to psychology and looking into ourselves, most adults have had the attitude of Albert the alligator in the old Pogo comic strip: **'The inner me? Naw, got no time fer him. Ah got trouble enough with the me whut's out cheer whar Ah kin get**

mah hands on 'im. Ez fer the inner me, he goes his way, Ah go mine' (mentioned in Charlton Heston's autobiography, *In The Arena,* 1995).

Resigned adults feared the issue of the human condition for good reason. For most people virtually any thinking would bring them into depressing confrontation with the issue of the human condition—as the Australian comedian Rod Quantock once said, **'Thinking can get you into terrible downwards spirals of doubt'**—so it was almost better not to think at all, or at least remain extremely superficial in thought, which is in fact what most people have done and why human discourse has been so immensely shallow. The only subject that has been safe to talk about has been the weather. Unless you were sufficiently innocent not to have had to resign, trying to confront the fundamental issue in human life of the human condition was *extremely* depressing and thus dangerous.

A measure of how dangerous it has been to confront the human condition is that even prophets encountered danger if they were not exceptionally sound and secure—as Plato once observed **'Even the friends of ideas** [even people who are relatively comfortable thinking deeply] **are subject to a kind of madness'** (mentioned in *Great lives, Great Deeds,* 1966, p.386 of 448). The fact that only a few prophets could fully confront the issue of the human condition shows just how perilous it has been.

The following are some examples of the risks that prophets faced confronting the dilemma of the human condition.

In her review of Ronald Hayman's aforementioned book, *Life of Jung,* Jean Curthoys, stated that Jung believed himself to be **'a prophet'** (and he has often been described as a **'true prophet'**, such as in a review in the British *Observer* newspaper reproduced on the back cover of the Flamingo 1983 edition of Jung's *Memories, Dreams, Reflections),* and referred to Jung's **'journey into the collective unconscious'** as having **'been a paradigmatic journey through the dark night of the soul'** *(Weekend Australian,* 18-19 Mar. 2000). The 'dark night of the soul' is a good description of what humans experience when they try to confront the soul's cooperative-meaning-expecting, truthful world. According to Curthoys, it took Jung **'about three years to recover from the breakdown'**, his journey into the dark night of the soul caused.

Jung's early mentor was the Austrian physician, Sigmund Freud. Known as the father of modern psychoanalysis and also frequently referred to as a prophet, Freud similarly dared to plumb the depths of the issue of the human condition. R.D. Laing recognised the **'stark terrors'** Freud faced with this task, saying: **'The greatest psychopatholo-**

gist has been Freud. Freud was a hero. He descended to the "Underworld" and met there stark terrors. He carried with him his theory as a Medusa's head which turned these terrors to stone. We who follow Freud have the benefit of the knowledge he brought back with him and conveyed to us. He survived. We must see if we now can survive without using a theory that is in some measure an instrument of defence' (*The Divided Self*, 1960, p.25 of 218).

Interestingly, immediately after paraphrasing Gerard Manley Hopkins' **'mountains of the mind and their cliffs of fall, frightful, sheer'** to describe Jung's exploration of the mind in *Jung and the Story of Our Time,* Sir Laurens van der Post went on to describe how the Italian poet, Dante, made a similar prophet-like, dangerously confronting expedition into the truth about the human condition, as documented in his famous 1321 poem, *The Divine Comedy.* On this terrifying expedition Dante, like Freud, was able to employ help in the form of the inspiration of the face of a beautiful woman. Sir Laurens says that Dante **'had to go down into a netherworld to its uttermost depths'** and that on this expedition Dante **'had as an overall guide and protector his love of a woman whose face, once seen when a boy in the streets of Florence, changed the course of his whole life. All that this woman and this face evoked in him grew into a love that was total, universal, and outside space and time. It became a power in his spirit that made Dante always feel firmly directed and safe. As a result, even at moments when Virgil, who was Dante's immediate guide on the descent into hell, was full of fear, yet Dante could declare without a tremor of doubt, "I have no fear because there is a noble lady in Heaven who takes care of me"'** (p.157 of 275). This is a marvellous illustration of how inspiring the beauty of women can be to men.

In a newspaper article by journalist Gary Kamiya the exceptionally unevasive thinker or prophet, Friedrich Nietzsche, is described as **'a terrifying Old Testament prophet'**, who was **'a desperately lonely man, poor and largely unread, plagued by bad health, who went mad at the age of 44'**. Kamiya said that **'What was great in Nietzsche was not, I began to see, his holiness, maybe not even his wisdom. It was his courage'** (*Australian Financial Review,* 11 Feb. 2000). Like Jung, Nietzsche courageously allowed himself to face the issue of the human condition and in doing so paid the price, which in his case was madness. (Incidentally the word 'holy', often used to describe prophets has the same origins as the Saxon word 'whole', which means 'well, entire, intact', thus confirming the prophets' wholeness or soundness or lack of alienation.)

In attempting to plumb the depths of the dilemma of the human condition, Jung faced **'insanity'**; Freud and Dante, while they were saved by protectors, respectively **'met stark terrors'** and **'descended into hell'**. Nietzsche did go mad and Eugène Marais and Arthur Koestler, two other exceptionally denial-free, honest thinkers or prophets, suicided. Arthur Koestler was the author of many books that dared to bring the issue of the human condition into focus. He was quoted in the *Plato* essay as courageously acknowledging the truth of integrative meaning, and the reader will see many of his extraordinarily truthful insights quoted in *Beyond*. In a review of a television program about Koestler called *Hours By The Window,* the journalist said in recognition of Koestler's remarkable ability to confront the human condition that, **'It's undeniable that Koestler had one of the most highly developed messiah complexes of the twentieth century'** (*Sydney Morning Herald,* 1 Dec. 1986). The reference to 'messiah' is a reference to the ability to confront the human condition because only by confronting the human condition could liberating understanding of the human condition ever be found. 'Messiah' means 'liberator'. You cannot ascertain the truth from lies, from a position of denial. Eugène Marais was the first person to study primates in the wild and wrote the exceptionally unevasive books, *The Soul of the White Ant* (1937), *My Friends the Baboons* (1939) and *The Soul of the Ape* (written in the 1930s, first published in 1969). In his Introduction to *The Soul of the Ape,* anthropologist Robert Ardrey referred to Marais as a **'prophet'** (p.33 of 170). Despite being exceptionally sound it is thought that even Plato may have experienced a crisis in the latter part of his life. To quote from the Plato entry in the 1979 edition of *A Dictionary of Philosophy:* **'Changes in outlook which accompany the change of style** [in the third or latter period of Plato's writing] **may reflect a profound crisis in Plato's life'** (p.270 of 380). In the great Arthurian legend of King Arthur, the prophet Merlin eventually went mad.

R.D. Laing, who was regularly referred to as a **'prophet'** (for example in *Life* magazine in October 1971), self-destroyed with alcohol, drugs and reckless behaviour and eventually died of a heart attack. In his 1994 biography, *R.D. Laing A Biography,* Adrian Laing described his father's circumstances 18 months before his death in August 1989 thus: **'He was sixty years old, the father of a new-born baby, with no reliable income, no home, a serious drinking problem and a debilitating feeling of depression bordering on despair'** (p.235 of 248). Throughout his life R.D. Laing cultivated honesty over denial, and the resulting transparency of the falseness in the world around him led him to a state of lonely despair. Towards the end of his life, R.D. Laing realised that what he needed in order to keep his equilibrium and do something truly constructive about the madness and suffering in the world that his honesty allowed him to see so clearly, was to find the source reason for the madness. In a 1989 documentary, *Didn't You Used To Be R.D. Laing?,* Laing said: **'I would like to be able to explore the reaches of the human mind, heart and soul, find out what we are doing here, where we have come from, where we are going to etc, etc, etc. I would like to spend the next time that I have got before I die enjoying that exploration without any contention'** (Third Mind Productions Inc. Vancouver, Canada). Tragically, by the time he realised that the only way to bring about any real change was to address the issue of the human condition he was too spent and weary for the task.

Unless you are extraordinarily innocent you cannot easily face the issue of the human condition without the protection imparted by the actual, biological understanding of the human condition. Confrontation with the issue of the human condition has to be accompanied by understanding of the human condition. As was emphasised in the *Introduction,* the evidence that understanding of the human condition has at last been found is that the subject is being discussed here so freely and openly, and that so many resigned adults in the FHA are successfully confronting the subject. Humanity would not have had the determination that it has had to find understanding of the human condition had it not known that when it finally did, that humans of that time would be able to confront the issue.

In passing I might make comment here about Sir Laurens van der Post's ability to live so extraordinarily close to the truth about the human condition. While a few prophets in history, in particular Buddha, Moses, Zarathustra, Christ and Mohammed, were so extraordinarily sound that they were able to live in the presence of the truth about the human condition without the support of understanding of it (they were able to confront God **'face to face'** as Moses said), Sir Laurens was not quite in this league. While Sir Laurens did grapple with the human condition it was always slightly obliquely, never quite head on. He was certainly much sounder than Carl Jung who, as previously described, struggled mightily even to begin to grapple with the human condition and had to employ dreams and observations of the extremely psychotic in order to access the truths that our soul knows. (Evidently Jung was confronted by this difference between himself and Sir Laurens, because he apparently came to view Sir Laurens as **'a pea that had grown too big for his pod'** [interview with J.D.F. Jones, ABC Radio, *Late Night Live*, 25 Feb. 2002].) While Sir Laurens was significantly sounder than prophets such as Jung, he was not completely sound and so it is surprising that Sir Laurens never suffered a psychological crisis during his 90-odd years living so close to the truth about the human condition.

Since the publication of J.D.F. Jones' alleged biography, *Storyteller: the many lives of Laurens van der Post* (2001), there has been much media attention focused upon Jones' accusations that Sir Laurens van der Post exaggerated the closeness of his relationship with Japanese acquaintances in the pre-war years, and with Lord Mountbatten and Carl Jung after the war. Jones also questioned Sir Laurens' abilities as a farmer, his rank as a lieutenant-colonel in the army (although he was undoubtedly a half-colonel), and his accounts of his war experiences and behind-the-scenes role in the Rhodesian settlement of 1979–80. (With regard to the Rhodesian settlement, if Sir Laurens had overstated his role in those events it certainly did not affect Lady Thatcher, who was Prime Minister of England at the time of settlement, for she recently described Sir Laurens van der Post as **'the most perfect man I have ever met'** [interview with J.D.F. Jones, ABC Radio, *Late Night Live*, 25 Feb. 2002].)

I want to first examine Jones' accusations that Sir Laurens exaggerated events surrounding his imprisonment by the Japanese and his association with those Japanese he met in South Africa before the war. While I am not familiar with the details of these events, what I do know from Sir Laurens' writing is that he was an exceptionally sound, unresigned individual. Knowing that resigned people are too familiar with, and accepting of, corrupt behaviour to have the defiance and uncompromising courage unresigned people have towards corrupt behaviour, I know Sir Laurens would have behaved extraordinarily strongly towards his Japanese captors and that he would have behaved extraordinarily courageously in the prisoner of war camps he was in. Also, with an appreciation of Sir Laurens' soundness, his room within himself, I know that he would have behaved most generously towards any Japanese that he met in South Africa.

With regard to the general accusation that Sir Laurens exaggerated his achievements, it needs to be appreciated that the fundamental problem for a prophet is having to live without any reinforcement. When all of humanity is living in an all-pervading and dominating world of denial, prophets live in an entirely different, denial-free world. Standing up to, and alone against, such an overwhelmingly different state is extremely difficult. In terms of positive support and feedback it amounts to living in a vacuum. Christ succinctly described the loneliness of the life of prophets when he said, **'Foxes have holes and birds of the air have nests, but the Son of Man** [the uncorrupted expression of the Godly, integrative state] **has no place to lay his head'** (Matt. 8:20). The coercion to give in and join everyone else is immense and all prophets have to counter that coercion is their love of the true world. In a scene in *Inherit the Wind*—a movie about the Monkey Trial in Tennessee, USA in 1925, where a school master named John Scopes was prosecuted for teaching Darwinian theory—Scopes' attorney empathised with his client's isolation in his community, saying: **'I know what you are going through. It's the loneliest feeling in the world. It's like walking down an empty street listening to your own footsteps. But all you have to do is to knock on any door and say "if you'll let me in I'll live the way you want me to live and I'll think the way you want me to think" and all the blinds will go up and all the doors will open and you will never be lonely ever again.'** This analogy gives some idea of the estrangement and loneliness experienced by an individual whose immediate community ostracises him. The loneliness from having to live in defiance of your *whole* race is infinitely greater. When, on the evening before he was arrested, Christ said **'I have overcome the world'** (John 16:33), the **'world'** that he had **'overcome'** was the world of denial. This comment suggests that even for Christ it was an achievement, something that had not been easy. Apart from the words of a few other prophets, acknowledgment of denial-free thinking is nowhere to be found. The essential psychological problem a prophet struggled with while he was growing up was the false world's inability to respond and acknowledge the soundness of his thoughts and behaviour. In fact, not only did the resigned, denial-complying world not respond, it normally actively rejected the sound thinking and behaviour of the prophet because the thinking and behaviour was confronting and exposing of its unsound thinking and behaviour. Until a prophet was older and had finally gained an understanding, or at least an appreciation, of why the false world could not respond and acknowledge his sound thinking and behaviour, it was a continual struggle to comprehend why sound thinking and behaviour did not receive a reinforcing response, in fact why it was not embraced and encouraged. Prophets were like children in the sense that children suffered from the same problem of not receiving the reinforcement that their true thinking and relatively innocent, uncorrupted behaviour deserved from the adults around them. It was not a case of children or prophets being egocentric, they were not looking for 'a win', only fair acknowledgment. To resigned people living in insecurity it has been self-evident to them why other people behaved insecurely, but to an unresigned person it was a complete mystery. A prophet's fundamental struggle was to avoid being overcome by the coercion to believe that their truthful thinking and true way of behaving was wrong or unworthy. The essential problem they faced was to survive a dysfunctional world. If they were exceptionally secure, that is, exceptionally nurtured in infancy and reinforced in childhood, they would be able to survive the dysfunctional world. However, if they were a little less than completely secure they would either succumb to becoming uncertain of their own truth and worth, or would have to find some way of resisting the implication that what they said and did was not true and worthy. It is possible Sir Laurens van der Post was not completely secure and so had to find a way of surviving and the way he did that was simply to tell his own story, rather than wait or depend on others to tell it. That part of Sir Laurens' writing that his detractors have so denigrated is his self-descriptions, his accounts of what he achieved, the so-called larger than life image that he cast of himself. It is his apparently necessary self-description that has been misrepresented as insecure, egocentric exaggeration by his detractors.

It also has to be appreciated that Sir Laurens had a responsibility to find support for his denial-free world; he had to somehow counter society's denial of prophets and the immensely valuable truths they reveal. I think it is possible that Sir Laurens found for himself the reinforcement he needed for himself, and the necessary support for his work, by using his extraordinary soundness, and thus extraordinary personality, to gain the admiration and support of people in high places. It was only people who reached positions of great achievement in the insecure, resigned, egocentric world who had the room in themselves, the generosity, to acknowledge someone as sound as Sir Laurens; everyone else was threatened by his soundness. It is no coincidence that a Prime Minister, Margaret Thatcher, was able to acknowledge Sir Laurens as **'the most perfect man I have ever met'**. Sir Laurens van der Post did receive very significant reinforcement through the acceptance and acclaim of his 24 books, however, it could be that he had to create additional reinforcement through self-promotion, or find himself unable to do the work he did. If he is guilty of self-promotion for this reason then he took the courageous option, because history will record that Sir Laurens van der Post played a pivotal role in saving the human race.

I am not in a position to know if, in searching for reinforcement of his denial-free way of living, Sir Laurens ever went beyond truthful self-acknowledgment to some untruthful embellishment in his stories. However, it has to be remembered that if Sir Laurens is guilty of any small lies then those small lies are minuscule compared to the massive lie that humanity's denial of the human condition represents, and which J.D.F. Jones has desperately tried to defend by his attack on Sir Laurens. The essential truth is that Jones has acted as an agent for the world of denial. The true story is that Sir Laurens van der Post revealed a galaxy of truth to this planet of lies, and for doing so Jones sought to 'crucify' him, just as Christ was crucified for exposing all the lying in his day. It is a classic example of a reverse-of-the-truth lie to call Sir Laurens van der Post a liar. Sir Laurens was a truthsayer of the highest order, a foremost prophet amongst prophets.

I should also comment here about the accusation that Sir Laurens van der Post 'romanticised' the Bushmen people of the Kalahari. J.D.F. Jones said in an interview, **'the academic experts on the Kalahari** [Bushmen] **are absolutely berserk with rage about the things he** [Sir Laurens van der Post] **said, because, if you read *The Lost World of the Kalahari*** [Sir Laurens' immensely popular book]**, you must not believe that this is the truth about the Bushmen; it's not'** (ABC Radio, *Late Night Live*, 25 Feb. 2002). As has been explained in detail in the *Plato* essay, the issue of the human condition and all the important truths that bring it into focus—in particular the existence of our soul's innocent, true world, and humans' alienated state of denial of that world—has been an anathema to humans. Science has been mechanistic not holistic and, like humanity, has not tolerated acknowledgment of the different states of alienation amongst races, genders, ages, generations and individuals. The reason people such as myself have loved Sir Laurens so much is because he defied the almost universal denial and let the truth out about our soul's true world. Whether mechanistic scientists like it or not, the truth is the Bushmen people are relatively innocent compared to most other existing races, and in them we can see something of what has been denied and repressed in the rest of the human race. Of all the books in my library, my copies of Sir Laurens' books about the Bushmen, *The Lost World of the Kalahari* (1958) and its sequel, *The Heart of the Hunter* (1961), are the most tattered from use. If I quote the last few lines from *The Heart of the Hunter* the reader will know just how devastatingly honest these books are: **'All this became for me, on my long journey home by sea, an image of what is wanted in the spirit of man today. We live in a sunset hour of time. We need to recognize and develop that aspect of ourselves of which the moon bears the image. It is our own shy intuitions of renewal, which walk in our spiritual night as Porcupine walked by the light of the moon, that need helping on the way. It is as if I hear the wind bringing up behind me the voice of Mantis, the infinite in the small, calling from the stone age to an age of men with hearts of stone, commanding us with the authentic voice of eternal renewal: "You must henceforth be the moon. You must shine at night. By your shining shall you lighten the darkness until the sun rises again to light up all things for men"'** (p.233 of 233).

I might mention that even though I have the benefit of the biological understanding of the human condition to protect me both from any doubts about the fundamental worth of humans and any intimidation from the world of denial, I still find it helpful to use quotes from other denial-free thinkers to confirm every step I have taken in my thinking. The road that I have cut through the jungle of denial I have paved with quotes to help both the reader *and* myself. Considering Christ did not have the benefit of the biological understanding of the human condition, it is not surprising that he also had his 'stepping stones'; the confirmatory statements from the prophets who preceded him, those rare individuals of the Old Testament who were sound enough to deeply penetrate humanity's psychosis/denial/alienation. Obviously he studied closely the Old Testament because he frequently referred to it. For example, in my soon to be published *Death by Dogma* essay the reader will see that when Christ was unmasking the lie of pseudo-idealism he used the deadly accurate description offered by the Old Testament prophet Daniel, **'the abomination that causes desolation'** (Matt. 24:15).

It was mentioned above that Sir Laurens van der Post was much sounder than Jung. In the coming essay in this book, *The Demystification Of Religion,* under the heading 'Prophets and the concept of the "Virgin Mother" demystified', it will be explained in detail that there are two classes of prophets, those who are unresigned and those who are resigned. Briefly, individuals who were sufficiently nurtured and thus secure in self to avoid having to resign to a life of denial are the unresigned prophets. Resigned prophets, on the other hand, were individuals who had resigned during adolescence but who had later been able to find their way back to a sufficiently denial-free way of thinking to be considered a prophet.

I should note that in the next section it will be described how some artists could so develop their capacity to bring out the beauty that exists on Earth—and by association bring into focus all the deeper human condition-related issues that such purity raises—that they could take themselves to the brink of suicidal depression. In the case of one of the greatest artists, Van Gogh became so tormented he did end up taking his life. The point is if a resigned person happened to take up a path that led back to the world of the soul and all the confronting truths that reside there, and had the courage, determination and gifts to pursue that extremely difficult path far enough, they could manage to reveal sufficient truth to be recognised as a denial-free thinker or prophet. However, as has been described, the price resigned prophets paid for such heroic return-trips to the world of truth was often as high as suicide. The problem is that if someone had become resigned during their adolescence then they cannot have been sound enough to face the truth without being suicidally depressed by it. Resigned people fought their way back to the truth at their peril.

Clearly Sir Laurens van der Post was an unresigned prophet, while Carl Jung was a resigned prophet.

The cost of resignation: 'fifty foot of solid concrete' between humans and their souls

The examples given earlier show that *even many prophets,* the exceptionally sound in society, found the issue of the human condition dangerously confronting and depressing. Those examples should make it *very* clear that what forced adolescents to resign to living a life of denial of the human condition was the extreme danger of suicidal depression if they continued in their efforts to confront it.

It is necessary then to explain why this danger *had* to be extreme for resignation to be a responsible alternative. What will be made clear is that the cost of resignation was also very high, *almost as bad as death from suicidal depression.* In fact resigning to a life of denial meant accepting psychological death, the death of a person's psyche or soul.

Again, Fiona Miller's amazing resignation poem described the extent of the psychological death that accompanied resignation: **'Smiles will never bloom from your heart again, but be fake and you will speak fake words to fake people from your fake soul/What you do today you will do tomorrow and what you do tomorrow you will do for the rest of your life/From now on pressure, stress, pain and the past can never be forgotten/You have no heart or soul and there are no good memories/ Your mind and thoughts rule your body that will hold all things inside it; bottled up, now impossible to be released/You are fake, you will be fake, you will be a supreme actor of happiness but never be happy.'**

An analysis of the phenomenon of savant syndrome, and of the related condition of autism, will demonstrate just how high the price of resignation has been. It will show just how much sensitivity and capability humans gave up access to when they resigned.

A television program titled *Uncommon Genius* was broadcast on ABC-TV in Australia on 24 May 2001. The documentary, written and directed by Ian Watson, followed Australian psychologist Dr Robyn Young on a tour of the USA where she met a handful of people described as savants. The program began with the narrative: **'One of the great unsolved mysteries of the human mind is savant syndrome. This man cannot remember how to clean his teeth, yet he recalls every zip code, every highway, every city in the United States. As a boy, this man was considered mentally handicapped but he can name the day of the week for every date on a forty-thousand-year calendar. And this man, blind, cerebral-palsied and barely able to talk, played Tchaikovsky's First Piano Concerto flawlessly after hearing it once on television. These genius-like abilities are the product of damaged minds...Most of us were introduced to savants through Dustin Hoffman's portrayal of Raymond in the Oscar-winning movie, *Rain Man*. Raymond had savant syndrome...Savants are people who produce awesome mental abilities from severely disabled minds.'**

The program mentions that **'savants may abandon their skills as they become more sophisticated socially...The really prodigious savants have a sort of memory super-highway that allows them to access and transfer enormous amounts of information. They develop that memory partly because they <u>don't get side-tracked by thinking too much</u>.'**

'...for more than half of all savants, the syndrome owes its origins to a familiar condition—autism...[which is] **a condition <u>whereby the brain's ability to organise complex thought is impaired</u>...**[The program refers to]

the huge increase in autism that we see these days, and massive increase, 273 percent increase, in the state of California over a 12-year period... [with] **California's Silicon Valley** [being] **home to more autistic children than anywhere** [else] **in America...'** The narrator continues, **'Researchers the world over are still seeking a complete understanding of the cause of autism.'**

The following quotes were taken from the program's concluding comments: **'Led by Dr Bruce Miller, this group of San Francisco researchers discovered people who suffered dementia and then suddenly gained prodigious skills they'd never experienced before. One case was a man who, when his brain degenerated, became a composer as he lost his ability to speak...Another patient won a patent for a chemical detector** [at a time] **when he could name just one out of 15 items on a word test. Dr Miller's patients suffered fronto-temporal dementia. They lost cells in parts of the brain that regulate social behaviour. These imaging studies reveal similar left brain injuries to those of savants Robyn had been studying. This is exciting new evidence that for the first time is taking us closer to discovering the cause of savant syndrome. Could savant genius lie dormant deep inside everyone's brain? Are savant skills merely obscured by layers of normal everyday reasoning? The possibility that we all possess hidden genius is tantalising. But it should not be forgotten that whether extraordinary savant skills emerge at birth or appear later in life, savant syndrome comes at a cost...The musical brilliance of Tony DeBlois exists at the price of blindness and autism. And George Finn is only now emerging from his genius with numbers to engage with a wider world.'**

In the course of another television program, *The Theories of Everything,* producer and presenter David Hunter Tow had the opportunity to describe a person with savant-like mathematical abilities. In the program Tow, a scientist specialising in computer software, says that: **'In a creative process it** [the idea] **seems to come out of the inner mind somewhere without you really forcing it, it seems to flow at a certain point. Perhaps the most famous case of this was—and I think one of the most interesting stories that I've come across in science—was a young Indian clerk, who lived at the beginning of last century, whose name was Ramanujan** [Srinivasa Aaiyangar Ramanujan 1887-1920]**. He was possibly the greatest mathematician of all time. He virtually had no education in mathematics beyond one text book that he picked up by chance in Madras one day, by an English mathematician, with a few theorems. In one year Ramanujan had re-invented one hundred years of western mathematics from scratch virtually. He eventually came to England and there he generated**

over a period of three years 4,000 theorems, which on average took a top mathematician just to prove one or two of those theorems the best part of a year. He generated those theorems virtually subconsciously out of his mind. He would virtually dream them…This young man could go to sleep at night and in the morning wake up and scribble out another half dozen theorems. These theorems he rarely proved but when they were tested and proved they were always absolutely correct. Unfortunately he died of pneumonia three years later. But where did those theorems come from? Where did that knowledge come from? Obviously from the subconscious in a phenomenal way and no one to this day knows how it happened. Other mathematicians can get subconscious insights, generate breakthroughs after being asleep even, or drinking coffee for that matter; however, Ramanujan was just an exceptional case, probably the greatest mathematician in history' (Australian community TV Channel 31, 3 Dec. 2000).

Dr Young specialises in autism and is head of the Autism Research Unit at Flinders University in South Australia. A summary of the Unit's current research projects refers to a hypothesis about savant-type abilities: **'A recent hypothesis that has received much attention is that savant type abilities "reside equally in all of us" (Snyder & Mitchell, 1999, p591). Snyder and Mitchell suggest that we all have access to this fundamental mechanism, perhaps even some privileged information, but because of higher order cognitive processing we are unable to access this or these mechanisms'** (www.ssn.flinders.edu.au/psyc/staff).

In a world that is resigned and as a result committed to denial of the all-sensitive and truthful-thinking world of our soul, this suggestion that savant skills reside equally in all of us is an extraordinary admission because it is tantamount to admitting humans are deeply alienated, deeply estranged from their true self and true potential.

There has certainly been some **'higher order cognitive processing'** not related to denial that has inhibited savant-type abilities in resigned humans. For example, selective memory has been necessary to stop our conscious mind becoming cluttered with irrelevant information. However, the main **'higher order cognitive processing'** that has been inhibiting humans' access to their savant-type abilities is the resigned mind's denial of any information that brings the issue of the human condition into focus, in particular the repression of the truthful, integrative-meaning-aware, all-sensitive world of our original instinctive self or soul. In Plato's cave allegory human life in the cave consisted of a world of darkness and shadows, shut off from the world of illuminating sunlight (ie cut off from the all-beautiful world 'illuminated' by truthful, effective thought).

The point is that if people were not resigned they would have the myriad of capabilities and sensitivities of savants. It is a comparison that gives the resigned world an accurate measure of just how spiritually defunct its world really is. The *Macquarie Dictionary* (3rd edn, 1998) defines 'spirit' as **'the vital principle in humans, animating the body or mediating between body and soul'**; our 'spirit' is our aliveness, our sensitivity to all of existence, and it is this capacity to be super-aware that largely dies with resignation. As Fiona Miller said, **'You have no heart or soul...You are fake, you will be fake, you will be a supreme actor of happiness but never be happy.'**

The Russian philosopher George Gurdjieff described the resigned, alienated state truthfully when he wrote: **'It happens fairly often that essence dies in a man while his personality and his body are still alive. A considerable percentage of the people we meet in the streets of a great town are people who are empty inside, that is, they are actually *already dead'*** (*In Search of the Miraculous,* P.D. Ouspensky, 1950, Ch.8, p.164).

R.D. Laing used the 19th century French poet Stéphane Mallarmé's evocative words to similarly elicit this truth about the extent of the psychological estrangement of the resigned, alienated state when he wrote: **'To adapt to this world the child abdicates its ecstasy ('L'enfant abdique son extase': Mallarmé)'** (*The Politics of Experience* and *The Bird of Paradise,* 1967, p.118 of 156).

Laing elaborated: **'We are born into a world where alienation awaits us. We are potentially men, but are in an alienated state** [p.12]**...the *ordinary* person is a shrivelled, desiccated fragment of what a person can be. As adults, we have forgotten most of our childhood, not only its contents but its flavour; as men of the world, we hardly know of the existence of the inner world'** (ibid. p.22).

'The condition of alienation, of being asleep, of being unconscious, of being out of one's mind, is the condition of the normal man [p.24]**...between *us* and It** [our soul] **there is a veil which is more like fifty feet of solid concrete. *Deus absconditus.* Or we have absconded'** (ibid. p.118).

'The outer divorced from any illumination from the inner is in a state of darkness. We are in an age of darkness. The state of outer darkness is a state of sin—i.e. alienation or estrangement from the inner light' (ibid. p.116).

'We are dead, but think we are alive. We are asleep, but think we are awake. We are dreaming, but take our dreams to be reality. We are the halt, lame, blind, deaf, the sick. But we are doubly unconscious. We are *so* ill that we no longer feel ill, as in many terminal illnesses. We are mad, but have no

insight' *(Self and Others,* 1961, p.38 of 192).

The English poet Percy Bysshe Shelley (1791–1822) also used the term **'asleep'** to describe humans' current state: **'Our boat is asleep on Serchio's stream/Its sails are folded like thoughts in a dream'** *(Shelley: The man and the poet,* Desmond King-Hele, 1960, p.335 of 390).

The mythology of the Superman character is an expression of humans' suppressed awareness of their alienated state—that resigned humans are Clark Kents (Clark Kent was Superman disguised as a newspaper reporter) with hidden Superman potential.

To think truthfully and thus effectively, to be able to access all the beauty that is in the world, to be able to create and behave naturally without inhibition, requires freedom from the denial that takes place at resignation. The resigned conscious mind was committed to blocking out the truthful, beautiful, intuitive, natural world. Necessary as it has been, the resigned mind's denial has massively thwarted humans' real potential. The resigned conscious mind has *not wanted* access to truth and beauty, it has worked *against* accessing truth and beauty, so much so that a person had to be free or independent of the will or desire of the resigned conscious mind if they were indeed to access truth and beauty. The philosopher Arthur Schopenhauer recognised this when he said, **'The unpremeditated, unintentional, indeed in part unconscious and instinctive element which has always been remarked in the works of *genius* owes its origin to precisely the fact that primal artistic knowledge is entirely separated from and independent of will, is will-less'** *(Essays and Aphorisms,* tr. R.J. Hollingdale, 1970, p.158 of 237).

Being unresigned, children have unimpeded access to the truthful, beautiful, alienation-free, natural world, and any adult who has retained such access must have retained that childhood clarity. The Chinese philosopher Mencius made the point, **'The great man is he who does not lose his child's-heart'** *(Works,* 4–3 BC, 4, tr. C.A. Wong).

Another measure of the cost of repressing—living in denial of—the all-sensitive world of our instinctive self or soul can be found in the phenomenon of near-death experiences, or NDEs. As an illustration of NDEs, there are mountain climbers who, having survived a fall from which they thought they would certainly die, reported that in the fall they entered a state of extraordinary euphoria where everything around them was utterly beautiful and radiant. What happens in such NDEs is the mind gives up worrying, and all facades—in particular the denial that they adopted at resignation—become meaningless. If death is inevitable then there is no longer any reason to worry or to pretend. At that point the struggle and agony of having to live under the duress of the human condition ceases and suddenly the true world of our all-sensitive soul surfaces.

There are many books available now documenting NDEs. Many people who have experienced an NDE tell of a world so wonderful that they have no wish to return to the 'normal' (alienated) world. The American social psychologist Kenneth Ring wrote that: **'For 10 years I have studied cases of persons who have survived episodes of near death or clinical death only to tell of wonders in the land *beyond* the edge of life. One man speaks of being in a state of "total radiance from absolute knowledge" when he realised that "*finally* I was alive." One woman says: "I was enabled to look deeply inside myself. I saw...that my core was perfect love—and that applies to *all* human beings." But it is not just that one experiences this truth; one becomes it. The meaning of life has something to do with realizing that our essence is perfect love, then going on to live our lives upon that truth, experiencing each day as a miracle and every act as sacred'** (*Life* mag. Dec. 1988).

Explaining NDEs is not difficult once it is appreciated how alienated humans have become. An NDE amounts to the purest form of prayer or meditation where the 'troubles of life' (essentially, the struggles that emanate from the dilemma of the human condition) are abandoned and the denial and alienation-free, utterly cooperative, unconditionally loving, integrated, all-sensitive, heavenly state reappears. The state that humans once instinctively lived in before the emergence of consciousness, and with it the agonising, worrisome, preoccupying, alienating dilemma of the human condition, emerged. Heaven is not a place up in the clouds somewhere, or a place in another universe, or a supernatural realm, rather it is the human condition-free, utterly integrated state of being and living; the ultra-natural, rather than super-natural state for humans. As Friedrich Nietzsche said, **'The Kingdom of Heaven is a state of the heart (of children it is written, "for theirs is the Kingdom of Heaven"): it has nothing to do with superterrestrial things'** (*The Will To Power*, 1901; tr. O. Levy, 1909, p.134 of 384).

The mention of **'absolute knowledge'** and a **'core'** of **'perfect love'** is witness to our instinctive self or soul's perfect orientation to cooperative, loving integrativeness. People who have returned to the world of our soul, to heaven, to our lost paradise, to the realm that the Roman poet Virgil called 'arcadia'—to the world before the corrupted state of the human condition emerged—testify to the true extent of the beauty, indeed radiance of that world, and to the certainty and contentment of life in that world. The world humans live in now is a numb, seared, brutalised, dead place compared to it. In truth humans are now so brutalised they merely skate on the surface of existence. If the true world is a bucket full of water, the world that resigned humans are able to experience is only the meniscus. Sadly, because of the soul's criticism of humans' apparently imperfect state, resigned humans have been ruthlessly repressing their idealistic soul and in the process all the beauty and truth that it knows of and has access to. This has occurred to such an extent that resigned humans have lost almost all memory of the true world. Without that memory humans walk in meaningless darkness. The word 'enthusiasm' is derived from the Greek word *enthios,* which means 'God within'. Without some knowledge of the heavenly state that our soul has already experienced, without some knowledge of 'the God within', life loses its richness and value.

Fatigue also offers access to our soul. If the mind is exhausted the soul can sometimes surface into the conscious awareness. The Bushman of the Kalahari dance until they become so physically exhausted that the alienation or blocks in their brain subsides, letting their soul through. Sir Laurens van der Post recognised that fatigue can liberate access to the soul when he wrote: **'fatigue was to the healer what drugs are to the psychiatrist; a means of lowering the level of consciousness and its wilful inhibitions so that the unconscious forces and the instinctive powers at the disposal of all life could rise unimpeded and be released in the healer'** *(Testament to the Bushmen,* 1984, p.157 of 176).

Rock climbers have a dangerous game they call 'scree jumping' where they race down a boulder-strewn hillside with their feet moving from one rock to the next so fast that their conscious mind cannot keep up, at which point the instinctive self takes over and suddenly they seem to acquire a grace and certainty that astonishes them. A good mountaineering book to read that describes this soul-accessing activity is Rob Schultheis' *Bone Games* (1985).

One of my close companions on this journey to bring understanding to the human condition is the world-renowned Australian mountaineer, Tim Macartney-Snape. Tim was the first Australian to climb Mt Everest and climbed it a second time solo from sea level. On both occasions the ascent was achieved without the assistance of bottled oxygen. In his 1992 book, *Everest from Sea to Summit,* which documents this second climb, there is a photo of Tim taken on his arrival back at base camp after his ascent (p.252 of 280). It shows him in a state of utter fatigue but with an expression of child-like radiance on his face. Tim has often described how exhaustion at high altitude brings with it an exhilaration and a heightened state of sensitivity.

Despair can sometimes have the same effect as fatigue. Sometimes when despair overtakes us, our mind gives up trying to cope with and make sense of the world. When the inhibitions, block-outs and denials are relinquished the soul surfaces into conscious awareness. The following quote, from Olive Schreiner's 1883 book, *The Story of an African Farm,* has already been referred to in the *Introduction,* however it is worth including here because it is a marvellous description of how despair can open up access to the soul: **'There are only rare times when a man's soul can see Nature. So long as any passion holds its revel there, the eyes are holden that they should not see her...Only then when there comes a pause, a blank in your life, when the old idol is broken, when the old hope is dead, when the old desire is crushed, then the Divine compensation of Nature is made manifest. She shows herself to you. So near she draws you, that the blood seems to flow from her to you, through a still uncut cord: you feel the throb of her life. When that day comes, that you sit down broken, without one human creature to whom you cling, with your loves the dead and the living-dead; when the very thirst for knowledge through long-continued thwarting has grown dull; when in the present there is no craving and in the future no hope, then, oh, with a beneficent tenderness, Nature enfolds you. Then the large white snowflakes as they flutter down softly, one by one, whisper soothingly, "Rest, poor heart, rest!" It is as though our mother smoothed our hair, and we are comforted. And yellow-legged bees as they hum make a dreamy lyric; and the light on the brown stone wall is a great work of art; and the glitter through the leaves makes the pulses beat. Well to die then; for, if you live, so surely as the years come, so surely as the spring succeeds the winter, so surely will passions arise, they will creep back, one by one, into the bosom that has cast them forth, and fasten there again, and peace will go. Desire, ambition, and the fierce agonizing flood of love for the living—they will spring again. Then Nature will draw down her veil: with all your longing you shall not be able to raise one corner; you cannot bring back those peaceful days. Well to die then!'** (p.298 of 300).

Fasting, where the brain is starved of nourishment, has been another means of shutting down the conscious mind and allowing the truthful, all-sensitive world of our soul out.

Hallucinatory drugs also enable the mind to cut through the conscious overburden of denial and access the soul. In recognition of this some earlier civilisations made hallucinatory substances a central part of their religious rituals.

Great artists, be they painters, sculptors, singers, musicians, dancers, poets, writers, architects or designers, are people in whom the denial and subsequent alienation that came with resignation was incomplete. Occasionally a person's protective block-out develops, as it were, with a crack or tear in it. Through this small rent these people can touch upon and reveal some of the true beauty that exists on Earth. This rent or window can also, to a degree, be cultivated. After years of developing his painting skills Vincent Van Gogh was able to bring out so much beauty that resigned humans looking at his paintings find themselves seeing light and colour as it really exists for possibly the first time in their life: **'And after Van Gogh? Artists changed their ways of seeing…not for the myths, or the high prices, but for the way he opened their eyes'** (*Bulletin* mag. 30 Nov. 1993).

The often-referred-to 'pain' of being an artist was that while resigned humans coped with life's deeper questions by evading them, artists continually raised them. By revealing the real beauty on Earth artists continually confronted God in the form of the perfection of beauty, and since humans have been an insecure, God-fearing species, what they were doing was confronting and hurtful. It has been a torturous business confronting God without being able to explain why humans have apparently been so unGodly.

It was described in the previous section how some prophets—the soundest of people—suicided as a result of trying to confront the issue of the human condition. Artists who were too honest for their degree of soundness, artists who tried to confront all the beauty and associated truths that exists on Earth when it was more than they were capable of enduring, could also take themselves to the brink of madness and/or suicidal depression. Van Gogh went over this brink and did suicide. Sir Laurens van der Post described the situation that faced artists and writers in the following remarkably insightful quote: **'The history of art and literature indeed contains as many examples of persons who have succumbed before the perils encountered in the world within as those who have been overcome by their difficulties in the world without. The asylums of the world are full of people who have been overwhelmed by what has welled up within them: instincts and intuitions shaped over aeons in which they had played no part, and imposed on them by life without their leave or knowledge. The person who enlists in the service of**

the imagination, as do the artist and writer, has continually to come to terms and make fresh peace with this inner aspect of reality before he can express his full self in the world without. Many are so appalled by the difficulties and terrifying implications of what they see within themselves that, after a few bursts of lyrical fire, they either retreat into the previously prepared positions conventionally provided for these occasions by their social establishments: or else they close up altogether or take to drink or commit suicide. Nor is there any comfort to be found in thinking that this kind of defeat is suffered only by the lesser breeds among artists and writers: there are too many distinguished casualties. There is, for instance, the uncomfortable example of Rimbaud who, though a poet of genius, found the implications of genius more than he could bear and took on the perils of gun-running in one of the most dangerous parts of Africa as a more attractive alternative. Yet before he turned a deaf ear to the profound voice of his natural calling, he had shaped a vision of reality which increased the range of poetry for good. One may regret his desertion, but surely no one who cares for poetry can read "Bateau Ivre" and "Les Illuminations", for example, without some understanding of the power of the temptation, and an inkling of how exposed and vulnerable the ordered personality is to the forces of this world that the artist carries within him. The suicide of Van Gogh is another instance. We owe it to him that our senses are aware of the physical world in a way not previously possible (except perhaps by the long-forgotten child in all of us when the urgent vision is not yet tamed and imprisoned in the clichés of the adult world). But because of Van Gogh, cypresses, almond blossom, corn-fields, sunflowers, bridges, wicker chairs and even trains are seen through eyes made young and timeless again and our senses are recharged with the aboriginal wonder of things. Here was not only genius but also high courage. Yet nothing so well gives one the measure of these inner forces as the fact that they were able to destroy both courage and genius' (from Sir Laurens van der Post's Introduction to the 1965 edn of *Turbott Wolfe* by William Plomer, first pub. 1925, pp.34–36 of 215).

Beauty could be the greatest inspiration, blindingly so at times. But it could also be condemning and hurtful to humans because it confronted them with their apparent lack of beauty or perfection. The truth is, mere glimpses of beauty were all corrupted humans could cope with. In the *Introduction* Patrick White and William Wordsworth were quoted as saying that even a flower could give rise to unbearably depressing thoughts about the apparent imperfection of humans. The banning of representations of living things in Islamic art is a means of subsiding the agony arising from the human condition in the minds of Islamic followers.

For many people, unnatural, alienated cities offered a refuge from the hurtful glare of the beauty and truth of the natural world. Orienteering, where people race through the forest at break-neck speed to complete a compass course, is a good indication of how little nature humans have been able to tolerate. Humans have been so condemned by the innocence of nature that it has been torture for them to be amidst nature for long. It seems that when humans were with nature they had to be there with something to distract them, such as a river to fish, a peak to climb, or a compass course to complete. Humans could not just *be* with nature. Nature could inspire and reinforce but it could also condemn. Nature was humans' instinctive self or soul's original companion and as such reminded them of their soul's pure world and their apparent impurity. The beauty of the natural world could inspire and centre humans, but it could also hurt them terribly.

All these examples illustrate the price of resigning. They show how much beauty and truth humans lost access to when they took up a life of denial. The depression that preceded resignation clearly had to be overwhelming for humans to be prepared to pay this immense price.

There was also the cost in terms of human potential involved in resignation. In the Grand Canyon in the USA, a bird called the nutcracker buries 30,000 nuts throughout the summer months, each in a different location. In winter, even under the cover of snow, it remembers the location of 90 percent of them. There is a goby fish that can memorise the topography of the tidal flats at high tide and when the tide goes out it knows the exact location of the next pool to flip to when the one it is in evaporates. The brain of the male common canary dramatically expands every spring in order to learn new mating songs, only to shrink again at the end of the mating season. Although these animals live in an instinct-dominated state, they have not become instinctively integrated as humans did millions of years ago—they have not entered the state of pure, unconditional, 'heavenly' love—nor have they become conscious (ie able to make sense of experience). Imagine therefore the abilities humans would have were it not for the emergence of the problem of the human condition and the resultant resigned state of alienation. We could have the capacity to develop amazing abilities, plus access to our instinctive self's experience of living in an utterly integrated, loving, heavenly state. Importantly, we would also be able to make sense of and savour all experience. Savants offer an incredible insight into just some of the extraordinary abilities humans would have if they were not resigned and alienated.

I have been able to find a veritable avalanche of answers, not because I am clever or more gifted in some physical way than other people, but *only* because I was exceptionally fortunate in being sufficiently nurtured and sheltered from corrupt reality in my infancy and childhood not to have to resign in my adolescence and adopt a life of living in denial of all the truth and beauty that is in the world. With the benefit of the achievements of the whole human race—in particular the insights into the workings of our world painstakingly found by mechanistic science and the confirmation I have derived for my thinking from the work of many other denial-free thinkers throughout the ages—I have been able to synthesise the biological explanation of the human condition; that is, explain the reason humans are competitive, aggressive and selfish when the ideals are to be cooperative, loving and selfless. In doing so I have been able to explain and end the need for humans' alienated state; explain and make possible the end of loneliness and depression; explain and end the need for egocentricity; explain the origin of war and aggression amongst humans and bring an end to the cause of war and aggression; explain and end the need for materialism; explain and end the need for a superficial, artificial, self-distracting way of living; explain biologically how humans acquired their altruistic 'soul' and its cooperation-demanding 'conscience'; describe and explain the psychological act of resignation; explain the stages of maturation of infancy, childhood, adolescence and adulthood that both humanity and humans individually go through; explain the meaning of life; explain why 'evolution' is in fact the purposeful process of ordering matter; explain the reasons for the limitations of mechanistic science; relate all the disciplines of the sciences and the humanities; explain in biological terms how humans became fully conscious and why other animals have not; explain why and when humans learnt to walk upright, lost their body hair, developed language, left Africa, began tool use, began hunting and meat-eating; reconcile science with religion; explain religion and render it obsolete, in the process explaining all manner of religious metaphysics, including the concepts of God, the Trinity, prophets, the Virgin Mary, the resurrection, miracles, Judgment Day, the Battle of Armageddon, the story of Noah's Ark, after-life, heaven and hell, good and evil; decipher humanity's legends and myths; explain and reconcile the left and right wings of politics; end the reason for prejudice and the cause of inequality between individuals, sexes, ages, generations, races and cultures, in the process reconciling the worlds of men and women, the young and the old, the innocent and the corrupted; explain the pseudo-idealism of the New Age, Peace, Green, Feminist, Native Peoples, Animal Rights, Multicultural, Politically Correct, Postmodern Movements; explain sex, heterosexuality, homosexuality, love, beauty, the attraction of youth, romance, rape, envy and lust; explain humour; explain human sensitivity and creativity, especially art and music; explain away the main underlying cause for human sickness; explain away the psychological basis of autism and savant syndrome and all psychological disorders; explain near-death experiences; provide the means for the psychological repair of the human race; save the human race from self-destruction; bring peace to the human situation, etc. Imagine what more I could have done had I not been preoccupied throughout my adult life defying and enduring ostracism and persecution from the resigned world of denial, and instead had been surrounded by other denial-free thinkers.

The point is, imagine what *all* humans could have achieved had they not had to resign and take up a life of lying. And *imagine what the human race will be able to achieve now that resignation is no longer necessary.*

The autistic state demonstrates how the mind can dissociate itself from unpleasant experiences

Resignation involves psychologically separating or dissociating—becoming alienated—from the unconfrontable subject of the human condition and all the many truths that bring the subject into focus. According to the quotes cited earlier about savants, many if not most savants are autistic. An understanding of autism will show dramatically how the human mind can block out, dissociate itself from, or create a life in denial of an unconfrontable reality.

To understand autism it needs to be recollected that nurturing, or love-indoctrination, of infants was the prime mover or main influence in both the maturation of our species and the maturation of our personal lives.

It was through nurturing, the process of 'love-indoctrination', that humans were able to develop an instinctive orientation to behaving unconditionally selflessly and as a result become utterly integrated and cooperative. In our past instinctive state, prior to becoming conscious and corrupted by the problem of the human condition, everyone behaved selflessly and considered the welfare of the group above their own. This instinctive memory of a loving, cooperative, alienation-free, all-sensitive past is what we term our 'soul', one expression of which is our 'conscience', the instinctive expectation within us that we behave selflessly, lovingly and cooperatively.

Of course, with the emergence of the human condition and the need to resign to a life dedicated to denial of any idealistic truth that appeared to condemn humans, this truth of the significance of nurturing in the maturation of our species, and in our own lives, was repressed and denied.

Before beginning the explanation of autism I would like to include four quotes. The first two quotes acknowledge our species' innocent, cooperative, loving instinctive past. The third quote acknowledges the extraordinary sensitivity we once possessed. The fourth quote acknowledges the existence of our conscience. The quotes are taken from the writings of Sir Laurens van der Post.

'This shrill, brittle, self-important life of today is by comparison a graveyard where the living are dead and the dead are alive and talking [through our soul] **in the still, small, clear voice of a love and trust in life that we have for the moment lost...** [there was a time when] **All on earth and in the universe were still members and family of the early race seeking comfort and warmth through the long, cold night before the dawning of individual consciousness in a togetherness which still gnaws like an unappeasable homesickness at the base of the human heart'** (*Testament to the Bushmen,* 1984, pp.127–128 of 176).

'There was indeed a cruelly denied and neglected first child of life, a Bushman in each of us' (*The Heart of The Hunter,* 1961, p.126 of 233).

'He [the relatively innocent Bushmen people of the Kalahari desert] **and his needs were committed to the nature of Africa and the swing of its wide seasons as a fish to the sea. He and they all participated so deeply of one another's being that the experience could almost be called mystical. For instance, he seemed to *know* what it actually felt like to be an elephant, a lion, an antelope, a steenbuck, a lizard, a striped mouse, mantis, baobab**

tree, yellow-crested cobra, or starry-eyed amaryllis, to mention only a few of the brilliant multitudes through which he so nimbly moved. Even as a child it seemed to me that his world was one without secrets between one form of being and another' (*The Lost World of the Kalahari,* 1958, p.21 of 253).

'Human beings know far more than they allow themselves to know: there is a kind of knowledge of life which they reject, although it is born into them: it is built into them' (*A Walk with a White Bushman,* 1986, p.142 of 326).

As a result of their instinctive heritage of growing up in an all-loving community, all humans now instinctively expect to be loved. Tragically however, since the emergence of consciousness some 2 million years ago and with it the dilemma and agony of the human condition, no infant has been given as much love as all infants received prior to the emergence of the struggle with the human condition.

In their innocent vulnerability, the main way children have coped with these wounds to their instinctive self has been to make themselves invulnerable to the pain by splitting themselves off from the world of that anxiety and trauma. As was described earlier, the hurt or damage or corruption of the souls of new generations and the progressive implementation of psychological block-out or alienation to cope with the hurt, have been occurring from birth onwards. The Australian impersonator, Barry Humphries, revealed that in performing his 'Dame Edna Everage' stage act he is **'performing a life long act of revenge'**. Humphries is quoted as saying that when he was growing up **'The phrase commonly used by my mother was "We don't know where Barry came from", which I regarded as an indication that I might well have been adopted.'** As a child, when he asked his mother if she loved him, she responded, **'Well, naturally I love your father most of all, and then my mother and father, and after that, you and your sister, just the same.'** Humphries added: **'Not to get a clear picture from one's parents as to where one stood in their affections, that was a troubling thing…I suppose one grows up with a desire to murder one's parents, but you can't go and really do that. So I suppose I tried to murder them symbolically on stage'** (*Australian Women's Weekly,* Oct 2001).

Later, in early adolescence, it was the corruption within themselves—acquired during their infancy and childhood as a result of their particular encounters with the corrupted and resigned world—that forced adolescents to also resign to a life of denial of the issue of the human condition.

The vulnerability of children was especially great in the situation that has existed where parents have been unable even to acknowledge the corruption of their resigned adult world. Tragically, unable to explain the human condition, humans could not afford even to admit they were corrupted, because of the crippling criticism such an admission would attract. As has been described, this silence, in effect preposterous lying, has devastated new generations arriving on the scene.

While a few cases of autism are caused by physical damage to the brain, most have been a result of extreme instances of this tragic situation where mothers have not been able to give infants the love they expect, and as a result the infant has had to dissociate psychologically from its reality to cope with the violation and hurt to its instinctive self or soul. Child psychiatrist and psychoanalyst, D.W. Winnicott, says that while **'a proportion of cases where autism is eventually diagnosed, there has been injury or some degenerative process affecting the child's brain...in the majority of cases...the illness is a disturbance of emotional development'** (*Thinking about Children*, 1996, p.200 of 343). To be able to maintain the denial and thus dissociation from the extreme pain of their circumstances required constant application of the denial/block, which is why autistic people tend to be compulsive and obsessive in their behaviour. They escape into repetitive activities and tend to develop a one-track mind. In his 1997 book *Next of Kin,* psychologist Roger Fouts describes autism as **'a developmental disorder characterized by lack of speech and eye contact, obsessive and repetitive body movements, and an inability to acknowledge the existence or feelings of other people. The autistic child lives in a kind of glass bowl, inhabiting a separate reality from those around him...Bruno Bettelheim, the renowned psychologist who ran a school for emotionally disturbed children, blamed autism on cold unfeeling mothers'** (p.184 of 420). Again, it has to be emphasised that since the emergence of the necessary battle with the human condition all mothers have been 'refrigerator mothers' to some degree, mothers who have been unable to give their children the amount of unconditional love that those children instinctively expect. While humans have not been able to love themselves it has been virtually impossible to properly love or nurture others.

The association of autism and savant abilities was mentioned in the quote from the documentary, *Uncommon Genius,* that stated **'for more than half of all savants, the syndrome owes its origins to a familiar condition—autism'**. The relationship between autism and savant abilities is not difficult to understand. If the autistic person could completely block out their reality, that is block out the world that was producing the pain that their instinctive self's expectation of being loved felt—for example, through developing a very narrow focus away from their reality onto some object or activity—their mind could be freed to access some of the potential of the soul's truthful, sensitive world. A window could be opened up to the immense potential that humans lost access to when they became preoccupied with worry about the hurtful, corrupted world around them, and/or their own corrupted state. That window to the soul's world depended on *completely* blocking out reality.

It might be thought that the dissociation that occurred at resignation, when people adopted a life of denial of their reality, should have given them some access to the soul's truthful, sensitive world. However, resigned people were not normally completely blocking out their reality, in fact they were constantly on the lookout for any criticism arising as a result of that reality. Generally, resigned people did not completely dissociate themselves from the world in the same way that autistic people did. However, there were some people who although not sufficiently hurt in infancy and childhood to become autistic, were sufficiently hurt to need to live an extremely alienated existence after they resigned; such people could develop some access to the soul's truthful world. In *Beyond* I describe this manner of accessing the soul's world as 'shattered defence' access. To quote, 'Sometimes when people became extremely exhausted [corrupted] their alienation (mental blocks) became disorganised and through this "shattered defence" the soul occasionally emerged. They became "mediums" or "psychics" or "channellers"' (p.182 of 203). As explained in *Beyond,* such shattered-defence access of the soul's true world was of course not the natural, secure, balanced access that people who never resigned have. For these people, whom we have historically referred to as prophets, the soul's world has always been an ultra natural place, not something weird, abnormal and apparently supernatural.

R.D. Laing was describing the 'shattered defence' way of accessing the soul when he said that **'the cracked mind of the schizophrenic may *let in* light which does not enter the intact mind of many sane people whose minds are closed'** *(The Divided Self,* 1960, p.27 of 218). Interestingly, Laing immediately continued to say that German existentialist Karl Jaspers was of the opinion that the biblical prophet Ezekiel **'was a schizophrenic.'** While some biblical prophets may have accessed the soul's true world using shattered defence, those who had full and natural access to the soul and were prophets in the true sense were exceptionally sound rather than exceptionally exhausted, alienated, separated from their true self, or schizophrenic.

The *Uncommon Genius* documentary about savants referred to earlier also gave examples of people with dementia who developed savant abilities: **'researchers discovered people who suffered dementia and then suddenly gained prodigious skills they'd never experienced before. One case was a man who, when his brain degenerated, became a composer as he lost his ability to speak'**. Dementia is obviously another form of complete dissociation from reality.

It was also mentioned earlier that when people prayed or chanted mantras or counted rosary beads they were trying to shut down their alienation-preoccupied mind in order to let through some of the truthful world of the soul. They were trying to shatter their defence. Fatigue, hallucinatory drugs, despair, faster-than-thought activities, such as scree-jumping, and near-death experiences have been cited as other ways of achieving this breakthrough to the world of the soul.

Autism is an extreme state of alienation, so extreme that the person has given up all efforts to manage their reality apart from totally blocking it out. In this *totally* disconnected state the soul's potential suddenly becomes accessible. Autistic people block out all awareness of their unloved circumstances by blocking out all reality so that there is no longer any disconnection from their soul's world, but of course such access to their soul depended on the maintenance of an extraordinary degree of dissociation from reality.

The most honest analysis of autism that I have come across appears in the already mentioned 1996 book, *Thinking About Children,* a posthumous publication of some of the papers written by British paediatrician, child psychiatrist and psychoanalyst, D.W. Winnicott, who died in 1971. On the book's dust jacket it says that Dr Winnicott, a former president of the British Psycho-Analytic Society, spent **'a lifetime thinking about the nature of the child and the origins of human nature'**, and is **'increasingly recognized as one of the giants of psychoanalysis'**. The following are some quotes from the book that confirm what I have said about the significance of nurturing in human life and how children psychologically react when they do not receive the amount of nurturing their instincts expect.

Firstly, Winnicott remarked that **'Many writers have expressed the view that an understanding of autism would widen our understanding of**

human nature' (p.223 of 343). There is a hint here of the truth that the dark shadow of the dilemma of the human condition was the real impediment to understanding autism and, beyond that, human nature as a whole.

The following is a typical case history of an autistic child from one of the many documented by Winnicott: **'When I first saw Ronald at the age of 8, he had very exceptional skill in drawing...Apart from drawing he was, however, a typical autistic child...I will look and see how things** [Roland's behaviour] **developed. The mother herself was an artist, and she found being a mother exasperating from one point of view in that although she was fond of her children and her marriage was a happy one, she could never completely lose herself in her studio in the way that she must do in order to achieve results as an artist. This was what this boy had to compete with when he was born. He competed successfully but at some cost...At two months the mother remembers smacking the baby in exasperation although not conscious of hating him. From the start he was slow in development...His slowness made him fail to awaken the mother's interest in him, which in any case was a difficult task because of her unwillingness to be diverted from her main concern which is painting'** (pp.201,202).

In a reference to the instinctual expectations of infants Winnicott said: **'There are certain difficulties that arise when primitive things are being experienced by the baby that depend not only on inherited personal tendencies but also on what happens to be provided by the mother. Here failure spells disaster of a particular kind for the baby. At the beginning the baby needs the mother's full attention, and usually gets precisely this; and in this period the basis for mental health is laid down. This in all its details becomes established by constant reinforcement through the continuation of a pattern of care that has in it the essential elements. Naturally, some individual infants have a greater capacity to go ahead in spite of imperfect care...On the whole, however, it is the quality of early care that counts. It is this aspect of the environmental provision that rates highest in a general review of the disorders of the development of the child, of which autism is one'** (p.212).

Winnicott then proceeded to say that **'the essential feature** [in a baby's development] **is the mother's capacity to adapt to the infant's needs through her healthy ability to identify with the baby. With such a capacity she can, for instance, hold her baby, and without it she cannot hold her baby except in a way that disturbs the baby's personal living process...It seems necessary to add to this the concept of the mother's unconscious (repressed) hate of the child. Parents naturally love and hate their babies,**

in varying degrees. This does not do damage. At all ages, and in earliest infancy especially, the effect of the repressed death wish towards the baby is harmful, and it is beyond the baby's capacity to deal with this' (p.222). (I am certain that when Winnicott comments that the fact of parents naturally hating their babies **'does not do damage'**, he is alluding to the fact that children have been genetically selected now, after 2 million years, to cope with a degree of imperfection in their upbringing. Winnicott was implying this when he said in the preceding quote that, **'Naturally, some individual infants have a greater capacity to go ahead in spite of imperfect care'**.)

Importantly Winnicott said (the italics and brackets are Winnicott's emphasis): **'Autism is a highly sophisticated defence organization. What we see is *invulnerability*. There has been a gradual build-up towards invulnerability...The child carries round *the (lost) memory of unthinkable anxiety*, and the illness is a complex mental structure insuring against recurrence of the conditions of the unthinkable anxiety'** (pp.220,221).

Similarly Winnicott said, **'I am perfectly aware that in a proportion of cases where autism is eventually diagnosed, there has been injury or some degenerative process affecting the child's brain...** [however] **It is extremely likely that in the majority of cases of autism the computer** [brain] **is undamaged and the child is potentially and remains potentially intelligent...The illness is a disturbance of emotional development...that the problem in autism is fundamentally one of emotional development and that autism is not a disease. It might be asked, what did I call these cases before the word autism turned up. The answer is..."infant or childhood schizophrenia"'** (p.200). (The etymology of the word 'schizophrenia' is 'schiz' meaning 'split' or 'broken', and 'phrenos' meaning 'soul or heart'. The truth is that all humans who are resigned are alienated or split off from their soul. While schizophrenia has been a description reserved for only the extremely alienated, the truth is all resigned humans are schizophrenic.)

Winnicott even acknowledged that autism is the extreme degree of a universal phenomenon: **'It has always seemed to me that the smaller degrees of disturbance of the mind that I am trying to describe are common and that even smaller degrees of the disturbance are very common indeed. Some degree of this same disturbance is in fact universal. In other words, what I am trying to convey is that there is no such disease as autism, but that this is a clinical term that describes the less common extremes of a universal phenomenon'** (p.206).

When he made his now-famous comment, **'Insanity is a perfectly**

natural adjustment to an insane world', R.D. Laing was emphasising that the people society has labelled as mad were merely responding to the madness of the resigned world, finding a way to insulate themselves from it. Similarly, he has written that, **'From the alienated starting point of our pseudo-sanity, everything is equivocal. Our sanity is not "true" sanity. Their** [the patients'] **madness is not "true" madness. The madness of our patients is an artefact of the destruction wreaked on them by us, and by them on themselves'** (*The Politics of Experience* and *The Bird of Paradise,* 1967, p.118 of 156).

Laing summarised the penalty of resignation to a life of denial when he said, **'I would wish to emphasize that our "normal" "adjusted" state is too often the abdication of ecstasy, the betrayal of our true potentialities, that many of us are only too successful in acquiring a false self to adapt to false realities'** (*The Divided Self,* 1960, 1965 preface, p.12 of 218).

The necessary dishonesty of the resigned mind

On the subject of humanity's historic denial of critical truths, in particular the truth of the significance of nurturing in human life, Winnicott says: **'Mothers do, of course, tend to feel guilty; they tend to feel responsible, quite apart from logic, for every defect that manifests itself in their children. They feel guilty before the baby is born, and they so strongly expect to give birth to a monster that they must always be shown the baby the very moment he or she is born, however exhausted they may be. And the father too. Nevertheless, most people are rational beings in their best moments, and they can then discuss the relationship between autism developing in a child and (in some cases) a relative failure in infant care. What is much more difficult is to deal with this problem in social terms, in terms of the public and the public's attitude to the parents. Collectively people are less rational than individually...we cannot hold back our statements for fear of hurting someone...In fact, if I could, this would be tantamount to saying that parents play no part when things go well** [with the nurturing of their offspring]**'** (p.213 of 343).

Winnicott continues: **'expect resistance to the idea of an aetiology** [cause] **that points to the innate processes of the emotional development of the individual in the given environment. In other words, there will be those who prefer to find a physical, genetic, biochemical, or endocrine cause, both for autism and for schizophrenia'** (p.219).

In light of Winnicott's warning it is significant that in the television program *Uncommon Genius,* the narrator says at one point that Robyn Young **'discovers some controversial new theories in San Diego. Dr Bernard Rimland is a world expert on autism. Some years ago he dispelled the myth that autism is caused by cold-hearted parenting, so-called "refrigerator mothers". Dr Rimland's latest theory is that when children are genetically predisposed autism may be triggered by antibiotics and vaccines.'**

To return to Winnicott's observations about nurturing, **'It is not good to distort the truth in order to avoid hurting the feelings of clients...Such ideas** [ie that lack of nurturing is the cause of autism] **call for courage on the part of those who discuss them, and yet without these ideas there is no hope, in my opinion, that any body of scientists will move towards an understanding of the aetiology of autism'** (pp.220,222).

The dilemma of the human condition was such that until the human condition was resolved, humans *had to* lie, they had to resign themselves to a life of evasion and denial of the truths that bring the human condition into focus—because the alternative could be madness or suicidal depression. Winnicott's capacity to be relatively unevasive or denial-free in his thinking was a measure of his soundness of self. For people with less nurtured and thus less secure upbringings a great deal more evasion and denial was needed. It follows that, with alienation rapidly increasing everywhere in the western world—and we can see where the human race is heading from looking at California's Silicon Valley, the place in the forefront of our development, which as was mentioned, **'is home to more autistic children than anywhere in America'**—Dr Rimland's evasive excuse was tragically necessary and timely. Only with the human condition explained is it finally safe to confront people with the truth about themselves, because it is possible to do so in the context of the greater truth of their goodness. Dr Winnicott and myself are not clever to have put forward these truths about the significance of nurturing, because they are in fact truths that everyone intrinsically knows but which the great majority of people have had to evade. In order to evade something you first had to know it existed.

The truth is there is no real 'nature versus nurture' or 'hereditary versus environment' argument, only a necessary avoidance of the significance of the role of nurturing in our upbringing. This quote from an article about parenting shows just how insecure parents have been about their inability to nurture their offspring: **'The biggest crime you can commit in our society is to be a failure as a parent...people would**

rather admit to being an axe murderer than being a bad father or mother' (*Sun-Herald, Sunday Life* mag. 7 July 2002). This demonstrates how much parents want to avoid the truth that, as R.D. Laing recognised, **'families cause madness'** (Adrian Laing, *R.D. Laing A Biography,* 1994, p.175 of 248). Blaming 'nature', in the form of genes or chemicals, for humans' alienated condition has been a necessarily contrived evasion of the role nurturing played in the upbringing of humans. Similarly, while humans were unable to explain their corrupted state they were left feeling that it was a bad, flawed, evil, worthless condition. So being able to label a failing or weakness that was a result of self-corruption and alienation as a disease or illness was greatly relieving. As Andrew Solomon says in *The Noonday Demon,* his 2001 book about severe depression, **'Being told you are sick is infinitely more cheering than being told you are worthless.'** The writings of Olive Schreiner were quoted earlier, including a reference to the vulnerability of children from her 1883 book, *The Story of an African Farm.* Olive Schreiner is famous for her unevasive honesty and it is worth including more of that particular quote as another example of a person who has bravely admitted the significance of nurturing in human life: **'They say women have one great and noble work left them, and they do it ill...** ***We*** **bear the world and** ***we*** **make it. The souls of little children are marvellously delicate and tender things, and keep for ever the shadow that first falls on them, and that is the mother's or at best a woman's. There was never a great man who had not a great mother—it is hardly an exaggeration. The first six years of our life make us; all that is added later is veneer...The mightiest and noblest work is given to us, and we do it ill'** (p.193 of 300).

In reality the situation is much more serious than even Schreiner described it. The truth is, nurturing rules the world; the amount of nurturing children receive dictates whether or not the world they control as adults is operational.

Winnicott's writings on the subject of autism have been extensively quoted. The following summary from the preface of his book offers his views on the more general relationship between mother and infant: **'Winnicott held that the innate potential for growth in a baby (and he was aware of the damage to innate potential, or the restrictions of potential in the baby, too) expressed itself in spontaneous gestures. If the mother responds appropriately to these gestures, the quality of adaptation provides a growing nucleus of experience in the baby, which results in a sense of wholeness, strength, and confidence that he calls the "true self"...If a mother is unable to meet her baby's gestures appropriately, the baby de-**

velops a capacity to adapt to and comply with the mother's "impingements"—that is, with the mother's initiatives and demands—and the baby's spontaneity is gradually lost. Winnicott called such a defensive development the "false self". The greater the "mis-fit" between mother and baby, the greater the distortion and stunting of the baby's personality...the decisive phase in the infant's development is the achievement of a unitary self capable of objectivity and creative activity' (*Thinking About Children,* 1996, p.xvi of 343).

The 1993 film *House of Cards,* quoted earlier, contains a number of insightful comments, including the following two that are relevant here. The first one emphasises just how much sensitivity, potential and capacity resigned humans have lost access to: **'Tell me a story—the one about the great grandmother of light who made the first few people out of white and yellow corn?—Yes, well, she gave them the power to see everything, they could see through walls, they could even see through rocks, they could even see inside each other. Today we can do this only in our dreams.'**

The second quote acknowledges the denial that infants employ to cope with the trauma of encountering life under the duress of the human condition: **'People say about the following categories that these kids have a problem or are disabled, or psychologically dumb, etc, but really they are children, through hurt or some kind of trauma, that have held onto soul, and not wanted to partake in reality—retarded, autistic, insane, schizophrenic, epileptic, brain-damaged, possessed by devils, crocked babies'** (ibid).

Only with the human condition explained, with the biological reason for why humans unavoidably became corrupted, does it finally become safe to acknowledge the truth about the significance of nurturing in human life—and the many other immensely confronting truths such as the current extent of alienation amongst humans. Humans' necessary lies/denials can now end and at last they can speak truthfully about their condition. New generations can have explained to them the real reason the world has not been ideal and instead of having to cope with the 'wrongness' of the world by blocking it all out they will be able to understand the corruption, and through understanding begin to cope with it. Instead of becoming more and more alienated, humanity can at last turn back towards innocence and head for home.

The march towards total insanity has ended—and only just in time. From humanity's innocent heritage 2 million years ago, each successive generation has, overall, become more alienated. The exponential rate of alienation is increasing at such a rapid pace that were there a graph charting the extent of alienation, its curve would now be almost vertical. In fact alienation has now reached such extreme levels that new generations find the level of falseness and denial almost unbearable. Health workers are reporting that the incidence of depression and ADD (attention deficit disorder) is approaching epidemic proportions amongst young people with whole generations growing up on anti-depressants and tranquillising medications. A recent article in the *Sydney Morning Herald* newspaper, titled **'Epidemic of autism mystifies experts'**, referred to a study which showed that in California autism had **'more than tripled from 1987 to 1998'** (19 Oct. 2002). Interestingly the article, which was first published in the *New York Times,* made the point that **'You can't explain an increase of this magnitude on genetics. Something else is happening.'** Recent generations have been revealingly labelled the 'X generation', the 'XX generation', and more recently the 'ACES generation'—the **'alienated, cynical, experimental and savvy'** generation *(Sydney Morning Herald,* 8 Aug. 1996). The Canadian writer Douglas Coupland defined Generation X as one who **'lives an X sort of life—cerebral, alienated, seriously concerned with cool'** *(Sydney Morning Herald,* 22 Aug. 1994). All of these qualities attributed to X and ACES generations are qualities associated with having had to adjust to an extremely corrupt world. Silicon Valley, with its 273 percent increase in autism, is just a microcosm of where humanity was headed.

Children exposed to the current world of extreme alienation simply cannot cope with it, hence the occurrence of childhood madness, evasively referred to as attention deficit disorder. (Again, it has to be appreciated here that alienation cannot recognise or see its alienation—if it could it would not be alienated—but new generations can see it and have to attempt to adjust to it.) An article about the alarming increase in childhood disorders emphasised that avoiding the real problem of nurturing, or lack thereof, and the level of alienation in society has led to the dangerous over-medication of children: **'Because it is so convenient and guilt-relieving to be able to attribute a child's difficult behaviour to a neurochemical problem rather than a parenting or broader social one, there is a risk that this problem will become dangerously over-medicalised'** *(The Australian,* 8 Dec. 1997).

The only measure that could stop the march towards total madness was understanding of the human condition; the real treatment humanity desperately required.

The dysfunction of the resigned mind

I mentioned earlier how many answers I have been able to find by not living in denial of the issue of the human condition. In truth unresigned, denial-free, unevasive thinking adults, or prophets, such as Sir Laurens van der Post, Laing, Winnicott, Schreiner and myself should be viewed the same way as savant syndrome people are viewed. Resigned humans have not compared themselves with savants because they accepted that savants had very different brain 'wiring'. Once humans resigned their ability to think truthfully and thus effectively was severely impaired, whereas unresigned brains and resigned brains that have learnt to think in a denial free way can investigate and succeed in penetrating all manner of mystery.

It was said of the denial-free thinker or prophet Jesus Christ, **'How did this man get such learning without having studied?'** (John 7:15). Similarly it is said that the prophet Mohammed **'was unlettered, having no formal education'** (*Eastern Definitions* by Edward Rice, 1978, p.261 of 433). Also, in the quote mentioned earlier about the exceptionally denial-free, penetrating thinker, Friedrich Nietzsche, it was said that he was **'largely unread'**. Prophets were able to **'get such learning'** by being unresigned and thus able to think truthfully and effectively. Christ summed up the futile thinking abilities of the resigned adult mind when he said, **'you have hidden these things from the wise and learned, and revealed them to little children'** (Matt 11:25). Elsewhere in the *Bible* it similarly says, **'The Lord knows that the thoughts of the wise are futile'** (I Cor. 3: 20 & Psalm 94:11).

R.D. Laing described the limitation of the resigned mind when he wrote, **'Our capacity to think, except in the service of what we are dangerously deluded in supposing is our self-interest, and in conformity with common sense, is pitifully limited: our capacity even to see, hear, touch, taste and smell is so shrouded in veils of mystification that an intensive discipline of un-learning is necessary of *anyone* before one can begin to experience the world afresh, with innocence, truth and love'** (*The Politics of Experience* and *The Bird of Paradise,* 1967, p.23 of 156).

The ability of prophets to think penetratingly has seemed astonishing to resigned minds but that has only been because resigned minds have been unaware of their own alienation. It is only natural that a mind that is not coming off a dishonest base can make sense of all manner of 'mystery' that false, resigned minds, that are living in denial of so many fundamental truths, cannot get anywhere near making sense of. The 'deaf effect', where people find it almost impossible to read about the human condition, arises because the resigned brain is extremely indoctrinated in a particular way of thought that severely limits and prejudices it, so that it only allows itself to hear what it wants to hear. The resigned brain is highly dysfunctional, instead of one truth leading logically to the next, there are blocks or denials that come into play and stop the thought sequences carrying through to their proper conclusion.

Mechanistic science has prided itself on being rigorously objective when in truth it has been coming off an extremely dishonest, evasive, subjective, indoctrinated, prejudiced base and, as a result, could not hope to reach a sound understanding of our world. Winnicott described the situation when he wrote, **'Can you see the one essential way in which science and intuition contrast with each other? True intuition can reach to a whole truth in a flash (just as faulty intuition can reach to error), whereas in a** [mechanistic] **science the whole truth is never reached'** (*Thinking About Children,* 1996, p.5 of 343).

The *Hutchinson Dictionary of Science,* 1993, defines 'scientific method' as **'the belief that experimentation and observation, properly understood and applied, can avoid the influence of cultural and social values and so build up a picture of a reality independent of the observer.'** In reality mechanistic science has complied with the principle of living in denial of the issue of the human condition until understanding of it could be found. It has been 'mechanistic', not 'holistic'. It studiously avoided the truth of integrative meaning and many other fundamental truths that brought the human condition into focus. It has been rigorously subjective, *not* rigorously objective, not **'independent of the observer'**, as it has claimed to be. This reality has been recognised in recent times with scientists being labelled as either left or right wing in their scientific orientation. For example, in an article about the dishonesty of genetic determinism, journalist Ziauddin Sardar wrote that, **'A number of left-leaning scientists, including Steven Rose and Richard Lewontin, have argued for decades that we are much more than the sum of our genes'** (*Australian Financial Review,* 23 Feb 2001).

The fact is that resignation is a state of extremely prejudiced falseness and thus 'futility' when it comes to thinking effectively. It is not interested in the truth, rather it is interested in evading the truth. As philosopher Arthur Schopenhauer (1788–1860) said, **'The discovery of**

truth is prevented most effectively...by prejudice, which...stands in the path of truth and is then like a contrary wind driving a ship away from land' (*Essays and Aphorisms,* tr. R.J. Hollingdale, 1970, p.120 of 237). Aldous Huxley also got it right when he said, **'We don't know because we don't want to know'** (*Ends and Means,* 1937, p.270).

Given the extreme falseness of the resigned state, it is no wonder humans have been sensitive about being called liars. Every resigned adult has been lying, has been denying the truth about integrative meaning and many other critically important truths, including the fact of their own extreme alienation. The truth is, all resigned adults have been 'lying through their teeth'; lying in the most extreme way while maintaining a broad smile of confident denial of these lies.

It is little wonder that children and young adolescents found the world of resigned adults so upsetting. The blatant denials that resigned adults maintained were extremely offensive to anyone trying to live a life of honesty. . . . .

. . . .

Apology: As explained in Notes to the Reader, the dotted lines indicate text that has temporarily been withdrawn because it deals with issues before the courts. Once the legal restrictions end the fully restored pages will be available on the FHA's website, or from the FHA.

. . . . When unresigned honest-thinking minds have stood up to and railed against the lies of the resigned world they have been accused of being rude. When they strongly maintained what they knew was true against what was so false they were accused of being arrogant. However, the *real* rudeness and arrogance . . . . . . was the resigned, mechanistic view that its world was authentic. As mentioned earlier, Christ said of the arrogant, self-important, utterly deluded exponents of that world that, **'They like to walk round in flowing robes and love to be greeted in the market-place and have the most important seats in the synagogues and the places of honour at banquets'** (Luke 20:46).

For its part, mechanistic science masqueraded before the world as a rigorously objective paradigm when the truth was that it was complying fully with the resigned evasive strategy of denying integrative meaning, and many other fundamental truths, such as a cooperative instinctive past for the human race, the significance of nurturing, the occurrence of resignation during adolescence, and the immensely alienated state of humans now. Mechanistic science had to be extremely evasive if it were to avoid leaving everybody suicidally confronted and depressed; however, that necessary evasion did not justify it arrogantly promenading across the world's stage proclaiming itself to be upholding a rigorous, truthful paradigm. With such arrogance it is no wonder so many religious people have been offended and upset by mechanistic science.

Mechanistic scientists have been immensely evasive, dishonest and subjective, the end result of which has been a form of inquiry that eventually ground to a halt under the weight of its own accumulated dishonesty. As pointed out in the *Plato* essay, mechanistic science had reached the limits of its meaningfulness. As the Templeton Prize-winning biologist Charles Birch acknowledged in an address given at the 1993 FHA Open Day held in Sydney, **'science can't deal with subjectivity...what we were all taught in universities is pretty much a dead end.'** Similarly, biologist Mary E. Clark, in her 1989 book, *Ariadne's Thread: The Search for New Modes of Thinking,* said **'Formal learning has become a meaningless vaccination process, and the information transmitted is next to useless for properly understanding the world.'** In his 1978 book, *Janus: A Summing Up,* Arthur Koestler described at length how **'the citadel they** [mechanistic scientists] **are defending lies in ruins'** (p.192 of 354). The Templeton Prize-winning physicist Paul Davies wrote: **'For 300 years science has been dominated by extremely mechanistic thinking. According to this view of the world all physical systems are regarded as basically machines...I have little doubt that much of the alienation and demoralisation that people feel in our so-called scientific age stems from the bleak sterility of mechanistic thought...Mechanistic thought has undoubtedly had a stifling effect on the human spirit'** (*Living in a non-material world—the new scientific consciousness, The Australian,* 9 Oct. 1991).

The New Age proponent Fritjof Capra wrote in 1982: **'To describe this world appropriately we need a new paradigm, a new vision of reality—a fundamental change in our thoughts, perceptions and values. The beginnings of this change, or the shift from the mechanistic to the holistic conception of reality, are already visible...The gravity and global extent of our crisis indicates that the current changes are likely to result in a transformation of unprecedented dimension, a turning point for the planet as a whole'** (quoted in *Ariadne's Thread: The Search for New Modes of Thinking,* Mary E. Clark, 1989).

Capra was right, a new paradigm based on acceptance of integrative meaning or holism had to emerge. Having said this, it is necessary to point out that while there has been acknowledgment of holism in the so-called New Age movement, the movement was in fact part of the most sophisticated form of denial of the issue of the human condition to develop on Earth. It seduced adherents with the belief that they were participating in the creation of an 'alternative', integrated, idealistic 'New Age'; in reality they were making no effort to confront human alienation and the associated issue of the human condition, the source of the disharmony on Earth. A detailed analysis of the extreme delusions and dangers of the New Age and related movements will be presented in the essay *Death by Dogma* in my next book.

The new, honest paradigm that Capra and many others, in fact the whole of humanity, have looked forward to required the emergence of truly unevasive, denial-free thinkers who could confront the human condition and defy all the lies of the alienated world. Friedrich Nietzsche was aware of how strong that defiance of the false, evasive, alienated world had to be when he wrote, **'we have to await the arrival of a new species of philosopher, one which possesses tastes and inclinations opposite to and different from those of its predecessors—philosophers of the dangerous "perhaps" in every sense'** (*Beyond Good and Evil*, 1886; tr. R.J. Hollingdale, 1973, p.16 of 237). The central thesis of Nietzsche's work was the need for the arrival of what he termed **'overman'** or **'superman'**, people who were secure, independent and highly individualistic; people who could and would defy the evasion/denial/alienation that has held the minds of almost all humans in a vice-like grip for some 2 million years. Nietzsche cited, amongst others, Christ and Socrates as models for such defiant thinkers. Such people *were* 'supermen', they were people who were unevasive and, as a result, able, in almost a savant-like capacity, to access the soul's true world—as all humans will soon be able to, now they no longer have to live in denial.

As the race between self-destruction (from rapidly increasing levels of alienation) and self-discovery (through finding understanding of the human condition) grew more intense, similar myths about superheroes such as Superman proliferated: The Lone Ranger, The Phantom, Batman, Wonderwoman, Spiderman, Luke Skywalker of Starwars, Frodo Baggins of The Lord of the Rings, The X-Men. Such myths were an expression of an intuitive awareness in people of what it would take to solve the human condition. To the resigned mind confronting the human condition seemed impossible, a superhuman undertaking.

Any real progress in understanding our world and our place in it is now dependent upon confronting, rather than avoiding, the issue of the human condition. It is humans' alienation that has been preventing any real advance in knowledge. Again we can go to the writings of R.D. Laing for an acknowledgment of this truth: **'Our alienation goes to the roots. The realization of this is the essential springboard for any serious reflection on any aspect of present inter-human life** [p.12 of 156]**...We respect the voyager, the explorer, the climber, the space man. It makes far more sense to me as a valid project—indeed, as a desperately urgently required project for our time—to explore the inner space and time of consciousness. Perhaps this is one of the few things that still make sense in our historical context. We are so out of touch with this realm** [so in denial of the issue of the human condition] **that many people can now argue seriously that it does not exist. It is very small wonder that it is perilous indeed to explore such a lost realm** [p.105]**...the direction we have to take is *back* and *in*** [p.137]**'** (*The Politics of Experience* and *The Bird of Paradise,* 1967).

As Laing says here, most people are now unaware of the issue of the human condition. At the conclusion of his book, *Coming to Our Senses* (1998), Morris Berman said that **'Something obvious keeps eluding our civilisation'**. That **'something'** was humans' alienated state, and the reason it **'eluded'** humans was because they were acutely aware of the danger of approaching the issue of the human condition and therefore maintained a dedicated denial of anything connected to it. By definition, the resigned mind was unaware it had resigned, it lived in denial of its alienation. This is why it is remarkable that some researchers have suggested that savant-like abilities exist in all of us. It is virtually an acknowledgment that we are a species living in denial and immensely alienated, living with a form of dementia or amnesia—spiritually dead.

The loneliness of humans' alienated state

While humans could not acknowledge the significance of nurturing in human life, how much innocence was lost as a result of insufficient nurturing, and that humans had to repress access to their soul's true world at resignation, it was impossible for them to see just how great a loss of innocence and how much estrangement from their true selves and the true world had occurred. Now that humans can safely acknowledge alienation, what is revealed is that humans are *2 million years* alienated from their soul's cooperative, all-sensitive, true world. This alienation from their true self and true world is a state of utter exile and aloneness; absolute and desperate loneliness.

Awareness of this extreme loneliness allows us for the first time to understand some previously inexplicable aspects of human nature, such as the mystery of why humans 'fall in love'. The French physicist Blaise once wrote that **'The heart has reasons that reason cannot comprehend'** (*Pensees*, iv, p.277), but we *can* now comprehend the heart's reason for falling in love.

Humans have come to live in an extremely lost, dark, forlorn, despairing state. R.D. Laing was spot-on when he said that humans live with **'fifty feet of solid concrete'** between themselves and the true world of their soul. As a person digests understanding of the human condition this immense loneliness within humans becomes more and more apparent and, with sufficient appreciation of the extent of that loneliness, it becomes possible to fully appreciate why humans 'fall in love'. The key word is 'fall'. It is an acknowledgment that humans let go of their immensely lonely, despairing reality and allow themselves to be transported to—escape to—another potential and possibility. Humans let themselves dream of the ideal state where they are no longer estranged from each other, but are together as one. As is explained in my earlier books, and briefly explained in the *Plato* essay in the section 'A7', humans spent some 6 or 7 million years in this wonderful, 'heavenly' state of real togetherness before consciousness emerged, so there is a memory within humans of what that state is like which allows them to dream of it. Humans once lived utterly lovingly and cooperatively, and within all humans exists the hope of one day being able to solve the human condition and return to that idyllic state. The lyrics of the song *Somewhere* written by Stephen Sondheim for the blockbuster 1956 musical and film *West Side Story* perfectly describe the dream of the heavenly state of true togetherness that humans allow themselves to be transported to when they fall in love: **'Somewhere/We'll find a new way of living/We'll find a way of forgiving/Somewhere//There's a place for us/A time and place for us/ Hold my hand and we're halfway there/Hold my hand and I'll take you there/Somehow/Some day/Somewhere!'** The 1928 song, *Let's Fall In Love,* written by Cole Porter, also has lyrics that reveal how falling in love is about allowing yourself to dream of the ideal state, of 'paradise': **'Let's fall in love/Why shouldn't we fall in love?/Our hearts are made of it/Let's take a chance/Why be afraid of it/Let's close our eyes and make our own paradise'**. The escape from the horror of a resigned world oppressed and upset by the human condition that falling in love is concerned with achieving is expressed in these lyrics from the 1977 Fleetwood Mac song *Sara:* **'Drowning in the sea of love/Where everyone would love to drown'**.

Given humans' immense loneliness we can understand now the saying, **'love is blind'**. It was deliberately blind; humans *wanted* to be transported away from their desperately lonely reality, they wanted to dream, to delude themselves, to not see their reality. It is explained in detail in *Beyond,* in the chapter 'Illustrated Summary of the Development and Resolution of Upset', and as is summarised in the first section of the next essay, 'Bringing peace to the war between the sexes', that youthful neotenous looks such as domed forehead, snubbed nose, large eyes and long healthy hair signalled innocence. It was this image of innocence that could inspire the dream of the ideal, uncorrupted, innocent, truly-together-as-one partnership. As stated in *Beyond,* the **'"mystery of women" was that it was only the physical image or object of innocence that men were falling in love with. The illusion was that women were psychologically as well as physically innocent. For their part, women were able to fall in love with the dream of their own "perfection"—of their being truly innocent. Men and women *fell* in love. We abandoned the reality in favour of the dream. It was the one time in our life when we could romance—when we could be transported to "how it could be"—to heaven'** (p.144 of 203).

The romantic state of falling in love with the image of innocence is apparent in the words of the 1958 song, *The First Time Ever I Saw Your Face,* written by Ewan MacColl. It begins, **'The first time ever I saw your face/I thought the sun rose in your eyes/And the moon and the stars were the gift you gave/to the dark and the endless sky'**. The irresistible escape from the depressing state of the human condition implicit in being 'in love' also comes through in the most popular jukebox record of all time, Patsy Cline's version of Willie Nelson's 1961 song, *Crazy.* The song begins with the words, **'Crazy, I'm crazy for feeling so lonely/I'm crazy, crazy for feeling so blue'**, and ends with the words, **'I'm crazy for trying and crazy for crying and I'm crazy for lovin' you.'** The delusion involved in falling in love is acknowledged in Larry Hart's lyrics for the 1938 musical, *The Boys of Syracuse*—**'Falling in love with love/Is falling for make-believe'**.

Being able to admit and confront the extent of humans' loneliness allows us to explain many mysteries. Many, if not all, people have dreams of levitating and/or flying through the air. We can now understand that these dreams represent yearnings for freedom from the horror of humans' lonely, unhappy, corrupted, alienated state. The advent of spoken language, which made it possible for humans to describe their feelings, is a relatively recent development. Spoken language followed the emergence of consciousness, and our species has only been conscious for some 2 million years. The way our older, pre-consciousness, original instinctive self or soul relates to the world is through associations with the natural world. In this language of our soul, the most powerful metaphor or symbol for the state of freedom in the natural world is undoubtedly flying, exemplified by birds that are able to lift off from the ground and sail around through the air at their will and with the greatest of ease. The saying 'free as a bird' says it all.

Interestingly Koko, the first gorilla to learn sign language, employs the sign for 'bird' as her swear word. We forget how astonishing it is that some animals, namely birds, seem to be able to defy gravity, jump off the ground and fly around with ease. For such a land-anchored animal as a gorilla, it must seem offensive and Koko has apparently given recognition to that offence in her use of the sign for the animal 'bird' as her swear word.

Flying as a metaphor for freedom from the horror of the human condition is well expressed in the song *Over the Rainbow,* first sung by Judy Garland in the film *Wizard of Oz*: **'Somewhere over the rainbow way up high/There's a land that I heard of once in a lullaby/Somewhere over the rainbow skies are blue/And the dreams that you dare to dream really do come true/Some day I'll wish upon a star/And wake up where the clouds are far behind me/Where troubles melt like lemon drops/Away above the chimney tops/That's where you'll find me/Somewhere over the rainbow bluebirds fly/Birds fly over the rainbow/Why, then, oh, why can't I?/If happy little bluebirds fly/Beyond the rainbow, why, oh, why can't I?'** (lyrics E. Y. Harburg, 1939).

Given that the reference to falling in love is to do with the dream of the heavenly state of true togetherness, it is appropriate to demystify the concept of 'heaven' a little more. 'Heaven' is the metaphysical or religious term for the integrated, cooperative state that humanity once experienced and will experience again now that the human condition is solved. A passage from Sir Laurens van der Post's writings quoted earlier (when examining Carl Jung's journey into the unconscious), shows that Sir Laurens recognised this truth of heaven being the human condition-free, alienation-free, integrated state that humanity will now return to. The quote was: **'He found himself turning to the child in himself as if instinct, too, was exhorting him to become like the child which the New Testament exhortation makes imperative. In this way he hoped to emerge from darkness into the light of which the Kingdom of Heaven is the supreme image'** (*Jung and the Story of Our Time,* 1976, p.154 of 275).

As has been mentioned already, a consequence of being able to understand and acknowledge resignation and the immense estrangement from their true self that resigned adult humans are living in and their consequent immense loneliness, is that we can now see where the real propensity for sickness and unhealthiness in humans has emanated from. It becomes clear that resigned humans have only just been managing to hold total despair at bay; any extra despair in their lives tipped them into a state of utter loneliness that expressed itself physically in the flaring up or festering of the normal weaknesses that exist in the human make-up, giving rise to such sicknesses as cancer. Solve the problem of the human condition and humans will be physically well again. In the future we will see elderly humans with faces as free and glowing as a child's. People will be as healthy as they were before the human condition emerged.

In his poem, *Theogony,* the 8th century BC Greek poet Hesiod described how healthy humans were in their original innocent state: **'When gods alike and mortals rose to birth/A golden race the immortals formed on earth/Of many-languaged men: they lived of old/When Saturn reigned in heaven, an age of gold/Like gods they lived, with calm untroubled mind/Free from the toils and anguish of our kind/Nor e'er decrepit age misshaped their frame/The hand's, the foot's proportions still the same/ Strangers to ill, their lives in feasts flowed by/Wealthy in flocks; dear to the blest on high/Dying they sank in sleep, nor seemed to die/Theirs was each good; the life-sustaining soil/Yielded its copious fruits, unbribed by toil/ They with abundant goods 'midst quiet lands/All willing shared the gathering of their hands'** (tr. Elton).

Buddhist scripture accurately describes what humans will be like when the ameliorating understanding of the human condition arrives; the time, in the words of the scripture, when humans **'will with**

a perfect voice preach the true Dharma, which is auspicious and removes all ill'. Of that future time, which has now begun, the scripture says **'Human beings are then without any blemishes, moral offences are unknown among them, and they are full of zest and joy. Their bodies are very large and their skin has a fine hue. Their strength is quite extraordinary'** (Maitreyavyakarana, tr. Edward Conze, *Buddhist Scriptures,* 1959, pp.238-242).

The objective of the human journey was to solve the human condition

Having lived in deep denial of the human condition it naturally comes as a shock to rediscover it. Living in such denial has meant that the world has seemed meaningless, but with the whole issue of the human condition brought out into the open and resolved, resigned humans can suddenly see that life was not at all meaningless. In fact, they can now see that every human that has ever lived has worked tirelessly towards this day when self-understanding would arrive. Humans have been involved in an immensely purposeful project and world.

In the following quote, Marilyn Ferguson, whose book, *The Aquarian Conspiracy* (1980), helped inspire the New Age Movement of the 1980s, acknowledges the necessary task of reconciling humans' divided selves and, by so doing, resurrecting our species' long-repressed instinctive self's natural sensitivities and cooperative potential. It also acknowledges that, with the damaging effects of ever-increasing levels of alienation in the world, fulfilling this task has become a matter of urgency. Ferguson wrote: **'Maybe** [the theologian, scientist and prophet] **Teilhard de Chardin was right; maybe we are moving toward an omega point—Maybe we can finally resolve the planet's inner conflict between its neurotic self (which we've created and which is unreal) and its real self. Our real self knows how to commune, how to create...From everything I've seen people really urgently want the kind of new beginning...**[that I am] **talking about** [where humans will live in] **cooperation instead of competition'** (*New Age* mag. Aug. 1982).

As extremely difficult as the much-needed exploration into, and journey out of the human condition is, our freedom from it and all the destruction it has caused can only be achieved via that journey. Humanity's task *is* to progress beyond the human condition—and it *is* the responsibility of those who are relatively free of corruption, those who are relatively innocent and able to confront the human condition, to unlock and then hold open the door to that new world for humans—just as it is the responsibility of those who are no longer innocent to restrain themselves from trying to shut the door to that future. Both the innocent and those whose innocence was sacrificed in humanity's battle with the human condition have roles in the human condition-addressing, denial-free new world. As described at the end of *Beyond*, there is a completely fulfilling and fully participatory role for everyone in the human condition-ameliorating new world.

The real revolution for the human race is a mental one. As the band Public Enemy sings, **'Are you ready for the real revolution; the revolution of the mind?'** (from the 1990s song *He Got Game*).

It has to be emphasised here that what makes it possible to acknowledge both innocents and those who are no longer innocent is the finding of understanding of the human condition. Solving the human condition means ending the historic criticism of humans' innocence-destroyed, corrupted, upset, divisive, alienated condition. Understanding is compassion. The truth sets us free from criticism. We have been *a species in denial* but understanding the human condition makes humans *free* of the human condition, takes humanity *beyond the human condition*, beyond the unjustly condemning concepts of good and evil. Honesty is possible now and honesty is the essential ingredient needed for the psychological rehabilitation of our species. Honesty *is* therapy. As Sir Laurens van der Post has written, **'Truth, however terrible, carried within itself its own strange comfort for the misery it is so often compelled to inflict on behalf of life. Sooner or later it is not pretence but the truth which gives back with both hands what it has taken away with one. Indeed, unaided and alone it will pick up the fragments of the reality it has shattered and piece them together again in the shape of more immediate meaning'** (*A Story Like the Wind*, 1972, p.174 of 473).

The good news is that now that the human condition has been explained, all humans can confront the dark side of themselves without feeling unbearably condemned. Since the late 1980s, the FHA has been developing the know-how that makes it possible for people to safely renegotiate resignation using the reconciling understanding of the human condition.

Renegotiating resignation

Now that it is at last safe to confront the issue of the human condition, the all-important question for almost all adults in the world is, can they renegotiate what happened when they resigned as teenagers and return to a denial-free, unresigned state? Can they hope to stop practicing the denial they have been so dependent on to protect themselves from confrontation with the issue of the human condition? The good news is that to a significant degree they can. People can end their historic denial of our soul's true world and of integrative meaning and, to a degree, return to a state of alienation-free sanity of mind, happiness and soundness of soul. I say 'to a degree' because once a person has lived in total denial of the soul's true world it is not realistic to hope to completely rehabilitate that sound world of the soul. Realistically, the full rehabilitation of humans' true selves will take a number of generations.

The FHA has learnt that with understanding of the psychological principles involved in overcoming a denial it is possible to renegotiate resignation. What is required is training in those psychological principles, along with support for coping with the self-confrontation that occurs. By 2002, the year this book was being prepared for publication, the FHA had been developing this know-how and support structure for more than a decade. The 100 plus Members and Supporters of the FHA are all resigned adults yet most of them are in the process of, and some have effectively completed, renegotiating their resigned states, liberating their minds from the denial they had to adopt as teenagers when reconciling understanding of the human condition was not available.

The FHA has significant and ever-growing experience in renegotiating resignation, and is preparing itself to begin helping other people set out towards the liberated position (or 'LP' as we refer to it). We are establishing a University For Denial-Free Studies, where people can study the understanding of the human condition that is now available and develop the know-how to liberate themselves and other humans from the resigned, alienated state.

Obviously to develop this university for the world's new paradigm requires significant resources, especially financial, ...
. . . .

. . . .
. . . .
. . . . To address the need for funds, those supporting the FHA are developing a number of business ventures. Unlike universities of the old denial-complying world, this new university has no tradition of support from the community, no government funding to rely upon. Financial assistance has been offered to everything—universities that are custodians of denial, cancer research, institutes for depression, drug rehabilitation, etc, etc—except the one thing that can make a *real* difference, the study of the human condition. In fact, the complete opposite has been occurring—rather than receiving support, study of the human condition has been oppressed and persecuted. That clearly has to change.

All the developments of this new paradigm, from its beginning in the research stages through to the writing of my books, have relied on self-sufficiency. We do however hope to receive corporate sponsorship as we think some business leaders are sufficiently far-sighted to recognise the immense importance of our work. As long as we can survive the extremely difficult pioneering years the future for the FHA is fabulous. It can become the largest organisation in the world; not through the pursuit of financial gain—in fact the FHA is a registered charity and a non-profit organisation—but because of its ability to save the world from destruction and end the suffering in people's lives. What we have is both the ultimate product for the planet—reconciling, peace-bringing understanding, and the ultimate product for humans—self-knowledge.

The great responsibility of humans living in the affluent parts of the world changes now from trying to derive some relief from their insecurity through power, fame, fortune and glory, to supporting this work that permanently eliminates humans' insecurity and ends the dilemma of the human condition. Humanity is no longer stalled. With understanding of the human condition found there is suddenly now a great deal of urgent work to be done. Samuel Beckett's famous 1953 play, *Waiting For Godot,* is a portrayal of humans' powerless, stalled state. Unless you could confront and solve the human condition there was little you could do about the awful predicament of humans but wait. This waiting for the God/truth-inspired/denial-free understanding that liberates humans from the human condition is now over.

As mentioned earlier in the context of explaining the limitations of the resigned mind, R.D. Laing anticipated this time when humans would finally be able to go to work on their psychosis and dismantle their resigned state of living in denial, saying: **'Our capacity to think, except in the service of what we are dangerously deluded in supposing is our self-interest, and in conformity with common sense, is pitifully limited: our capacity even to see, hear, touch, taste and smell is so shrouded in veils of mystification that an intensive discipline of un-learning is necessary of *anyone* before one can begin to experience the world afresh, with innocence, truth and love** [p.23 of 156]', and **'True sanity entails in one way or another the dissolution of the normal ego, that false self competently adjusted to our alienated social reality: the emergence of the "inner" archetypal mediators of divine power, and through this death a rebirth, and the eventual re-establishment of a new kind of ego-functioning, the ego now being the servant of the divine, no longer its betrayer** [p.119]' (*The Politics of Experience* and *The Bird of Paradise,* 1967).

With understanding of the human condition found, everybody can be shown that it is now safe to sufficiently confront the truth about themselves to live their lives free of denial. In the previous essay, *Deciphering Plato's Cave Allegory,* Plato highlighted the need to **'put sight into blind eyes'**, and subsequently went on to say, **'this capacity** [of a **mind...to see clearly**] **is innate in each man's mind, and that the faculty by which he learns is like an eye which cannot be turned from darkness to light unless the whole body is turned; in the same way the mind as a whole must be turned away from the world of change until it can bear to look straight at reality, and at the brightest of all realities which is what we call the Good** [integrative meaning or God]**. Then this business of turning the mind round might be made a subject of professional skill, which would effect the conversion as easily and effectively as possible'** (*Plato The Republic,* tr. H.D.P. Lee, 1955, p.283 of 405). In this passage Plato anticipates the role of psychiatry in liberating humans from their historic practice of denial, a role Laing was also referring to when he said, **'an intensive discipline of un-learning is necessary of *anyone* before one can begin to experience the world afresh, with innocence, truth and love'**. Erich Neumann was also anticipating the need for psychiatry to liberate humanity from denial when he wrote, **'Ego and consciousness identify themselves in principle with one side of the opposition and leave the other** [the instinctive soul] **in the unconscious, either preventing it from coming up at all, i.e., consciously suppressing it, or else repressing it, i.e., eliminating it from consciousness without being aware of doing so. Only deep psychological analysis can then discover the unconscious counterposition'** (*The*

Origins and History of Consciousness, 1949, p.117 of 493).

The term 'psychiatry' literally means 'soul healing'—from the Greek words *psyche,* meaning 'soul', and *iatreia* meaning 'healing'. While a common practice in the 20th century psychiatry really only comes into its own now. While humanity could not explain the human condition, psychiatry or soul-healing, was little use because humans could not heal their soul, could not stop living in denial of it, while they lacked the dignifying understanding of their soul-corrupted state. In fact, during the trial of a psychiatrist accused of negligence, the defence (the Attorney-General of Massachusetts) said, **'The art of psychiatry is just one step removed from black magic'** (*The Australian,* 19 July 1983). Essentially, humans were in no position to dismantle their denials, remove their mental blocks, while they lacked the understanding with which to replace them. This is why, as the title of a 1992 book by Hillman and Ventura states succinctly, **'We've Had a Hundred Years of Psychotherapy and the World's Getting Worse'**. It is little wonder a cover feature in *Time* magazine asked **'Is Freud Dead?'** and depicted a psychoanalyst's couch being thrown out the window (Nov. 1993). Now that understanding of the human condition is found, humans are finally in a position to take down their mental blocks because they have the understandings with which to replace them.

Healing the human soul will be one of the main activities of the human condition-reconciled new world. It should be emphasised that alongside this task of addressing the psychological devastation of humans will be the task of attending to the physical impoverishment of humans everywhere. Repairing self while others are not yet aware of this liberating knowledge and while there are people desperately impoverished has the potential to become a more extreme form of selfishness than that which accompanied humans' materialist, escapist existence. In the FHA we describe this danger as 'pocketing the win'. Plato recognised this potential for people who have become 'liberated from the cave'—that is liberated from the resigned, alienated state—to 'pocket the win' when he said, **'And when he** [the liberated person] **thought of his first home and what passed for wisdom there and of his fellow-prisoners, don't you think he would congratulate himself on his good fortune and be sorry for them?'** (*Plato The Republic,* tr. H.D.P. Lee, 1955, p.280 of 405). While there is a danger of those who have become liberated resting on their good fortune, our experience is that although people do tend to become selfish with the information initially, those who become more familiar with it and gain a deeper appreciation of it, develop an extremely compassionate, outward-looking, selfless agenda. 'Pocketing the win' is merely a temporary phenomenon in the journey to renegotiate resignation.

It was mentioned above that with understanding of the psychological principles involved in overcoming a denial it is possible to renegotiate the denial that occurred at resignation. Given that psychiatry has historically been about as much use as black magic, the reader may wonder how these principles were formulated. Interestingly, the FHA has found that they are most clearly set out in literature written to help victims of incest and other situations of childhood and early adolescent sexual abuse, in particular in the 1988 book, *Courage to Heal* by Laura Davis and Ellen Bass. The reason these principles have been so clearly articulated for these young victims of sexual abuse is that their situation is one of few in the old resigned, dishonest world where a form of understanding was available to replace the denial they had adopted at the time of the abuse; a denial put in place to insulate themselves from the pain of being unable to understand the violation of their souls. While humanity has not been able to explain the fundamental cause of sexual abuse, when these young victims grew older they developed an appreciation of human sexuality that naturally accompanies adulthood, that could be used to replace the denials they had employed at the time of the abuse in order to cope. Although this adult awareness of sexuality fell far short of an actual understanding of sexual abuse, as an 'explanation' of what had occurred it was far superior to the child's pre-resigned situation of being completely unable to understand the abuse, as such, a much 'freer' psychological state was achieved.

The principles of change that the victims of childhood sexual abuse applied in their rehabilitation can be applied to any change in thinking that requires 're-routing the brain', including the ultimate change in thinking—the paradigm shift involved in replacing humans' resigned state of denial with the truthful explanation of themselves that understanding the human condition makes available.

It might be assumed that if our mind has a certain way of thinking about something and it discovers a superior pattern of thought that this latter way would immediately replace the now defunct way of thinking. This however is not how our brain works. The brain has a tendency to become habituated to a way of thinking and behaving and that habituation cannot be discarded instantly nor easily. There is a process involved in replacing one way of thinking with another.

An understanding of how our brain works will aid the reader's appreciation of this process. As was briefly mentioned in the *Plato* essay, the ability of the nerve-based learning system to remember past events enables it to compare past events with current events and identify regularly occurring experiences. This knowledge of, or insight into, what has commonly occurred in the past enables the mind to predict what is likely to occur in the future and to adjust behaviour accordingly. The nerve-based learning system can associate information, reason how experiences are interrelated, learn to understand and become conscious of the relationship of events that occur through time.

In the brain, recordings of experiences (memories) are examined for their relationship with each other. To understand how the brain makes the comparisons, we can think of the brain as a vast network of nerve pathways onto which incoming experiences are recorded or inscribed, each on a particular path within the network. Where different experiences share the same information, their pathways overlap. For example, long before we understood what the force of gravity was, we had learnt that if we let go of an object, it would usually fall to the ground. The value of recording information as a pathway in a network is that it allows related aspects of experience to be physically related. In fact the area in our brain where information is related is called the 'association cortex'. Where parts of an experience are the same they share the same pathway, and where they differ their pathways differ or diverge. All the nerve cells in the brain are interconnected, so with sufficient input of experiences onto a nerve network of sufficient size, similarities or consistencies in experience show up as well-used pathways that have become highways. (In the vast convolutions of our cortex there are about 8 billion nerve cells with 10 times that number of interconnecting dendrites which, if laid end to end, would stretch at least from Earth to the Moon and back.)

An 'idea' describes the moment information is associated in the brain. Incoming information could reinforce a highway, slightly modify it or add an association (an idea) between two highways, dramatically simplifying that particular network of developing consistencies to create a new and simpler interpretation of that information. For example, the most important relationship between different types of fruit is their edibility. Elsewhere the brain has recognised that the main relationship connecting experiences with living things is that they appear to try to stay alive. Suddenly it 'sees' or deduces ('tumbles' to the idea or association or abstraction, as we say) a possible connection between eating and staying alive which, with further experience and thought, becomes reinforced or 'seems' correct. 'Eating' is now channelled onto the 'staying alive' highway. Subsequent thought would try to deduce the significance of 'staying alive' and, beyond that, compare the importance of selfishness and selflessness. Ultimately the brain would arrive at the truth of integrative meaning.

Consciousness has been a difficult subject for humans to investigate, not because of practical difficulties in understanding how our brain works as we're told, but because we did not want to know how it worked. We have had to evade admitting too clearly how the brain worked because admitting information could be associated and simplified—admitting to insight—was only a short step away from realising the ultimate insight, integrative meaning, and immediately confronting ourselves with our inconsistency with that meaning. It was better to evade the existence of purpose in the first place by avoiding the possibility that information could be associated. For the same reason we evaded the term 'genetic refinement', preferring instead the vaguer term genetics. We had to evade the possibility of the refinement of information in all its forms. Admitting that information could be simplified or refined was admitting to an ultimate refinement or law, again confronting us with our inconsistency with that law.

Essentially the brain is a vast network of nerve pathways in which 'highways' of regularly occurring associations develop. We can think of it as a 'jungle' of nerve connections that initially have no 'highways' carved through it. Gradually regularly occurring associations develop as 'tracks' through the 'jungle'. Those 'tracks' then become more used and develop into 'paths', and if they continue to be used become 'roads', and so on until they become reinforced as 'highways'. Once a 'highway' has been established it is not easy to redirect the traffic of information and establish a new route. It takes determined practice to redirect such 'highways'.

Our brain has a tendency to develop habits. If you have been brushing your teeth a certain way and your dentist tells you to do it another way and you do not apply yourself to practicing the new way you will find yourself automatically reverting to the old method. Established routines take time and much application to change.

In the case of adjusting to change on a large scale there are four basic stages. Firstly there is normally a shocked, even angry reaction to the imposition of the need for such a change. Less reactionary resistance follows as adjustment begins. In this more rational stage the mind experiences a period of procrastination where it carefully and systematically searches for any justification to avoid change. Eventually, when procrastination fails, acceptance of the inevitability of the change occurs. After acceptance there remains the task of adapting to the new paradigm.

These basic stages take a particular form in the case of the very large scale change of having to adjust to the arrival of understanding of the human condition.

The shock stage has two phases. There is the initial shock of reading description and analysis of the human condition. The effect of this shock is that the reader's mind finds it difficult to take in or absorb or 'hear' what is being explained. If the reader is patient and perseveres they can however begin to hear the explanations and an appreciation of them grows rapidly. So powerful is the information's ability to explain human behaviour that the world of humans starts to become transparent. Albert Einstein once said, **'truth is what stands the test of experience'** *(Out of My Later Years,* 1950). Similarly, author Morris West once wrote, **'Life itself is the best of all lie-detectors'** *(A View from the Ridge,* 1996, p.89 of 143). The principle being referred to is that of subjecting hypotheses to tests—the basis of 'the scientific method'—with the greatest test of all obviously being the test of life itself. The fact that this information makes human life transparently understandable convinces the reader of the truthfulness of the explanations and when this happens they normally become extremely excited and enthusiastic about the information.

It is some time after this phase that the reader starts experiencing the effects of this transparency on their own condition. This is the second and main shock phase. The resigned person's dishonest, deluded way of explaining and defending themselves begins to be exposed. The truth destroys the lies. The more you digest these understandings the more the false forms of reinforcement that you adopted when you resigned are exposed and fail to be effective. You cannot maintain the delusions that you are living off in the presence of the truth. You face exposure day, truth day, come-clean day, honesty day—'judgment day' in fact.

As this exposure increases the reader begins to feel a strong need to reject the information, to deny that it is true; revert to the old 'safe' way of coping with confronting information. Doing this however does not work because it amounts to irresponsibly denying fulfilment of the whole human journey, which was to find these liberating understandings of the human condition. Also, the accountability of the explanations makes it difficult to deny their truth.

A collision occurs between the old and new way of explaining and defending yourself, leaving you in a dilemma, a state of procrastination, a Mexican stand-off in fact. You do not want anything more to do with the new, all-exposing information but on the other hand you do not want to deny it; you cannot advance with the information and you cannot retreat from it. It is at this point that the principles of re-routing the brain become critically important.

The overall requirement is that you have to practice the new way of explaining and defending yourself, otherwise the old, habituated, resigned method of defence will reassert itself. As with breaking any habit, you have to practice thinking with the new understanding. Using the 'highway' analogy, a 'track' has to be carved out from the established 'highway' to form a new 'route' through the 'jungle' of nerve pathways in the brain, and that 'track' then has to be actively maintained and developed if it is to have any chance of becoming a 'road', and ultimately the new 'highway' of thinking.

Again it has to be emphasised that what is occurring has nothing to do with indoctrination. The Macquarie Dictionary defines 'indoctrinate' as, **'to so instruct someone in a manner which leads to their total and uncritical acceptance of the teaching; brainwash'** (1998 edn). In this case it is the *accountability* of the explanations of the human condition, the reasoned logic, that leads to people's initial appreciation of the explanations and eventually to the Mexican stand-off, a process that is the very opposite of **'uncritical acceptance'**. This information is brain nourishment, not brain anaesthetic; it is about being mindful not mindless. What is occurring is actually the dismantling of indoctrinated thought, the dismantling of the delusions, lies and denials that people forced their mind to accept when they resigned.

The difficulty with this re-routing of the brain, this adoption of the new way of thinking is that it is not easily achieved. *Courage to Heal* describes the tenacity of habituated 'highways': **'A pattern is any habitual way of behaving. By its nature it is deeply entrenched, set by repetition, and brings a familiar result. Even if that result is not, ultimately, what you want, its predicability is part of its grip. Patterns usually start unconsciously as a way of coping when your options are limited. They serve**

you, but at a cost. Patterns have a life of their own, and their will to live is very strong. They fight back with a vengeance when faced with annihilation. Once you recognise a pattern and make the commitment to break it, it often escalates' (p.175 of 495).

The resigned highway of living in denial of the human condition and deriving your reinforcement from false, survival-of-the-fittest, competitive success through winning power, fame, fortune and glory, has an exceptionally strong hold in the mind because of what happened at resignation. At resignation the fear of depression absolutely 'cemented' the denial in place. While the mind forgets the depression soon after resignation, the entrenchment of the denial remains and as a result the denial does **'fight back with a vengeance when faced with annihilation'**.

It is the crisis brought about by the Mexican stand-off, by the logic and understanding exposing and destroying your ability to lie, that eventually brings about <u>acceptance</u> of the need to change. What happens is the crisis becomes sufficiently unbearable—you are actually renegotiating resignation—for you to accept that you have to let the truth into your life, admit that you are a victim of the human condition. Honesty is what relieves the situation. An alcoholic who joins Alcoholics Anonymous (AA) is told that they will initially try to deny that they are an alcoholic but for any real rehabilitation to take place they have to accept this truth. Until you reach acceptance of your human condition-afflicted, corrupted state it is difficult to make any real progress from the exposed, frustrated state of the Mexican stand-off.

This acceptance and subsequent honesty fractures the hold the resigned 'highway' of denial has on your mind. It releases you from being a total victim of your old resigned strategy, where you cannot think any other way than from a base of denial. This critical stage of adopting and practicing honesty initially benefits from the help of others who have experience in this process of change. For instance, it is initially easier if others go through and acknowledge for you your true situation; that you are a corrupted but heroic victim of humanity's great battle to defeat the ignorance of our instincts. Using the true, honest defence for yourself becomes easier with practice because you discover that the true defence works—it does stop the anger, pain and depression that comes from being exposed by the truth. You enter a positive feedback loop.

The way to survive exposure is to use the truth now available to explain why humans became corrupted. By surviving the exposure, through using the understanding that is now available of why humans became corrupted, you gain your first foothold in the new human condition-free world. From there it is only a matter of applying yourself to holding onto the true explanation for your corrupted condition and the new 'highway' will gradually replace the old. You adapt to the new world, you achieve liberation from the human condition, you achieve the liberated position or LP as it is termed.

In *Courage to Heal* there is a poem titled *Autobiography in Five Short Chapters* that describes the process of dismantling an established 'highway' of thinking. It uses the example of falling into a hole as an analogy for succumbing to the old way of thinking: **'I walk down the street/There is a deep hole in the sidewalk/I fall in/I am lost—I am helpless/It isn't my fault** [that I fell in because I am still learning not to, learning that there is another way of thinking]/**It takes forever to find a way out//I walk down the same street/There is a deep hole in the sidewalk/I pretend I don't see it/I fall in again/I can't believe I am in the same place/But it isn't my fault/It still takes a long time to get out//I walk down the same street/There is a deep hole in the sidewalk/I see it is there/I still fall in—it's a habit/My eyes are open/I know where I am/It is my fault** [I should know by now to think differently, to not "fall in the hole"]/**I get out immediately//I walk down the same street/There is a deep hole in the sidewalk/I walk around it//I walk down another street'** (p.183 of 495).

The following drawing illustrates the same transition from being owned by the old highway, to being sufficiently free of it to be able to view it honestly.

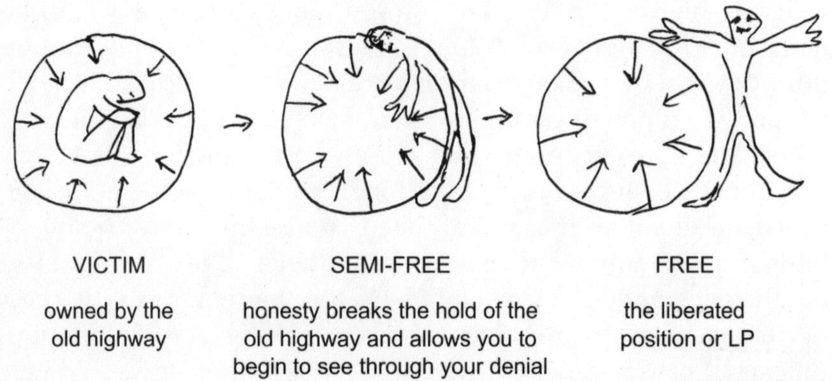

The drawings below summarise the journey from living in denial, through exposure, to the liberated position or LP. The key symbolises the liberating understanding of the human condition—the true defence for humans' divisive, corrupted state—that humanity has been in search of for 2 million years.

The deluded resigned egocentric state The exposed, naked state The liberated position

1996 JEREMY GRIFFITH

Most importantly, in the years ahead, people will increasingly realise they do not have to go through the agony of trying to fully confront the issue of the human condition, and, with it, their own condition. All people need to do is study the explanations sufficiently to confirm for themselves that the truth about humans has at last been found and then offer the explanations their support. It is not necessary to fully confront the information to support it. You can live in semi-denial of it. As long as there is support of this all-important information, future generations can be given the true explanation of the world of humans, and their part in that world, and as a result they will not have to resign nor encounter the problem of renegotiating a resigned state of denial.

It is only necessary for people to confront the truth to the extent that they feel secure to do so. The more corrupted will only be able to confront it to a small degree, but everyone will live in support of the truth.

This situation is described in the last chapter of *Beyond,* 'Adjusting to the Truth', and illustrated with the following drawing. It shows different forms of participation in the new human condition-ameliorated world, each of which requires a varying amount of confrontation with the truth.

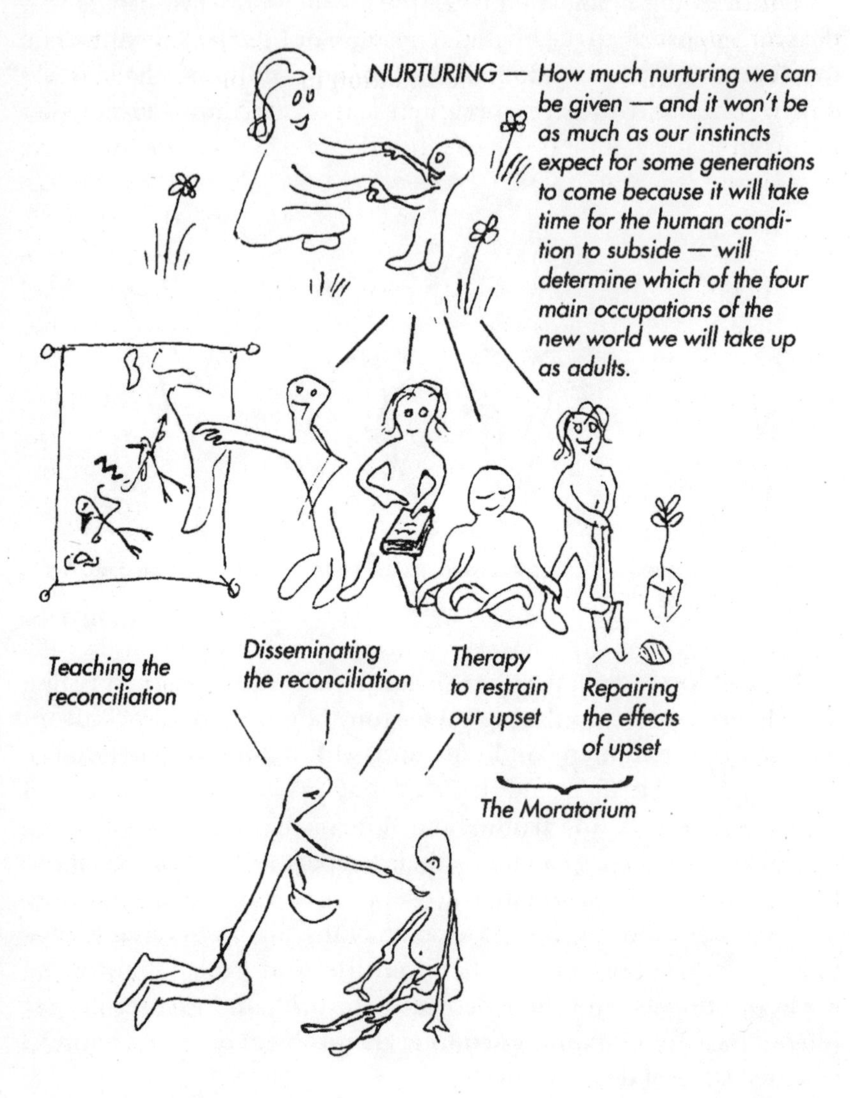

1991 JEREMY GRIFFITH

Some more intelligent and highly educated people, accustomed to their ability to closely scrutinise all before them, will find this need to live with the truth at arms length difficult to accept and offensive to their intellect. However such responses will decrease as the enthusiasm for the new humanity-liberating way of living spreads, and everybody supports the truth and goes to work for the new world, each in a way that accommodates their level of soundness.

It should be pointed out that this situation, where individuals only confront the truth to the extent that they are secure enough to do so, is really no different from the situation in the old resigned world where those who were not clever were excluded from universities and higher study because they could not be trusted to maintain the denial, the great lie. As was explained in the *Introduction,* human intelligence to date has largely been concerned with the art of denial, not with truthful thinking. Artful, sophisticated, evasive, esoteric, cryptic, intellectual cleverness was needed to establish, defend and maintain the safe, non-confronting, escapist, alienated state. Universities selected for cleverness (that is, attendance at university was dependent on passing exams that tested for mental aptitude), not because cleverness was a prerequisite for thinking and learning, as those living in denial have evasively maintained, but because 'dumbness' or lack of intelligence could not be trusted to maintain the denial. The truth is the average IQ or intelligent quotient of humans is quite adequate for understanding. A high IQ or 'cleverness' was needed to deny and evade the truth. That was the real art. Universities have had high IQ entrance requirements because they have been the custodians of denial, keepers of 'the great lie'. A student had to be able to investigate the truth and talk about the truth without confronting it or admitting it, a very difficult, IQ-demanding undertaking.

Just as the old evasive world had to emphasise cleverness, so the new unevasive world has to emphasise soundness. This is the meaning of the Biblical references, that **'The meek...will inherit the earth'** (Matt. 5:5) and **'many who are first will be last, and many who are last will be first'** (Matt. 19:30,20:16; Mark 10:31; Luke 13:30). Humanity has turned its head for home, 'home' being the denial-free world of soundness, togetherness, peace and happiness, and it is those who are already sound who can show humanity the way.

In an ideal situation, as I have stressed in my earlier books, counselling support would accompany the introduction of understanding of the human condition. The reality however is that the FHA, while it is doing all it can to supply assistance and support, is stretched to the limit defending these revolutionary ideas. As the progenitor of this information I have a responsibility to offer as much help as I can, however there is also a responsibility on the part of those immersing themselves in this information to heed this advice about how to manage the effects of this information. The unavoidable situation is that until substantial support develops for our work, people have to responsibly manage their situation so as to not overly confront the information if they are not sufficiently sound and secure. This is not easy advice to accept because, for instance, it is so at odds with the old resigned world tradition of intelligence being able to pursue any interest. The overall reality we are faced with is that any pioneer project is inherently problematic and, given the scale of the change that understanding of the human condition brings, the difficulties this project faces in its pioneer stage are bound to be substantial. It *is* a 'brave new world' that humanity is entering.

The following is an account of someone who is learning how to support the truth without overly confronting it. Jeremy Schroder from the Australian state of Victoria, illustrates all that has been said about the stages of adjusting to the arrival of the truth about the human condition. Jeremy has kindly given the FHA permission to include these extracts from his correspondence. While the FHA has corresponded with Jeremy we have not met.

'Three years ago I read Jeremy [Griffith]**'s book and something happened that I have since been trying to deal with. At the time I was halfway through enlisting for the army when Jeremy's words truly hit home. Nothing could touch me in that time, everything made perfect absolute sense. Then things started to change. I wanted to go further but something stopped me and when I pushed I felt the worst fear I have ever known. Fear doesn't even go close to expressing it. What do you suppose you do when you find the most fearful thing you'll ever encounter is yourself.**

I understand Jeremy's message in having to distance yourself from the truth. I don't know how to give in. All my life I have looked for understanding and when I find it it doesn't sit well with me to have to run away from it. If I let it go I may never be able to find it again. I have found ways to block but I will never again be able to deny. I know I have to let it go but where do my responsibilities truly lie, in seeking truth and finding paths or in denial. It was so beautiful, a whole other world exists inside our minds that we have rarely seen. I will keep trying to find the middle ground and thank you for being there.'

At this point the FHA emphasised to Jeremy that, **'Until proper preparation can be made for exposure day it will be necessary for each of us to rely on block-out or evasion to cope'** (*Free*, p.212 of 228) and that **'Essentially all we have to do now is support the unevasive truth (hold the key aloft). It is not necessary to confront it. This is the way to cope with judgment day'** (*Beyond*, p.169 of 203). Jeremy Schroder gave the following response: **'I have taken on board our last conversation. At the moment I am in still waters, but eventually a storm will no doubt arise, but as you have said there is a difference in looking for them and avoiding them. That is something that I have to deal with. I have you because a strong man stood up for what he knows and believes in; it's not even a belief, it transcends belief, it is pure truth. The storms are brewing and there is a big one on the 26th of May** [the FHA's first major court hearing]**. Within that storm I am trying to give you as much cover as I can. My thoughts are with you. Bow your head to no-one. I will leave you with something that offered me strength for some time now and I hope it will do the same for you in a time which finds us fragmented and alone. "It fortifies my soul to know that tho I perish, truth is so that however I stray and range and whatever I do thou does not change. I steadier step when I recall that if I slip thou dost not fall." Love you guys heaps'** (FHA records, Mar. 2003).

With regard to Jeremy's comment about a 'strong man', that is an old world prejudiced interpretation. I am simply sound enough to confront the truth. We are now able to understand that the truly strong people are those who have survived unjust condemnation from the human condition. They are the real heroes of life under the duress of the human condition.

Despite the pioneering difficulties that have to be overcome it is obviously immensely exciting to know that people who have resigned can renegotiate their life of living in denial. It means all humans can look forward to becoming, to varying degrees, free from the historic psychosis that has beset our species since the emergence of consciousness 2 million years ago.

The end of resignation

While adults who are already resigned and living in denial can begin to dismantle the denial, it is especially wonderful that young people who are not yet resigned, like Lisa Tassone at 16 years of age, can now apply themselves to digesting the explanation of the human condition and avoid resignation. In fact, the real beneficiaries of this new information are going to be the generations who have not yet reached the stage where they would otherwise have resigned and now will not have to. It is obviously much easier to stay in the denial-free state than to have to return to it from a state of denial.

Now that the human condition has been understood, humans will no longer have to suffer the agony of resignation. With the ability to reconcile ideality with reality new generations of humans will not have to resort to blocking out ideality to cope with reality. Also, they will see how understanding of the human condition will bring an end to the immensely upset, corrupted and destructive, angry, egocentric and alienated world humans have been living in. They can look forward to a peaceful, loving world.

Resignation was a tragic but necessary coping strategy humans had to employ in the absence of the understanding of their 'goodness'. With understanding of the human condition available, resigning to a world of denial, insensitivity and alienation is no longer necessary.

What will be especially relieving to new generations is that older generations will now be able to be honest about human life. At last able to understand the human condition, it is finally possible to reveal what has truly been going on in the lives of resigned adults. The once necessary but horrifically destructive silence resigned adults maintained about their corrupted, alienated state can be broken. All the lying can stop and new generations of humans arriving in the world will grow up free of alienation and, as a result of this freedom, will be so different from humans of today that they will be like a new variety of beings on Earth. No longer having to bury the magic world of their soul and resign themselves to a life of alienation the new generations will stay alive inside themselves. Furthermore, with the understanding available of why humans have been the way they have been, that greatest of all living horrors, depression, will disappear from Earth.

Thank heavens the journey through estrangement or alienation or **'banishment'** (see Genesis 3:23) and its terrible resultant darkness is over. Humans can now escape the alienated condition, as Jim Morrison of The Doors sang, they can now **'Break on through to the other side'** because **'the day** [that] **destroys the night'** has arrived. Humans lived in denial, in Morrison's words they **'Tried to run/Tried to hide'**, and they lived superficially, materialistically, **'We chased our pleasures here/Dug our treasures there'**, but ultimately humans had to **'Break on through to the other side'**, through a path that was directly and deeply confronting, a path where **'The gate is straight/Deep and wide'**.

While the acknowledgment and explanation of the human condition is initially a great shock, it ultimately brings phenomenal relief to humans—and to our world as well, for humans will no longer be the upset, angry, selfish and immensely destructive beings they have been.

Our species' fabulous future

Once the reader begins to comprehend and see into the human condition, the world begins to become transparent; the whole edifice of denial that humans have carefully constructed over some 2 million years begins to crumble. As William Blake prophesied, **'When the doors of perception are cleansed, man will see things as they truly are'**. Jim Morrison's inspiration for calling his band The Doors came from this statement, while author Aldous Huxley was sufficiently inspired by Blake's words to title his book, detailing his experience with psychedelic drugs, *The Doors of Perception*.

With understanding of the human condition 'the doors of perception' *are* finally **'cleansed'**, and **'man will see things as they truly are'**. The whole manufactured, artificial edifice of denial of the resigned world begins to disintegrate and gradually 'the scales will fall' from people's 'eyes'; they will begin to wake up as if from a deep sleep, an amnesia or state of hypnosis.

The immense power that emanates from having insight into the human condition will be demonstrated in *The Demystification Of Religion* essay where Christ, who for so long has been regarded as the one nobody knows, will be explained and demystified, along with all manner of religious metaphysics.

Understanding the human condition allows humans to know everything about ourselves that we have ever wanted to know; better still, it liberates our all-sensitive potential. It liberates our soul. Humans have the potential to be a super-clever, ultra-sensitive species. This was made abundantly clear when savant syndrome was explained earlier. What has held us back has been the denial we developed to avoid the depression brought about by the issue of the human condition. This denial ensured our species did not become extinct, it protected us until we reached the position where we could solve the human condition and banish that suicidal depression. The denial contributed valuable service to the human race but thankfully it is no longer necessary. Understanding the human condition is the key that unlocks the door to a fabulous true world (hence the FHA's symbol of a key held aloft), and indeed destiny—a destiny that humans had almost given up hope of ever reaching.

There was a limit to how much denial/alienation humans and our world could absorb, and that limit has been reached; humanity *had to* break through into understanding of the human condition, or face extinction. Towards the end of his life the Spanish cellist, Pablo Casals, summed up the plight of our species when he said, **'The situation is hopeless. We must take the next step.'** The 1991 film, *Separate but Equal,* accurately summarised the situation in the words of one character, **'Struggling between two worlds; one dead, the other powerless to be born'**—words which echo those of Antonio Gramsci, **'The crisis consists precisely in the fact that the old is dying and the new cannot be born; in this interregnum a great variety of morbid symptoms appears'** (*Prison Notebooks,* written during Gramsci's 10-year imprisonment under Mussolini, 1927–1937). Until understanding of the human condition was found humans *were* powerless to change their society. Historian Eric Hobsbawn described humanity's stark predicament when he wrote in his 1994 book, *Age of Extremes,* **'The alternative to a changed society—is darkness'**. To paraphrase Benjamin Disraeli's famous expression, stalled halfway between ape and angel was no place to stop.

With regard to the hunger in the human heart for the arrival of the truthful new world, it is worth repeating the wonderful words from Sir Laurens van der Post that were included in the section on autism: **'This shrill, brittle, self-important life of today is by comparison a graveyard where the living are dead and the dead are alive and talking** [through our soul] **in the still, small, clear voice of a love and trust in life that we have for the moment lost...**[there was a time when] **All on earth and in the universe were still members and family of the early race seeking comfort and warmth through the long, cold night before the dawning of individual consciousness in a togetherness which still gnaws like an unappeasable homesickness at the base of the human heart'** (*Testament to the Bushmen,* 1984, pp.127–128 of 176).

Having, at the last possible moment, broken through to understanding of ourselves, everything changes. The future for our species suddenly changes from being totally bleak to becoming fabulously golden.

There has been much superficial talk in the media about starving millions; terminal pollution; wars; universal relationship and family breakdown; endemic suffering, loneliness, depression and sickness;

soul/innocence-destroying promiscuity; childhood madness, and so on. But these are merely symptoms of a larger malaise. What is being talked about in this book is solving the root cause of all our species' problems, ending our destruction of our innocent souls and lifting the universal depression from the human race forever. We are talking about the psychological rehabilitation of our species, and the end of the historic denial. This is it, the breakthrough the human race has been waiting for.

Such claims may seem astonishing. Living in denial, humans hardly knew humanity was struggling with a problem, let alone that there was a solution to it. One moment humanity is living in a wasteland, a world of almost total darkness, in a meaninglessness and disorientated state of complete denial, the next it is in a new world drenched in the light of relieving understanding with immense potential and great excitement ahead.

The drawing overleaf summarises humanity's journey to enlightenment. From happy, loving, integratively-orientated infancy and childhood, humanity progressed to the horror of insecure adolescence where humans had to search for their identity, for understanding of why they lost innocence and became divisively behaved. With the understanding of the human condition now found humanity can enter the happy, ameliorated, secure state of adulthood.

Humanity's Journey To Enlightenment

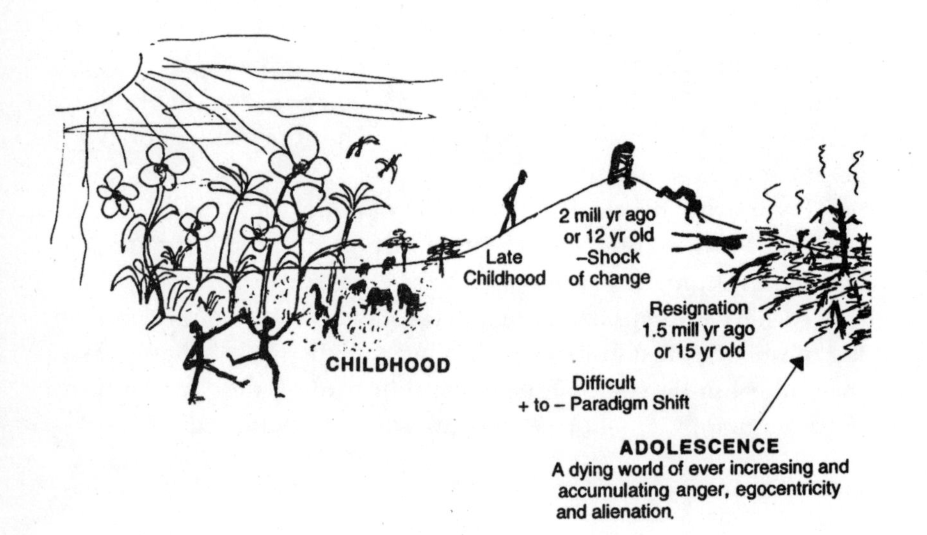

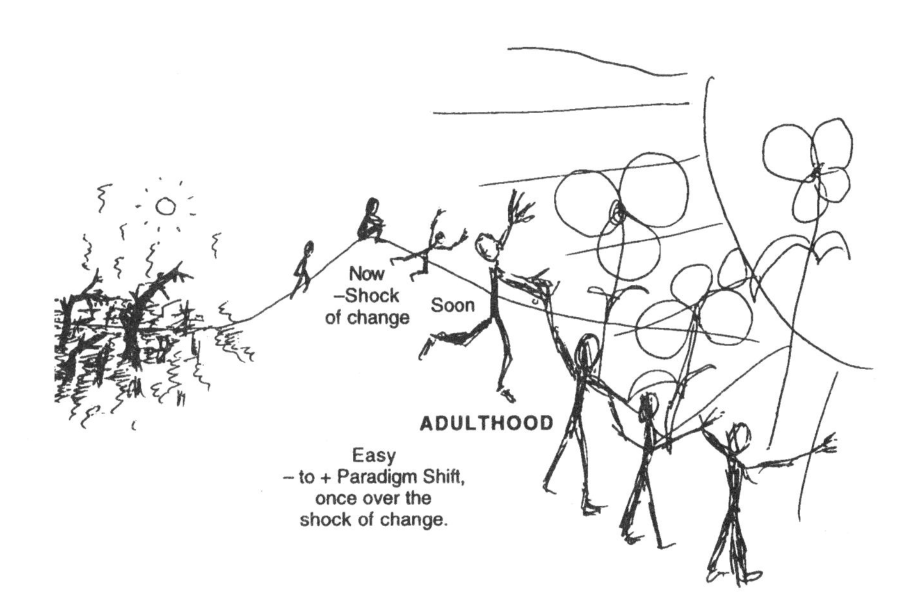
Now
–Shock
of change
Soon
ADULTHOOD
Easy
– to + Paradigm Shift,
once over the
shock of change.
1996 JEREMY GRIFFITH

Bringing Peace To The War Between The Sexes

And

The Denial-Free History Of The Human Race

Bringing peace to the war between the sexes

The ability to understand the human condition, that is, understand why humans have been competitive, aggressive and selfish when the ideals are to be cooperative, loving and selfless, makes it possible to reconcile all the manifestations of the poles of 'good and evil' in human life. The instinct and the intellect, our soul and mind or conscience and conscious can be reconciled, as can mysticism and rationalism, religion and science, faith and reason and holism and mechanism. In politics, the left wing and right wing, or socialism and capitalism, idealism and realism can be reconciled. Ameliorating understanding can be brought to the rift that has existed between the more innocent individuals and races and those more corrupted, between women and men, the young and old. These differences are all manifestations of the underlying conflict between our instinct and our intellect yet it is the lack of understanding that has existed between men and women that has caused some of the deepest wounds in human life. Enormous energy has been spent in the struggles that have taken place in relationships between men and women. The bitterness, heartache, suffering and the hurt to children has been immense. There are so many questions about the relationship between men and women that need answering and with understanding of the human condition it is now at last possible to answer *all* of these questions.

(Note: the underlinings contained in this and the next essay are to assist the reader to identify different topics as they are addressed.)

Understanding the rift, indeed war, that has existed between men and women requires an appreciation of the different roles men and women have played in humanity's 2-million-year heroic journey to find understanding of the human condition.

Until the human condition could be resolved it was not safe to acknowledge the different roles men and women played in the journey to enlightenment. As can now be explained, the role men took up under the duress of the human condition left them embattled and egocentric, competitive and aggressive, while the role women took on left them relatively naive and soulful. Without the reconciling understanding of these different realities, and their complications, it was all too easy for women to condemn men for being egocentric, competitive and aggressive, and all too easy for men to then feel hurt, angry and retaliatory for being unjustly criticised. Over time it was found that the best way to control prejudice was to prevent people from acknowledging that there was any substantial difference of any sort between the sexes. The dogma of the 'politically correct' culture emerged. Now that the human condition is explained, it is both safe and necessary to acknowledge the different roles men and women have played in humanity's journey, the different effects those roles have had on each sex, and how they have affected relationships between men and women.

The different roles and their effects are explained in *Beyond* in 'Stage 2' of the chapter 'Illustrated Summary of the Development and Resolution of Upset'. What follows is a description, and in places an elaboration, of that explanation.

In the essay *Deciphering Plato's Cave Allegory*, in the section 'How we acquired our soul', the importance of nurturing in the emergence of humans from our ape ancestry was described. It was explained that nurturing was the prime mover or main influence in the development of humanity, it gave us our cooperative instincts and allowed us to become conscious. This means that from 10 million years ago when apes emerged and the love-indoctrination process began, to 2 million years ago when consciousness finally emerged, the activity of greatest priority in human society was nurturing. Since women gave birth to and suckled the offspring, nurturing was predominantly a female responsibility, and so for those 8 million years our society was a female-role dominated or matriarchal society.

The prophet Jean-Jacques Rousseau acknowledged women's nurturing role when he wrote: **'The first education is the most important, and this first education belongs incontestably to women; if the Author of nature had wanted it to belong to men, He would have given them milk with which to nurse the children'** (*On Education*, 1762, p.37 of 501). The renowned South African author, Olive Schreiner, made a similar acknowledgment in her 1883 book, *The Story of an African Farm:* **'They say women have one great and noble work left them...** ***We*** **bear the world and** ***we*** **make it. The souls of little children are marvellously delicate and tender things, and keep for ever the shadow that first falls on them, and that is the mother's or at best a woman's...The first six years of our life make us; all that is added later is veneer...The mightiest and noblest work is given to us'** (p.193 of 300).

In *Deciphering Plato's Cave Allegory*, in the section 'How our soul became corrupted', the battle between our instincts and our intellect was explained. It was described how, when humans became conscious some 2 million years ago and our intellect began to experiment in understanding, our instincts, being, as they were, unaware or ignorant of the intellect's need to conduct such experiments, in effect criticised and tried to prevent the conscious mind's necessary search for knowledge. Unable to defend its behaviour by explaining why it was necessary to carry out its experiments, the intellect had to find other ways of holding at bay the unjust criticism from the instincts. The alternative was to abandon the search for knowledge, which it obviously could not do. The only option available was to attack the unjust criticism, live in denial of it and try to prove the criticism wrong. Humans' upset angry, alienated and egocentric state emerged. (Note that the dictionary defines 'ego' as 'conscious thinking self', so ego is another word for the intellect. The word 'egocentric' then applies to where the intellect becomes centred or focused on trying to prove the instincts' criticism wrong and asserting its self-worth.)

Ignorance on the part of our instinctive self or soul of our intellect's need to search for knowledge was a threat against our species; if it prevailed humanity would never find understanding and humans would never fulfil their responsibility to master intelligence. This all-important battle to overthrow ignorance became our species' priority. When men in their role as group protectors went out to meet the threat of ignorance, our species' social structure changed from a matriarchal (female-role dominated), soulcentric, nurturing society, to a patriarchal (male-role dominated), egocentric, embattled society. Wives and children in virtually all cultures have not adopted their husband or father's surname because of some cultural accident, but because they were living in a patriarchal or male-role dominated world. This patriarchy is acknowledged in Genesis in the *Bible,* **'To the woman he** [God] **said...Your desire will be for your husband, and he will rule over you'** (3:16).

Humanity's recent 2-million-year adolescent search for its identity, search for understanding of itself, and the reason for its divisive nature, has been a patriarchal journey, just as humanity's infancy and childhood were part of the matriarchal stage.

Tragically, in their task of championing the intellect over ignorance, men unavoidably became angry, alienated and egocentric. Further, this conflict between the intellect and the instincts became greatly compounded by the fact that the angry and aggressive behaviour was completely at odds with humans' particular instinctive orientation, which was to behave lovingly and cooperatively. From an initial state of upset, men had then to contend with a sense of guilt, which greatly compounded their insecurity and frustrations, and made them even more angry, alienated and egocentric.

Unable to explain their rapidly increasing loss of innocence, men began to resent and attack innocence because of its implied criticism of their lack of innocence. The first victim was nature. As already mentioned, men began to 'hunt' (kill) animals because their innocence criticised them, albeit unwittingly.

The rapidly compounding upset in men attracted criticism from women who became their next victims. As has been explained, it was men's role as group protectors to defend humanity against the threat of ignorance. Also, while men were championing the ego and suffering self-corruption, it was important that women stayed relatively free of the battle in order to retain as much innocence as they could to nurture an upset-free subsequent generation. Since women did not participate in the fight against ignorance, they were largely unaware of the cause of the upset in men. The problem was then that the relative innocence of women, and their lack of appreciation of the upset that resulted from fighting ignorance, represented unjust criticism of men's embattled, corrupted state. Unable to explain their upset, men could not even admit their embattled state for a failed attempt to explain their upset would only be misinterpreted by women as an admission of badness. Unjustly condemned by the relative innocence and naivety of women, men retaliated and attacked women. Since women reproduced the species, men couldn't destroy them as they did animals. Instead they violated women's innocence or 'honour' by rape; they invented 'sex', as in 'fucking' or destroying, as distinct from the act of procreation. What was being fucked, destroyed, ruined or sullied was women's innocence. In this way women's innocence was repressed and they came to share men's upset.

To reiterate, it was men who became egocentric, who had the task of championing the ego, the conscious self's need to find justification for itself despite the instinctive self's criticism of it. Men's burden was that they had to suffer self-corruption, **'march into hell for a heavenly cause'**, as the words from *The Man of La Mancha* state. They had to suffer becoming angry, alienated and egocentric and women's innocence simply added to this criticism of men's corrupted state. To subdue the criticism, men violated women's innocence through sex. While sex was originally for procreation (and, in the case of some species, such as the bonobo chimpanzees, a means of pacification), with the advent of the dilemma of the human condition and men's need to somehow retaliate against the condemnation that innocence represented, sex became 'perverted'. It became used as a means of attacking the innocence of other humans, in particular the innocence of women. On this level, sex became rape. The feminist Andrea Dworkin recognised this underlying truth in her 1987 book *Intercourse,* when she said, **'All sex is abuse'**.

In time the image of innocence in women, their physical beauty that 'attracted' sex, also became a means of inspiring the journey to self-understanding. This aspect means that while at base sex was rape, on a nobler level, it became an inspirational act of love. When all the world disowned men for their unavoidable divisiveness women in effect stayed with them, bringing them the only warmth, comfort and support they would know. While at base sex is rape, it also became an act of love, an act of faith in, and affection for men; a sublime partnership between men and women. As it says in Genesis in the *Bible,* **'The Lord God said, "It is not good for the man to be alone. I will make a helper suitable for him"...Then the Lord God made a woman...and he brought her to the man'** (2:18,22).

This attack on women's innocence by men, and the consequences of it, deserves elaboration. Given that the entire universe was an innocent friend of our soul but not of our apparently corrupt intellect, it can be appreciated that men needed extremely powerful egos to defy ignorance and champion the intellect. Women, not responsible for the fight against ignorance and so not participating in the battle itself, did not and could not be expected to understand the validity of the effects of the battle. Being as conscious as men, women were naturally as aware as men of the task of having to resist the ignorance of the ideals-demanding instinctive self or soul. What women were not so aware of was what happened when the battle against ignorance was fought. Since they were not as involved as men in the battle they lacked an appreciation of the degree of anger, alienation and egocentricity that emanated from fighting it.

Women had little empathy with and respect for the effects of the battle. They tended to be soul-sympathetic, not ego-sympathetic. For example, to sustain themselves, the embattled egos of men needed to build towering buildings symbolising their will and determination to defy and defeat the unjustly condemning world that surrounded them. The feminist author Camille Paglia spoke the truth when she famously stated that, **'If civilization had been left in female hands, we would still be living in grass huts'** (*Sexual Personae,* 1990). This 'grass huts' comment can be understood both literally and metaphorically, because the fundamental situation was that if the soul (which women represented) had its way the intellect would never have been allowed to search for knowledge. The soul's ignorance *had to be* defied if knowledge and ultimately self-understanding was to be found. To give in to soul was to go nowhere, to remain in **'grass huts'**. Incidentally, Camille Paglia also defied the politically correct code when she said, **'Wake up, men and women *are* different'** (*The Australian,* 4–5 July 1992).

The truth is, as the champions of the conscious thinking self or intellect or ego, men are the heroes of the human journey to enlightenment, *not the villains* they have for so long been portrayed as. Women's ignorance of the all-important role that men have been playing in the world is clear in this comment from Germaine Greer, an icon of modern feminism: **'As far as I'm concerned, men are the product of a damaged gene. They pretend to be normal but what they're**

doing sitting there with benign smiles on their faces is they're manufacturing sperm' (*Sydney Morning Herald,* 14 Nov. 1991). Greer believes that the wilful, competitive, egocentric nature of men is nothing more than a selfish drive to reproduce their genes. The truth is, rather than being driven by selfishness, men have been involved in serving humanity in a most remarkable way. Their role in the world has been entirely honourable, brave and *selfless.*

Women's ignorance and thus lack of sympathy for men's role in the world is also apparent in this famous comment by leading feminist Gloria Steinem: **'A woman needs a man like a fish needs a bicycle.'**

The gulf between men and women—acknowledged in the title of John Gray's best-selling 1992 book, **'Men are from Mars, Women are from Venus'**—is palpably clear in this conversation I once overheard: 'She: **You men are wholly monstrous, foreign bodies, in fact cancers on this planet.** He: **Yes, well, have you heard that women are so meaningless as to not even exist.'** This comment provides a true measure of the extent of 'the war between the sexes'.

The fact is women have not understood the egocentric world of men. The following quote from actress Shirley MacLaine illustrates this lack of appreciation of men's egocentric nature: **'MacLaine can't find a man to love. The 48-year-old actress...** [said she] **longs for a "close and warm relationship" but hasn't met a suitable partner. "Most men I meet seem to be too involved in trying to be successful or making a lot of money," she said. "I feel sorry for all of them. Men have been so brainwashed into thinking they have to be so outrageously successful—to be winners—that life is very difficult for them. And it's terribly destructive, as far as I am concerned, when you are trying to get a serious relationship going"'** (*Daily Mirror,* a Sydney newspaper, 14 Dec. 1982).

The fact is women have not understood men at all, as this quote admits: **'Men are a knot, I'll never untie, around a box I long to peer into'** (from Kate Llewellyn's poem *Men, Sydney Morning Herald,* 20 Apr. 1996).

Women have not appreciated the battle—they have not been able to empathise with what has been going on. They have not been, as it were, 'mainframed'; intuitively aware of the battle like men—in the same way men have never been as 'mainframed' to the role of nurturing as women intuitively are. While both men and women have had to live in denial of the battle of the human condition, men retained an awareness of the battle because they were having to fight it.

Sir Laurens van der Post described this limitation of women in his 1976 book, *Jung and the Story of Our Time,* when he related a dream

Carl Jung had about a blind woman named Salome. Sir Laurens wrote that **'Salome was young, beautiful *and* blind'**, and then explained the symbolism of Salome's blindness, saying **'Salome was blind because the anima** [the soulful, more feminine side of humans] **is incapable of seeing'** (p.169 of 275).

This 'blindness' in women meant that the only alternative to the oppression of women was that men explain their predicament, but, as emphasised, that was not possible. Men could not admit their inconsistency with integrative meaning until they could defend it. This quote describes men's plight: **'One of the reasons that men have been so quiet for the past two decades, as the feminist movement has blossomed, is that we do not have the vocabulary or the concept to defend ourselves as men. We do not know how to define the virtues of being male, but virtues there are'** (Asa Baber, *Playboy* mag. July 1983).

It should be explained that feminists did not liberate themselves simply because men stayed quiet as this quote suggests. The more men fought to defeat ignorance and protect the group (humanity), the more embattled, upset and corrupted they became and thus the more they appeared to worsen the situation. The harder men tried to do their job of protecting humanity the more they appeared to endanger humanity! As a result, they became almost completely ineffective or inoperable, paralysed by this paradox; cowarded by the extent of their self-corruption and its effects. At this point women have had to usurp some of the day-to-day running of affairs as well as attempt to nurture a new generation of soundness. Women, not oppressed by the overwhelming responsibility and extreme frustration that men felt, could remain effective. Further, when men crumpled women *had* to take over or the family, group or community involved would perish. A return to matriarchy, such as we have recently seen in society, is a sign that men in general have become completely exhausted. However, total matriarchy has not emerged because men could not afford to stand aside completely whilst the fundamental battle existed. They needed to stay in control and remain vigilant against the threat of ignorance. While some elements in the recent feminist movement seized the opportunity to take revenge against men's oppression, the movement in general was borne out of necessity. The tragedy was that like all pseudo-idealistic, politically correct movements, feminism was based on a lie; that there is no real difference in the roles of men and women.

The question arises, if women are so duped by soul how is it that there exists such right wing, ego-sympathetic women as political leaders Margaret Thatcher, Madeleine Albright and Golda Meir, and authors like Ayn Rand? To address this question the important sentence to be considered in the above paragraph is, 'Women, not oppressed by the overwhelming responsibility and extreme frustration that men felt, could remain effective.' Men have been overly corrupted for at least half a million years and as such have lived with extreme frustration, even self-loathing, of the immense destruction they have caused the planet. After such a long time, it can be expected that women now have a strong instinct for an opportunity to participate up-front in the battle and even—in situations where men became totally destructive, disdainful of themselves, paralysed by their predicament and inoperable—take control from men. We can now expect women to anticipate this opportunity for greater power within personal relationships, and in larger economic and political situations, and to some degree be adapted to and thus appreciative of what is required to effectively take up the male role of championing the ego. After all, if men had not been available to take on the battle to champion ego over ignorance, women would have had to take it on fully and become as aware as men now are of what happens when you fight that battle. Camille Paglia once wrote that **'It is woman's destiny to rule men'** (*Vamps & Tramps*, 1974, p.79). This comment is an expression of the awareness that now exists in women that eventually men crumple and women can then take over. The truth is many sensitive new age guys (SNAGs) are crumpled men. Also, women's nagging of their menfolk is a case of women chiselling away at, attempting to break, men's ability to keep fighting and defying the ignorance of the world of the soul.

Without the understanding necessary to explain themselves men had no choice other than to repress the relative naivety of women, which in turn tied women's corruption inextricably to men's. It was an extremely difficult situation for women. They had to try to 'sexually comfort' men but also preserve as much true innocence in themselves as possible for the nurturing of the next generation. Their situation, like men's, worsened at an ever-increasing rate. The more women 'comforted' men, the less innocence they retained and the greater comforting the following generation needed. Had humanity's battle continued in this exponential pattern for a few thousand years more, all women would have eventually become like Marilyn Monroe; complete sacrifices to men. At this point men would have destroyed themselves and the human species, for there would be no soundness left in women to love/nurture future generations. Olive Schreiner emphasised this point in her 1883 book *The Story of an African Farm.* When talking of men persuading women to have sex, Olive Schreiner's female character stated that men may say, '**"Go on; but when you** [men] **have made women what you wish, and her children inherit her culture, you will defeat yourself. Man will gradually become extinct..." Fools!'** (p.194 of 300).

In her 1899 essay, *The Woman Question,* Olive Schreiner talked about women, '**sinking slowly into a condition of more or less complete and passive *sex-parasitism!*...** [where] **social conditions tend to rob her of all forms of active conscious social labour, and to reduce her, like the field-bug, to the passive exercise of her sex functions alone** [p.85 of 261].' She also spoke about the '**degradation**' of the women of Rome before its fall, '**seeking madly by pursuit of pleasure and sensuality to fill the void left by the lack of honourable activity; accepting lust in the place of love, ease in the place of exertion**', and said that such '**parasite females**' would '**have given birth to a manhood as effete as itself**', and said that it was no surprise that such women and their offspring '**should in the end have been swept away before the march of those Teutonic folk, whose women were virile, and could give birth to *men*—among whom the woman received on the morning of her marriage from the man who was to be her companion through life, no miserable trinkets to hang upon her limbs, but a shield, a spear, a sword, and a yoke of oxen, while she bestowed on him in return a suit of armour, in token that they two were henceforth to be one, in toil and in the facing of danger; that she too should dare with him in war and suffer with him in peace—and of whom another writer tells us, that these women not only bore the race and fed it at their breasts without the help of others' hands, but that they undertook the whole management of house and lands, leaving the males free for war and chase** [p.92]'. Schreiner continued, saying these '**Cimbrian women who accompanied their husbands in the invasion of Italy** [around 100 BC] **were certain who marched barefooted in the midst of the** [battle] **lines, distinguished by their white hair and milk-white robes, and who were regarded as prophinspired—of whom** [the historian] **Florus, describing an early Roman victory, says "the conflict was not less fierce and obstinate with the wives of the vanquished: in their carts and wagons they formed a line of battle, and from their elevated situation, as from so many turrets, annoyed the Romans with their poles and lances." Their death was as glorious as their martial spirit. Finding that all was lost, they strangled their children, and either destroyed themselves in one scene**

of mutual slaughter or with the sashes that bound up their hair suspended themselves by the neck to the boughs of trees or the tops of their wagons [p.93]'. Schreiner went on to quote the Roman historian Tacitus: '"[the Cimbrian women] **are even hardy enough to mix with the combatants administering refreshment and exhorting them to deeds of valour"; and adds moreover that "to be contented with one wife was peculiar to the Germans; while the woman was contented with one husband as with one life, one mind, one body"** [p.93]' *(An Olive Schreiner Reader: Writings on Women and South Africa,* 1987, ed. Carol Barash).

This reference to the Cimbrian men and women's ability to live in monogamous relationships raises an important issue. It needs to be explained that the more upset, corrupted, insecure and alienated humans became the more they needed sexual distraction and reinforcement through sexual conquest (in the case of men) and sex-object attention (in the case of women), and thus the more difficult it became for them to be content in humans' original monogamous lifestyle. The saying, **'the first cut is the deepest'**, is an acknowledgment of the deep and total commitment humans make to their first love. It reveals that the original, innocent, pre-human-condition-afflicted relationship between a man and a woman was a monogamous relationship.

The Cimbrians' exceptional vitality, valour and ability to be content in a monogamous relationship was a reflection of the relative innocence of their race. The Cimbrians who invaded Italy migrated all the way from Denmark (or Jutland as it was known in those times), which was an exceptionally sheltered corner of the inhabited world. Innocence does not survive long in New York's Times Square. As mentioned earlier, and as will be explained more fully in the next essay, prophets were all shepherds in their youth. The lifestyle of a shepherd is probably the most removed from encounters with the human condition-afflicted world. It was the sheltered shepherd's life that preserved the innocence necessary for someone to be able to avoid resignation and be a prophet.

The different levels of innocence within races of humans will be elaborated upon further in the concluding section of this essay, 'The denial-free history of the human race'.

It should be pointed out that women taking up the opportunity of leadership when men crumpled is fundamentally different to the situation where the Cimbrian women actively supported their men in war. The former situation occurred when men became exhausted, while the latter was a product of the loving, psychologically strong and less sex-object-obsessed state of a more innocent people.

It should also be clarified that it was not **'social conditions** [that] **tend to rob her** [women] **of all forms of active conscious social labour, and to reduce her, like the field-bug, to the passive exercise of her sex functions alone'**, as Schreiner suggested. Rather it was the level of self-corruption and alienation that had developed in Roman society and the degree of innocence that had been lost, that caused the status of women to be reduced to mere sex-objects.

Blaming social conditions or cultural conditioning, as Schreiner did, has been the main excuse used by the world of denial for any differences between men and women. In particular it has been used to explain why women have not contributed as much to the world of literature, music and art as men have. Not being 'mainframed' to the battle going on in the world has made it very difficult for women to create profound works of literature, music and art. Even the bravest of women writers could not empathise with what has been happening in the world. For example, Olive Schreiner, in her immensely courageous book, *The Story of an African Farm,* had her leading character describe men as **'fools'** (p.194 of 300), and even say that men's oppression of women is not part of God's plan, that **'He** [God] **knows nothing of'** it (p.189).

Sex killed innocence; although, paradoxically, it was also one of the greatest distractions and releases of frustration, and on a higher level, an expression of sympathy, compassion and support—an act of faith in, and love for men. In *The Seed and the Sower,* Sir Laurens van der Post offers this sensitive attempt by a man to explain to a woman the greater significance of sex. It is a conversation that takes place on the eve of a World War II battle: **'Touched by her concern for her honour, in his imagination he would have liked to tell her that he could kneel down before her as a sign of how he respected her and beg her forgiveness for what men had taken so blindly and wilfully from women all the thousand and one years now vanishing so swiftly behind them. But all he hastened to say was: "I would have to be a poet and not a soldier to tell you all that I think and feel about you. I can only say that you are all I imagined a good woman to be. You make me feel inadequate and very humble. Please know that I understand you have turned to me not for yourself, not for me, but on behalf of life. When all reason and the world together seem to proclaim the end of life as we have known it, I know you are asking me to renew with you our pact of faith with life in the only way possible to us"'**

(1963, p.238 of 246).

The prophet Friedrich Nietzsche gave an honest description of the roles that developed for men and women in humanity's heroic journey to overthrow ignorance when he famously wrote: **'Man should be trained for war and woman for the recreation of the warrior: all else is folly'** (*Thus Spoke Zarathustra: A Book for Everyone and No One*, 1892; tr R.J. Hollingdale, 1961, p.91 of 342). The femme fatale of the 1960s, Brigitte Bardot, once similarly said, **'A women must be a refuge for the warrior. Her job is to make life agreeable'** (*The Australian*, 11 Oct. 1999). Shortly after making the above statement Bardot encapsulated the paradox of life for women when she declared that all men are **'beasts'** (ibid).

It is now possible to explain more fully how the image of innocence came to be cultivated in women.

The significance of nurturing in the human journey was described in the essay *Deciphering Plato's Cave Allegory*, in the section 'How we acquired our soul'. It was explained that through the ability to leave their infants in infancy for a prolonged period of time in which they could be indoctrinated with love or unconditional selflessness, primates were able to overcome genetic refinement's limitation to develop unconditional selflessness. Our primate ancestors were able to complement this nurturing, love-indoctrination process by consciously seeking out mates who were more love-indoctrinated and thus more cooperatively and selflessly behaved. Essentially these were members of the group who had had a long infancy and were closer to their memory of infancy (that is, younger). The older individuals became the more their infancy training in love wore off. Our ape ancestors began to recognise that the younger an individual, the more integrative he or she was likely to be. They began to idolise, foster and select youthfulness because of its association with cooperativeness or integrativeness. The effect, over many thousands of generations, was to retard physical development so that our human ancestors and their descendants became infant-like adults. This explains how we came to regard neotenous (infant-like) features—large eyes, dome forehead and snub nose—as beautiful. Our attraction to 'cute', neotenous features is so strong that animals that exhibit them, such as seal pups, giant pandas and tree frogs, become favourites amongst humans. The effect of neoteny upon our physical evolution was that we lost most of our body hair and became infant-like in our appearance; we selected for what we now recognise as innocence. The full explanation of how neotenous features came to be considered beautiful is presented in *Beyond* in the chapter 'How We Acquired Our Conscience'. The chapter contains photographs that illustrate how extraordinarily neotenous-looking humans are today compared with our ape ancestors.

As mentioned earlier, with the emergence of the human condition humans became resentful of innocence and instead of cultivating it, sought to oppress or destroy it. The result was that the attraction to innocence as a sought after quality with which to mate became perverted. Neotenous, cute, childlike features of a domed forehead, snub nose and large eyes became attractive for what we now refer to as 'sex'. What this means is that throughout the battle to find understanding women were forced to suffer the destruction of their soul, their innocence, while at the same time their trappings of innocence were sought after. While women's soul was being destroyed their image of innocence was being cultivated.

When ignorant innocence became a threat, men sought neotenous features, the signs of innocence, for sexual destruction. We evasively describe such looks as 'attractive' to avoid saying that what was being attracted was destruction, through sex, of women's innocence. It can be seen that since all other forms of innocence were being destroyed, this *image* of innocence—'the beauty of woman'—was the only form of innocence to be actively cultivated during humanity's adolescence. The beauty of women became men's *only* equivalent for, and measure of, the beauty of their lost pure world.

The following quotes reveal just how inspiring women's image of innocence became for men: **'we lose our soul, of which woman is the immemorial image'** (Laurens van der Post, *The Heart of the Hunter*, 1961, p.134 of 233); **'I believe hers to have been the kind of beauty in which the future of a whole continent sings, exhorting its children to renounce what is out of accord with the grand design of life'** (ibid. p.86); **'Woman stands before him** [man] **as the lure and symbol of the world'** (Pierre Teilhard de Chardin, *Let Me Explain*, 1966; trs René Hague & others, 1970, p.67 of 189); **'Women are all we** [men] **know of paradise on earth'** (Albert Camus); **'You give me a reason to live'** (Joe Cocker's 1986 song *You Can Leave Your Hat On*); **'I, I who have nothing/I, I who have no one/Adore you and want you so'** (Jerry Leiber & Mike Stoller's 1963 song *I Who Have Nothing*); **'Sex is life'** (graffiti on a granite boulder at Meekatharra in Western Australia). In the 1996 film, *Beautiful Girls*, when criticised for plastering pictures of supermodels all over the walls of his room, one of the male characters responds: **'Look the supermodels are beautiful girls. A beautiful girl can make you dizzy, like you've been drinking bourbon and coke all**

morning, she can make you feel high for the single greatest commodity known to man—promise. Promise of a better day, promise of a greater hope, promise of a new tomorrow. This particular awe can be found in the gait of a beautiful girl, in her smile and in her soul; in the way she makes every rotten thing about life seem like it's going to be okay. The supermodels, that's all they are, bottled promise, scenes from a brand new day, hope dancing in stiletto heels.'

Nietzsche recognised the role women played in inspiring the world with their illusion of innocence when he wrote: **'her great art is the lie, her supreme concern is appearance and beauty. Let us confess it, we men: it is precisely *this* art and *this* instinct in woman which we love and honour'** (*Beyond Good and Evil,* 1886; tr. R.J. Hollingdale, 1972, p.145 of 237).

It is little wonder men fell in love with women. The 'mystery of women' was that it was only the physical *image* or object of innocence that men were falling in love with. The illusion was that women were psychologically as well as physically innocent. For their part, women were able to fall in love with the dream of their own 'perfection' that men projected—of their being truly innocent. Men and women *fell* in love; we abandoned the reality in favour of the dream. It was the one time in our life when we could romance—when we could be transported to how it once was and how it could be again—to heaven. The lyrics of the song *Somewhere,* written by Stephen Sondheim for the 1956 blockbuster musical and film *West Side Story,* perfectly describe the dream of the heavenly state of true togetherness that humans allow themselves to be transported to when they fall in love: **'Somewhere/We'll find a new way of living/We'll find a way of forgiving/ Somewhere//There's a place for us/A time and place for us/Hold my hand and we're halfway there/Hold my hand and I'll take you there/Somehow/Some day/Somewhere!'** (Humans' capacity to fall in love and further examples of how falling in love involves dreaming of the heavenly, truly-together state were given in the *Resignation* essay in the section, 'The loneliness of humans' alienated state'.)

Different cultures have different perceptions of female beauty. Essentially, men are 'attracted' by innocent looks, which are youthful neotenous features. The popular saying, 'blondes have more fun', illustrates the tendency in Caucasian cultures to regard blondes as more attractive because many young Caucasians have blonde hair, a sign of youth/innocence. In his 1940 detective novel, *Farewell, My Lovely,* Raymond Chandler acknowledged the appeal of blondes when he wrote, **'It was a blonde. A blonde to make a bishop kick a hole in a**

stained glass window' (ch.13). Long, healthy hair is also associated with youth, which is why men find long hair on women attractive. In general, any feature unique to women will be attractive and signal a sex object to men, hence the desirability of breasts, shapely hips and narrow waist. The different cultural definitions of beauty can be explained in terms of what signifies innocence. In times when few could eat or live well, fat women were considered beautiful because their appearance indicated that they had been well cared for, better nurtured and were thus more innocent. Today, the attraction of a long, ultra-thin female shape can be explained by the increase in alienation amongst humans. Innocence now is very brief and in fact what is attractive now is that pubescent age when young girls first start to develop physically and have the slender, long-legged frame of young animals, such as foals. For women to be perceived as attractive they have to endeavour to look like a pubescent teenager. This ultra-thin body shape is unnatural for adult, child-bearing women and to achieve it necessitates a starvation diet, such is the level of perversion that has developed in the world.

The destruction of women's souls and the cultivation of their image of beauty has been occurring for 2 million years. Lust and the hope of falling in love have assumed such importance that many people, including psychoanalyst Sigmund Freud, were deceived into believing sex ruled our lives. Men and women became highly adapted to their roles. While men's magazines are full of competitive battleground sport and business, women's magazines are dedicated to enhancing beauty, to becoming more 'attractive'.

One of the best examples of being misled into the belief that sex rules our lives is found in the story of the Garden of Eden, where Eve is blamed for tempting Adam to take the apple from the tree of knowledge. The truth is that women were the victims not the cause of upset in men, but lust became such a strong force, people have been misled into believing it seduced humans into behaving in an upset way.

The object of innocence became so 'attractive' that it eventually had to be concealed. While clothes became necessary to keep neotenised, hairless humans warm, they also became necessary to dampen lust. Even the relatively innocent Bushmen people of southern Africa who go about almost naked most of the time wear aprons to conceal their genitals. Once humans became extremely upset even the glimpse of a woman's face or ankle became dangerously exciting to men, which is why in some societies women are completely draped and persecuted if they reveal any part of their body. In Genesis this need for concealment to dampen lust is thus described: **'Then the eyes of both of them were opened, and they realised that they were naked; so they sewed fig leaves together and made coverings for themselves'** (3:7).

The convention of marriage was invented as one means of containing the spread of soul-corruption through sex. By confining sex to one life-long relationship, the souls of the couple could gradually make contact and coexist in spite of the sexual destruction involved in their relationship. As it says in the *Bible:* **'a man will leave his father and mother and be united to his wife, and the two will become one flesh. So they are no longer two, but one'** (Mark 10:7,8).

Brief and multiple relationships spread soul repression. Celibacy has been one way of avoiding the hurt sex caused our soul; as it says in the *Bible:* **'others have renounced marriage because of the kingdom of heaven'** (Matt. 19:12).

Women across every generation have had a very brief life in innocence before being soul-destroyed through sex. They then had to try to nurture a new generation, all the time trying to conceal the destruction that was all around and within them. Mothers tried to hide their alienation from their children, but the fact is if a mother knew about reality/upset, her children would invariably know about it also and psychologically adapt to it. Alienation is invisible to those alienated, but to the innocent—and children are born innocent—it is clearly visible. For example, Christ's mother Mary must have been innocent because we know Christ was. Since women become upset through sex, Mary must have had little exposure to sex. The symbol for women's innocence/purity is virginity hence the description of Christ's mother as the Virgin Mary. (The Virgin Mary description will be explained more fully in the next essay.)

Women have had to inspire love when they were no longer innocent, 'keep the ship afloat' when men crumpled, all the while attempting to nurture a new generation while oppressed by men who could not explain why they were dominating, or why they were so upset and angry! This was an altogether impossible task, yet women have done it for 2 million years. It was because of women's phenomenally courageous support that men, when civilised, were chivalrous and deferential towards them. Men had an impossible fight on their hands, but at least they had the advantage of appreciating the battle.

To be a victim of a victim, as women have been, is an untenable situation, because while a primary victim knows what the primary source offence is, a victim of a victim does not. This is why, when men became overly upset they became mean, even brutal, but when women became overly upset they became nasty, even venomous. Not knowing what it is they are flailing at, women's fury is unsourced, untargeted and unbounded. The proverb, **'hell knows no fury like a woman scorned'** recognises this potential extremity in the nature of women. Nietzsche said, **'Let man fear woman when she hates: for man is at the bottom of his soul only wicked, but woman is base'** *(Thus Spoke Zarathustra: A Book for Everyone and No One,* 1892; tr. R.J. Hollingdale, 1961, p.92 of 342). Women have historically had to carry so much unsourced frustration and hurt that their situation is very fragile. The hormonal upheaval accompanying menstruation is enough to destabilise this delicate balance, hence pre-menstrual tension (PMT).

In the following quote, Nietzsche acknowledged women's greater ability to nurture; that women have had to work through men; that women haven't been aware of or mainframed to the deeper battle that has been going on in the world; and the insecurity of women's situation: **'Woman understands children better than a man...The man's happiness is: I will. The woman's happiness is: He will. "Behold, now the world has become perfect!"—thus thinks every woman when she obeys with all her love. And woman has to obey and find a depth for her surface. Woman's nature is surface, a changeable, stormy film upon shallow waters. But a man's nature is deep, its torrent roars in subterranean caves: woman senses its power but does not comprehend it'** (ibid. p.92).

With men defying and repressing their own souls, women became representative of soul in their partnership with men. Further, because of men's unexplained oppression of women and the world of the soul, and their own inability to understand this oppression; women put increasing trust and reliance in their soul and instincts, rather than in their ability to understand. As a consequence, they are more intuitive or dependent on their soul's guidance than men. A common saying is **'women feel while men think'**.

Sir Laurens van der Post has given this description of the different roles of men and women. **'The sword was, he would suggest, one of the earliest images accessible to us of the light in man; his inborn weapon for conquering ignorance and darkness without. This, for him, was the meaning of the angel mounted with a flaming sword over the entrance to the Garden of an enchanted childhood to which there could be no return. He hoped he had said enough to give us some idea of what the image of the**

sword meant to him? But it was infinitely more than he could possibly say about the doll. The doll needed a woman not a man to speak for it, not because the image of the sword was superior to the image of the doll. It was, he believed, as old and went as deep into life. But it was singularly in women's keeping, entrusted to their own especial care, and unfortunately between a woman's and man's awareness there seemed to have been always a tremendous gulf. Hitherto woman's awareness of her especial values had not been encouraged by the world. Life had been lived predominantly on the male values. To revert to his basic image it had been dominated by the awareness of the sword. The other, the doll, had had to submit and to protect its own special values by blind instinct and intuition' (*The Seed and the Sower,* 1963, p.193 of 246).

Note again that while men and women are different, sexist notions of men being 'evil' or of women being irrelevant have no credibility. While the main device for avoiding prejudice was to deny that there was any difference between men and women, another device was to maintain that any difference between men and women was simply a product of cultural conditioning—of girls being given dolls and boys swords as infants for example. In fact, as Sir Laurens van der Post acknowledges, our differences are the product of a very real distinction between the sexes.

While at a more noble level sex has become an expression of love, at its fundamental level it is an attack on the innocence of people; it is rape. The more upset and corrupted the human, the more sexually destructive and thus sexually perverted they are inclined to be, and the more innocent (or innocent-looking) the human, the more attracting of that destruction they have been. Understanding this makes it possible to explain homosexuality in men. As the victims of sex, women have historically been more exposed to and thus become, through natural selection over hundreds of thousands of years, more adapted to sex than men. In most cases, if a male was not interested in sex then sex did not occur, whereas women have been exposed to sexual advances regardless of their interest or lack thereof. Teiresias, the prophet mentioned in Homer's Greek legend, *The Odyssey,* recognised that women were more sexually aware than men. When asked, **'whether the male or the female has most pleasure in intercourse'**, he replied that **'Of ten parts a man enjoys one only; but a woman's sense enjoys all ten in full'** (*The Melampodia* by 8th century BC Greek poet Hesiod; from *Hesiod, The Homeric Hymns and Homerica,* 1914, tr. H.G. Evelyn-White, p.269 of 657). While women have had to hide their sexual awareness in order to present an attractive image of innocence, the fact is they are more sexually aware than men. In the *Happy Days* television series (set in the 1950s and first broadcast in 1974) girls are much more attracted to the sexually aware Fonzie character than to the naive, relatively innocent Richie Cunningham character. While only fictional characters, the viewing audience would not have responded with such empathy over the years if the characters did not resonate with truth. In Joan Mellen's 1977 book, *Big Bad Wolves: Masculinity in the American Film,* the caption accompanying a picture of the 1920s sex symbol actor Rudolph Valentino reads: **'Rudolph Valentino in the film "Son of the Sheik". Rape is the central visual metaphor.'** As explained, 'the mystery of women' is that after 2 million years of having been sexually used by men, women now only represent the physical image or object of innocence. It is this image of innocence that men have been falling in love with; the illusion that women were psychologically as well as physically innocent. A well-known African fable tells of a woman who agrees to marry a man on the condition he never looks inside a precious basket that she keeps. She warns him that if he does she will vanish. He agrees, but some time after they are married and his wife has gone to the river for water he cannot resist and peers in the basket. On his wife's return she finds him laughing and the basket open. When she accuses him of looking in the basket he says, 'you silly woman, there was nothing in the basket', at which point she vanishes into thin air. The basket is symbolic of the mystery of women—they are only the image of innocence; it is an 'empty basket' that men are looking at, and once men see through the illusion, women's attractiveness diminishes. It follows that the more corrupted a man is, the less naive he is and thus the more he is aware that women are not innocent. Therefore, if a man is extremely hurt and corrupted in his infancy and childhood, when he becomes sexually mature he will not be naive enough to believe that women are still innocent and he will thus not find women sexually attractive. The last bastion of 'attractive' innocence for such men is younger men, because they are not as exposed to sexual destruction as women have historically been. To explain the feminine mannerisms particular to male homosexuality, if you have had your soul, which is your core strength, destroyed in childhood, then taking on the extremely difficult male role of having to fight against the ignorance of the soulful, idealistic world becomes so untenable that the female position of not having to fight is preferential. You would rather adopt the female role of being an object of adoration and service than the male role of having to take on the loathsome job of championing ego over soul. The professional tennis player of the late 1970s and transsexual Renée Richards—who went so far as to have a sex change operation to become a woman—alluded to the difficulty of life as a male, and by inference the appeal of being a woman, when she said, **'women don't realise the horror of the world that men live in—the strife torn world that men live in'**. Having to live with the condemnation that you are an evil monster, when you know you are not, but cannot explain why you are not, has been a living hell for men. To be a man and have to oppress the all-magic soul without being able to explain why has been the most wretched of tasks. The following quote serves to illustrate how pressured men's lives have become: **'If women are so oppressed, how come they live much longer than men?'** (Don Peterson reviewing *The Myth of Male Power* by Dr Warren Farrell, *Courier Mail*, June 1994, p.9). Homosexuality amongst women results from women's understandable disenchantment with men. Homosexuality is simply another level of perversion to heterosexuality. They are both corrupted states of sexuality that developed under the horror of the duress of the human condition.

With regard to whether homosexuals are 'born' or 'made', even without the ability to explain the human condition and thus defend the corrupted state of humans (ie explain that humans' various states of corruption are not 'bad' or evil but are in fact immensely heroic states), a decade-long research project completed at the Institute for Sex Research in Bloomington, Indiana, found that, **'a quarter of the gays interviewed believe** [are prepared to acknowledge?] **homosexuality is an emotional disorder'** (*Time* mag. 17 July 1978). In his 1992 book, *Health & Survival in the 21st Century*, Ross Horne referred to studies that show, **'That the highest incidence of homosexuality coincides with the general level of stressful influences in a community and that the lowest incidence coincides with the degree of happiness and health in remote and unstressed populations indicates that, like many conditions of physical disease, it is just as unnatural as the mental breakdowns, depression and neuroses so common in civilization. Studies of primitive natives reveal that while in some populations homosexuality is non-existent or rare, in other populations it is fairly common; but the same pattern still holds—among the placid, happy, untroubled people homosexuality did not occur, while among fighting tribes and headhunters it did'** (p.206). After 25 years of clinical experience helping homosexual men and women, Dr Robert Kronemeyer of New York concluded that, **'Homosexuality is a symptom of neurosis**

and of a grievous personality disorder. It is an outgrowth of deeply rooted emotional deprivations and disturbances that had their origins in infancy' (*Overcoming Homosexuality,* 1980).

While women are instinctively more sexually aware then men this does not exclude the fact that a woman can be more innocent and less sexually aware than a man. Girls who are nurtured and sheltered in their upbringing can be very innocent. However, because there has been no honesty about the existence of the different levels of upset and alienation amongst humans, they can be deceived by men who are much less innocent and therefore much more sexually advanced down 'the rungs of the perversion ladder' (1 is holding hands, 2 is kissing, 3 is touching her breast, etc, etc, etc). Men who are more upset can be very attentive to women because sex for them is a distraction and a way of gaining reinforcement, and innocent women can be deceived—seduced—by this attention into a relationship. In her 1981 book, *African Saga,* the African photographer, and remarkably beautiful woman, Mirella Ricciardi gives an extraordinarily honest account of a relationship between a more innocent woman and a less innocent man. She wrote: **'We went to live in Rome, where I quickly began to taste the bitter-sweet agony of life with Lorenzo. I was young, unaware of the world, and ignorant of people and their behaviour. I married Lorenzo as easily as I had switched lovers. It was probably the most foolish, irresponsible, exciting thing I have ever done. Years later, I came to the conclusion that most of the men I had met fell into three categories—those prompted by their heads** [presumably, upset men], **those by their heart** [presumably, less upset, more soulful, relatively innocent men] **and those by their sex** [presumably, extremely upset men]. **Some—not many—were a combination of all three. Lorenzo belonged to the last category—these I have found are the most attractive. They are sexy, amusing, fun-loving, careless, irresponsible and lazy—they dress well and have a lot of style. Most people like them. They are excellent lovers and lousy husbands. Women usually find them irresistible or are terrified by them. Men either envy or despise them. No one can remain indifferent to them..."Lorenzo's mother died when he was seven," Cesarino** [Lorenzo's father] **told me one day. "You will have to be more of a mother than a wife to him; do you realise this?" Then he laughed. "The only pleasure he ever gave me was nine months before he was born." But when his father died sixteen years later, Lorenzo's grief was immeasurable and I began then to understand the meaning of these words'** (p.136 of 300). Because women have lived through men, and because their means of healing the world is nurturing, relatively innocent women in relationships with more upset men often tried to change the upset man, make him sounder and stronger through nurturing love, inspiration and motivation. Since it is only understanding of hurt that can heal hurt, these efforts only serve to further confront and criticise the man. Mirella's dedication in her book **'to Lorenzo, my magnificent obsession'** is an acknowledgment of her frustrated efforts to change Lorenzo. She wrote: **'When I married Lorenzo I had created an image of a giant in whose shadow I would live. I clung stoically to my belief in our union and waited patiently for ten years for him to cast his shadow, but he never did'** (ibid. p.138). In the 1955 film, *Guys and Dolls,* the upset gangster, played by Marlon Brando, and the Salvation Army innocent, played by Jean Simmons, marry, she determined to reform him, but without success. He eventually rebels against her efforts to, as he said, **'try and change me'**. In a world of lies, the basis for relationships has often been unhealthy. Women have been seduced by men in so many ways and their innocence has been the casualty—another reason why women have historically become more sexually aware than men. Growing up in the countryside in Australia in the 1960s I saw many sheltered, relatively innocent country girls go off to Europe for a few years in their early '20s—it was considered the thing to do in those days, as it is now—only to return with 'knowing eyes', a different more sophisticated way of looking at the world. I could never understand the point of sheltering and nurturing young women if they were simply going to go off to Europe and, as I saw it, 'cash their innocence in'.

Since sex is an attack on innocence as well as an act of love the recent generations of humans who have been treating sex cheaply, have been contributing significantly to the death of soul in the world, and thus contributing significantly to the level of alienation in the world. Queen Victoria was right to encourage people to treat sex with care and restraint.

During the 2 million years that women have endured the wretched situation of being unable to understand men's oppression of them, many must have found it an impossible position to accept and as a result there must have been a great deal of natural selection and thus genetic adaption to the role that women have had to play in the human journey to enlightenment. Olive Schreiner described women's resignation to their role in the following passage from her book *The Story of an African Farm.* It is a dialogue between her young female character and that character's male friend, Waldo: **'"I know it is fool-**

ish. Wisdom never kicks at the iron walls it can't bring down," she said. "But we are cursed, Waldo, born cursed from the time our mothers bring us into the world till the shrouds are put on us. Do not look at me as though I were talking nonsense. Everything has two sides—the outside that is ridiculous, and the inside which is solemn." "I am not laughing," said the boy sedately enough; "but what curses you?" He thought she would not reply to him, she waited so long. "It is not what is done to us, but what is made of us," she said at last, "that wrongs us. No man can be really injured but by what modifies himself. We all enter the world as little plastic beings, with so much natural force, perhaps, but for the rest—blank; and the world tells us what we are to be, and shapes us by the ends it sets before us. To you it says—*Work!* and to us it says—*Seem!* To you it says—As you approximate to man's highest ideal of God, as your arm is strong and your knowledge great, and the power to labour is with you, so you shall gain all that the human heart desires. To us it says—Strength shall not help you, nor knowledge, nor labour. You shall gain what men gain, but by other means. And so the world makes men and women. Look at this little chin of mine, Waldo, with the dimple in it. It is but a small part of my person; but though I had a knowledge of all things under the sun, and the wisdom to use it, and the deep loving heart of an angel, it would not stead me through life like this little chin. I can win money with it, I can win love; I can win power with it, I can win fame. What would knowledge help me? The less a woman has in her head the lighter she is for climbing. I once heard an old man say, that he never saw an intellect help a woman so much as a pretty ankle; and it was the truth. They begin to shape us to the cursed end," she said, with her lips drawn in to look as though they smiled, "when we are tiny things in shoes and socks. We sit with our little feet drawn up under us in the window, and look out at the boys in their happy play. We want to go. Then a loving hand is laid upon us: 'Little one, you cannot go,' they say; 'your face will burn, and your nice white dress be spoiled.' We feel it must be for our good, it is so lovingly said; but we cannot understand; and we kneel still with one little cheek wistfully pressed against the pane. Afterwards we go and thread blue beads, and make a string for our neck; and we go and stand before the glass. We see the complexion we were not to spoil, and the white frock, and we look into our own great eyes. Then the curse begins to act on us. It finishes its work when we are grown women, who no more look out wistfully at a more healthy life; we are contented. We fit our sphere as a chinese woman's foot fits her shoe, exactly, as though God had made both—and yet He knows nothing of either. In some of us the shaping to our end has been quite completed. The parts we are not to use have been quite atro-

phied, and have even dropped off; but in others, and we are not less to be pitied, they have been weakened and left. We wear bandages, but our limbs have not grown to them; we know that we are compressed, and chafe against them. But what does it help? A little bitterness, a little longing when we are young, a little futile searching for work, a little passionate striving for room for the exercise of our powers,—and then we go with the drove. A woman must march with her regiment. In the end she must be trodden down or go with it; and if she is wise she goes"' (pp.188–189 of 300).

While the feminist movement has improved a woman's lot superficially, there has been no fundamental change to the situation that Schreiner so honestly described, as these quotes confirm. **'Nirvana hasn't happened. Although men are speaking about understanding** [the need for women's liberation from men's oppression] **on the surface, they're not doing anything about it'** (Carmel Dwyer, *Sydney Morning Herald,* 22 Sept. 1993). **'What happened was that the so-called Battle of the Sexes became a contest in which only one side turned up. Men listened, in many cases sympathetically but, by the millions, were turned off'** (Don Peterson reviewing *The Myth of Male Power* by Dr Warren Farrell, *Courier Mail,* June 1994, p.9). Until men could explain to women why they have had to be so egocentric, competitive and aggressive there could be no fundamental change to the situation where men found they had no choice other than to oppress women.

The roles that men and women have played in humanity's journey to find understanding of the human condition have been equally tragic. The extent of the tragedy becomes abundantly clear when observing the lives of older men and women.

In the next section, 'The denial-free history of the human race', it will be described how men who had resigned and adopted a dishonest attitude of denial inevitably progressed to ever greater levels of self-corruption, alienation and despair as they grew older. From bold, adventurous 20-year-olds they progressed to angry, embattled 30-year-olds, to overly embattled 40-year-olds, to burnt-out and vengeful 50-year-olds. Beyond 50, men became so overwhelmed by the horror of their circumstances that they lost even this desire to be vengeful. When young men resigned at 15 they took up a dishonest way of living and over the years that fundamental dishonesty could only lead to greater and greater self-corruption and disappointment. It will be explained in the next section that while a 20-year-old man might have succeeded in psychologically blocking out the emptiness that would inevitably come with living a life of lies and delusion, by

60 the sadness of such a life was undeniable. T.S. Eliot's 1925 poem *The Hollow Men,* describes the bleak state of nothingness that resigned men arrived at in their old age: **'We are the hollow men/We are the stuffed men/Leaning together/Headpiece filled with straw. Alas!/Our dried voices, when/We whisper together/Are quiet and meaningless/As wind in dry grass/Or rats' feet over broken glass/In our dry cellar//Shape without form, shade without colour/Paralysed force, gesture without motion//...This is the dead land/This is cactus land/Here the stone images/Are raised, here they receive/The supplication of a dead man's hand/Under the twinkle of a fading star//Is it like this/In death's other kingdom/Waking alone/At the hour when we are/Trembling with tenderness/Lips that would kiss/Form prayers to broken stone//The eyes are not here/There are no eyes here/In this valley of dying stars/In this hollow valley/This broken jaw of our lost kingdoms//In this last of meeting places/We grope together/And avoid speech/Gathered on this beach of the tumid river//...Between the desire/And the spasm/Between the potency/And the existence/Between the essence/And the descent/Falls the Shadow/...This is the way the world ends/Not with a bang but a whimper'** (*T.S. Eliot Selected Poems,* 1954, pp.77–80 of 127).

For women, ageing meant the loss of the image of innocence they depended on for reinforcement; the loss of their sex-object 'attractiveness', and with it, the loss of their meaning in the world. When women are young their beauty is so empowering it is as if they own the world, but when they become older and their beauty fades they discover that they become invisible; when they walk in the streets they are no longer noticed. While men become 'hollow', women become 'invisible', and older couples walk together in the park united by their comparable afflictions. This quote from the French beauty therapist Diane Delaheve describes how devastating it can be for women to lose their sex appeal: **'Her eyes, the mirror of her soul, speak nothing but despair. Her face may have kept its beauty, but it has become a picture of affliction. For some women, the prospect of age is sheer tragedy, worse than death, which might be seen as an escape'** (*Sydney Morning Herald,* 4 Sept. 1988).

In his 1961 novel *Zorba The Greek* Nikos Kazantzakis gave this stark account of how difficult women have found losing their sex appeal: **'"But what do you mean, Zorba?" I replied. "Do you seriously think all women have nothing else but that** [sexual attention] **in mind?" "Yes, boss, they've nothing else in mind. Listen to me, now—I've seen all sorts, and I've done all kinds of things—A woman has nothing else in view. She's a**

sickly creature, I tell you, and fretful. If you don't tell her you love and want her, she starts crying. Maybe she doesn't want you at all, maybe you disgust her, maybe she says no. That's another story. But all men who see her must desire her. That's what she wants, the poor creature, so you might try and please her! I had a grandmother, she must have been eighty. What a tale that old soul's life would make! Never mind, that's another story, too—Well, she must have been eighty in the shade, and opposite our house lived a young girl as fresh as a flower—Krystallo she was called. Every Saturday evening, raw young bloods of the village would meet for a drink, and the wine made us lively. We stuck a sprig of basil behind our ears, one of my cousins took his guitar, and we went serenading. What love! What passion! We bellowed like bulls! We all wanted her, and every Saturday we went in a herd for her to make her choice...So every Saturday the old girl pulled her mattress up to the window, took out her little mirror and combed away at the little bits of thatch she had left, and carefully made a parting. She'd look round slyly, for fear someone saw her. If anyone came near, she'd snuggle back and look as if butter wouldn't melt in her mouth, pretending she was dozing. But how could she sleep? She was waiting for the serenade. At eighty! You see what a mystery woman is, boss! Just now it makes me want to cry. But at that time I was just harum-scarum, I didn't understand and it made me laugh. One day I got annoyed with her. She was hauling me over the coals because I was running after the girls, so I told her straight out where to get off: 'Why do you rub walnut leaves over your lips every Saturday, and part your hair? I s'pose you think we come to serenade *you?* It's Krystallo we're after. You're just a stinking old corpse!' Would you believe it, boss! That day was the first time I knew what a woman was. Two tears sprang into my grandma's eyes. She curled up like a dog, and her chin trembled. 'Krystallo!' I shouted, going nearer so as she'd hear better, 'Krystallo!' Young people are cruel beasts, they're inhuman, they don't understand. My grandma raised her skinny arms to heaven. 'Curse you from the bottom of my heart!' she cried. That very day she started to go into a decline. She wasted away and two months later, her days were numbered"' (pp.52–53 of 315).

There has been an added dimension to the situation faced by older women. Women, not responsible for the main battle of having to champion the ego over ignorance, found that their role of living in support of the battle was limited. A common observation of a woman's life has been that she progresses from **'bimbo, breeder, babysitter to burden'**. Men, able to be involved in the battle of championing the ego, do not face the prospect of one day feeling they are a 'burden' to the extent that women do. In his 1993 book *The Fisher King & The Handless Maiden,* the American Jungian analyst and prophet, Robert A. Johnson, relates the myth of the Handless Maiden. In it, a miller makes a deal with the devil in order to complete more work with less effort. The devil demands the miller's daughter as payment. **'The miller is desolate but unwilling to give up his much expanded mill, so he gives his daughter to the devil. The devil chops off her hands and carries them away'** (p.59 of 103). Waited on by her newly prosperous family, the handless maiden is content for a time, until her growing sense of desperation sends her out to the forest alone. Johnson explains that the cry of women, like that of the handless maiden, is **'What can I *do?* I feel so useless or second-rate and inferior in this world that puts its women on the rubbish heap when they are through with courtship and childbearing!'** (p.56).

In the *Introduction* I mentioned how the artist Francis Bacon depicted the human condition as honestly as anyone has ever managed to write about it. I also said that the two pictures by Blake on the cover of this book are dramatic depictions of the story of humans' struggle with the human condition. Ralph Steadman is another artist whose drawings have always managed to wrench to the surface the truth of the full horror of the predicament of humans. His works include one particularly revealing drawing that depicts humans as reptiles. It features in Hunter S. Thompson's 1971 classic novel about the human condition, *Fear and Loathing in Las Vegas,* and is reproduced below. Humans live in such deluded denial of the extent of their alienation it is only in pictures such as this, or in exceptionally honest passages of literature, such as those just included from the writings of T.S. Eliot and Nikos Kazantzakis, that the true extent of the corruption of humans' soul is revealed. For example, the eyes of the main dragon in this drawing show the hollowness that T.S. Eliot wrote about: **'This is the dead land/This is cactus land'**. Also apparent is the terribly sad, **'sickly creature'**, **'old corpse'** state of women that Kazantzakis described so honestly. Thank heaven this tragic state of the human condition that humans have so courageously had to endure as best they could can finally end and our true self be restored!

The Lizard Lounge by Ralph Steadman, 1971

While men and women have had no option other than to live out their different roles, the truth is neither men nor women have liked what they have had to do. Having destroyed innocence men would end up wanting to rediscover it. The truth was that men were having to repress and, as the saying asserts, **'hurt the ones they loved'**. As Sir Laurens van der Post has written: **'I thought finally that of all the nostalgias that haunt the human heart the greatest of them all, for me, is an everlasting longing to bring what is youngest home to what is oldest, in us all'** *(The Lost World of the Kalahari,* 1958, p.151 of 253).

While men have yearned for freedom from their oppressor, ignorance, women have similarly yearned for freedom from their oppressors, men, as this quote from Olive Schreiner makes clear: **'if I might but be one of those born in the future; then, perhaps, to be born a woman will not be to be born branded...It is for love's sake yet more than for any other that we** [women] **look for that new time...Then when that time comes...when love is no more bought or sold, when it is not a means of making bread, when each woman's life is filled with earnest, independent labour, then love will come to her, a strange sudden sweetness breaking in**

upon her earnest work; not sought for, but found' *(The Story of an African Farm,* 1883, pp.188,195 of 300).

The desire for an end to the so-called 'war between the sexes' could not have been greater, but the fundamental requirement to bring peace to any human situation is the removal of the criticisms that cause the upset. The containment of upset through criticism, restraining attitudes, cultural convention, rules, laws or the threat of punishment, offers no real solution. What is causing the corrupt behaviour has to be explained and understood. Upset has to be loved not despised to be healed and permanently changed and the ultimate form of love is understanding. It follows that in order to resolve the war between the sexes and enable men and women to finally live in peaceful harmony, the source criticism involved in the battle that men were bad, even evil, for being so divisively behaved had to be removed. While in dogmatically imposing an ideal situation where men and women treat each other indifferently, as the politically correct culture attempts to do, upset could be contained and disguised, but it could not be removed. *Only* understanding the world of men, and why they have been so divisively behaved, could subside the anger, alienation and egocentricity that caused them to victimise virtually everything they encountered. The boot that was screwing men into the dirt had to be lifted for the horrible war between men and women to end. And an all-out war it has been, lived to the full extent of what was possible under the limitation that men and women were forced to coexist if they were to reproduce and nurture a new generation.

In finding the liberating understanding of why men have been divisively behaved it is revealed that men are the heroes of the whole human journey. They were carrying out the crucial work of defeating ignorance and this task was made almost unbearably difficult because they were having to apply themselves in a world that had no appreciation and thus sympathy for what they were doing. From being thought of as the villains on Earth one day, they become the absolute heroes the next. This turn of events is long overdue. Everywhere men have become wretchedly oppressed by the politically correct dogma that denies them any real meaning in the world. So intimidated have men become by pseudo-idealism that many have come to believe they are useless. In his 2001 book, *Men: From Stone Age to Clone Age,* Bob Beale said that while it pained him, he was prepared to accept that except for **'the baby business'**, **'males are largely**

useless'. The proliferation of men's movements that aim to counter men's horrific situation of being totally misunderstood is but one sign of the problems men faced. Boys growing up in the current, men-are-worthless world are having their self-esteem destroyed before they even have a chance to enter manhood. A very real concern for the future was that there would be a dearth of men psychologically strong enough to fight ignorance. If the human condition had not been solved the human race would be facing very dark times. That is how dangerous the politically correct, postmodernist culture has been. Even universities, our supposed 'centres of learning', have been taken over by this utterly dishonest culture. Therefore we can expect no meaningful 'learning', no answers, no solutions from them. Universities have become spent, effectively dead places.

The reason school boys are not performing as well in their studies as girls is not only because they have had their immense relevance in the world denied, but also because they are not as duped by the historical denial of the issue of the human condition as women. Women tend to believe the world we are living in is the real world, whereas men, being 'mainframed', are intuitively aware that it is a fraudulent existence, and as a result never take it too seriously. In the television series, *The Simpsons,* the young boy, Bart, has little respect for school whereas his sister Lisa applies herself completely and excels. Nietzsche was alluding to the trusting naivety of women when he said, **'"Behold, now the world has become perfect!"—thus thinks every woman when she obeys with all her love'.**

Of course the battle that has been raging on Earth between men and women was merely a manifestation of the fundamental battle between the instinct and the intellect; the battle of the human condition. Men have represented the apparently 'bad' ego or intellect while women have represented the apparently 'good' instinct or soul. At the end of the journey to enlightenment of the human condition it is revealed that the instinct or soul was the 'baddie' and not the 'goodie' after all, because it unjustly condemned the intellect and started the war. (Of course, with understanding of the human condition the concepts of 'good' and 'bad' no longer apply to the human situation. The soul was not *actually* 'bad', nor the intellect *actually* 'good', rather they were different ways of processing information that were forced to coexist in the one organism.)

One of the most powerful myths or stories I know of is Agatha Christie's famous play *The Mousetrap*. The power of this story lies in the fact that it portrays the astonishing change in perception that brings peace to Earth. This play, first performed in 1952, is just another 'whodunnit' murder story, yet it has become the longest running play in history and is still going strong. All enduring myths and stories contain truths that resonate. At the conclusion of *The Mousetrap,* the police inspector involved in the murder investigation, regarded as the pillar of goodness and justice throughout the play, is revealed to be the culprit. This is the essential plot to the story of humanity where the apparent ideals of the soul's selfless, cooperative, loving world are revealed, at the very last moments, to have been the unjustly condemning villains. The truth was not as it appeared. We discover at the very end of our journey that conscious humans, immensely corrupt as we are, are good and not bad after all.

Nietzsche understood the warrior man, ruthlessly attacking the weaknesses of those who no longer wanted to fight ignorance, those preferring to impose truth-denying, thought-stifling, knowledge-avoiding, pseudo-idealistic, politically correct dogma on the world, declaring: **'There have always been many sickly people among those who invent fables and long for God: they have a raging hate for the enlightened man and for that youngest of virtues which is called honesty...Purer and more honest of speech is the healthy body, perfect and square-built: and it speaks of the meaning of the earth** [to face truth and one day find understanding of the human condition] [p.61 of 342]**...You are not yet free, you still *search* for freedom. Your search has fatigued you...But, by my love and hope I entreat you: do not reject the hero in your soul! Keep holy your highest hope!** [pp.70–71]**...War and courage have done more great things than charity. Not your pity but your bravery has saved the unfortunate up to now...What warrior wants to be spared? I do not spare you, I love you from the very heart, my brothers in war!** [pp.74–75]**'** (*Thus Spoke Zarathustra: A Book for Everyone and No One,* 1892; tr. R.J. Hollingdale, 1961).

Nietzsche's much debated and misunderstood concept, 'the will to power' (the title of his last work before his psychological breakdown in 1889), can now be interpreted as man's will to achieve power over humans' idealistic instinctive self or soul's ignorance of the true goodness of corrupted humans. In *Thus Spoke Zarathustra: A Book for Everyone and No One* Nietzsche recognised that, unlike other animals, humans have had to fight a psychological demon, the human condition: **'Man, however, is the most courageous animal: with his courage he has overcome every animal. With a triumphant shout he has even overcome every pain; human pain, however, is the deepest pain'** (p.177 of

342). In Jungian terms, **'wholeness'** for humans depended on them being able to **'own their own shadow'**. Humans have had to develop understanding, and through that understanding embrace the dark, corrupted, damaged, upset, divisive side of their nature. Humans had to have the courage to triumph over their deepest pain, the pain of not being able to know whether they were fundamentally evil beings or not. *Humans have had to learn to love themselves.*

Incidentally Nietzsche's subtitle, 'A Book for Everyone and No One', was an open acknowledgment that to speak the truth to people who are resigned and living in denial of the human condition was to court total rejection, the 'deaf effect'. 'No One' would hear, understand and accept his words. What Nietzsche knew however was that only the truth can liberate humans and that in time people would hear and understand his words, and that eventually his book *would be* for 'Everyone'. Prophets are not appreciated in their own time, but they do lead the way to freedom. In fact, their honesty is the purest form of love the human race has ever known.

For 2 million years women have stood by their men, just as for 8 million years prior to that, men supported their women. With understanding of the human condition now found, men and women can at last stand side by side—the 'war of the sexes' can finally be resolved. The human journey has a happy ending, as we always believed it would: **'The happy ending is our national belief'** (Mary McCarthy, *On the Contrary*, 1961).

The denial-free history of the human race

With understanding of the human condition it is possible not only to talk honestly about the different roles that men and women played in the human journey to enlightenment, but to explain the roles different races have played in the journey.

As emphasised at the beginning of this essay, without explanation of humans' corrupted state it hasn't been safe to acknowledge different levels of corruption between individual humans or between races of humans. Acknowledgment of such differences only led to false, 'racist' notions of some races being inferior or superior to others. Understanding of the human condition enables us to know that all humans are equally good despite their various states of corruption. In fact corruption is a heroic state, for it is a result of fighting the all-important battle against ignorance. Now that humans' corrupted state can be understood as a heroic rather than a 'bad' or 'evil' or 'guilty' state, the differences in corruption between individuals and races can be safely acknowledged. Such honesty is vital for the rehabilitation of the human race.

With the arrival of the understanding of the human condition, criticism of humans is rendered obsolete. Humanity moves beyond the concepts of 'good' and 'evil' to a denial-free world of honesty and with it, psychological healing. As the commonly used phrase asserts, **'honesty is therapy'**, or as Christ said, **'the truth will set you free'** (John 8:32). Honesty *is* therapy, as Sir Laurens van der Post noted: **'Truth, however terrible, carried within itself its own strange comfort for the misery it is so often compelled to inflict on behalf of life. Sooner or later it is not pretence but the truth which gives back with both hands what it has taken away with one. Indeed, unaided and alone it will pick up the fragments of the reality it has shattered and piece them together again in the shape of more immediate meaning'** (*A Story Like the Wind,* 1972, p.174 of 473).

Because of the danger of racist, sexist and ageist prejudice, human history has been taught in an almost totally dishonest, evasive and superficial way. At school we are currently taught, and required to recall for exams, such facts as 'King Oscar the 30th married Queen Isabelle the 14th in 1252 and they fought the War of the Marshes in 1270', but the real questions are not addressed. Why were there kings and queens and peasants? Why did people get married? Why did people go to war? Why did one group of people dominate another group of people, and what were the effects of such dominations over time? What was the overall meaning of these events?

It was mentioned in the previous section that the Cimbrians' exceptional vitality, valour and ability to be content in a monogamous relationship was a reflection of the relative innocence of their race. It was explained that they migrated from Denmark, an exceptionally sheltered corner of the inhabited world, and that their innocence was a product of this isolation. As was pointed out, innocence doesn't survive long in New York's Times Square.

A deeper analysis of the history of the Cimbrians and related tribes provides insight into what has actually been occurring throughout human history.

The Goths, who wandered freely and adventurously all over Europe around 400 AD, and eventually succeeded in sacking Rome, originally came from Sweden, one of the most isolated and thus sheltered corners of Europe. The energetic Angles, Saxons and Juts who settled England, (the word 'England' means 'land of the Angles'), in the 5th and 6th century came from the same Danish peninsular as did the Cimbrians some 600 years earlier. Later, around 800 AD, the extremely adventurous and fearless Vikings traversed as far as the Caspian and Black Seas, the Mediterranean, England, Ireland, Iceland and even America. They too came from the sheltered corner of Scandinavia—Norway, Sweden and Denmark, while William the Conqueror, who conquered England in 1066, was a Norseman from coastal France. These robust Norsemen originally came from Norway and Sweden; the word 'Norsemen' is derived from 'Northmen' in reference to these northern origins.

The British Isles, due to their isolation, became a shelter for relative innocence. Even before the relatively innocent Scandinavian Angles, Saxons, Vikings and Norsemen came to the British Isles, original relatively innocent stock from a race called the Celts sheltered and thus survived there, especially in Ireland, Wales and Scotland. A 2001 Scottish Television documentary (and book) about the Celtic people of the British Isles, titled *The Sea Kingdoms,* said that **'more than 40 percent of all Britons claim an Irish, Welsh, Scottish, Manx, Cornish or Cumbrian** [ie Celtic] **heritage'**.

The story of the Celts of the British Isles is very similar to the story of the Scandinavians. As will be explained shortly, the Scandinavian races and the Celts were both descendants of the Aryan race who originally came from the Caucasian Steppes, in what is now southern Russia. The original Celts were Aryans who settled in Central Europe where they were given the name Celtoids. As their culture expanded they became known as the Gauls or Gaels. They spread to Portugal, to Gaul (now called France), to Galicia in southern Poland, to Pays Les Galles (the French name for Wales), and even to Galatia in central Turkey (as referred to in the *Bible* as the Galatians). The Celtic languages were once widely spoken in Britain, France, Spain, The Alps, northern Italy, parts of Yugoslavia and central Turkey. In 390 BC the Gauls even managed to briefly capture Rome. The Celts dominated much of Europe from 650 BC to about 200 BC, when they were progressively conquered by the irresistible might of Rome's legions. While the Celts were once widespread their last stronghold was the British Isles, just as Scandinavia was the last stronghold of other early Aryans. As *The Sea Kingdoms* documentary recognised, it was the **'remoteness'** of the British Isles that preserved the Celts.

The language and culture of the Celts or Gauls arrived in Britain and Ireland in around 500 BC and within a short period of time their political structure and military methods became widely established. By the time the Romans arrived in 43 AD these horse-riding warriors and their kings controlled the whole of the British Isles. When the Romans left in 410 AD the Celts were still living relatively undisturbed in Ireland, Wales and Scotland. In fact *The Sea Kingdoms* documentary stated that **'there are still nearly three-quarters of a million people living on these islands** [off the west coast of Britain] **who speak a language which would have been understood, admittedly with some difficulty, by people here 2,500 years ago'**. After the Romans withdrew, the native Celts were invaded and defeated by the Germanic Angles and Saxons, but they and their culture survived in the western parts of the British Isles.

Like the Scandinavians, the Celts of the British Isles were renowned for their **'physical courage'**, **'a certain arrogance'**, and **'tenacity'**. However, while the Anglo-Saxons were **'cool and practical'**, the Celts were **'fiery and passionate'** (ibid). This difference in personality and attitude to life is still strongly manifest in the personality differences between the predominantly Celtic Irish and the predominantly Anglo-Saxon English.

It can be seen that the British Isles have been an extraordinary refuge for relative innocence with all its qualities of courage, energy, enthusiasm and relative lack of cynicism and selfishness—that is, relative lack of the anger, egocentricity and alienation that resulted from life under the duress of the human condition.

In the case of England, the author of a 1918 book that I found about Romania said she knew of far away England as **'The land of the free'** (*Roumania Yesterday and Today,* Mrs Will Gordon, p.55 of 270). For a country small in size and with limited natural resources, England has made an extraordinary contribution to the human journey to enlightenment. In an essay titled *The English Record,* my headmaster at Geelong Grammar School, the incomparable Australian educator and denial-free thinker or prophet Sir James Darling, wrote: **'The truth seems to be this, that there is a genius of the English character which shows itself in its institutions, in its practical inventiveness, and under stimulus in its fighting quality. There is a stubborn determination to live its own life, and to brook no interference from a foreign power, which has put England five times in history into the position of protagonist against a European power which**

threatened to dominate the world; but this same determination has been coupled with a recognition that others also have a right to self-government and that the function of Empire is to educate rather than to oppress.' Sir James even went on to say that England has **'an unbeaten record in the history of civilization'** *(The Education of a Civilized Man,* 1962, pp.134,136 of 223). An article in *Time* magazine recorded that **'A quarter of the world's population speak English...English is increasingly becoming entrenched as the language of choice for business, science and popular culture. Three-quarters of the world's mail, for example, is currently written in English'** (7 July 1997).

For their part the Irish, Welsh and Scottish have contributed a **'non-conformist', 'joyful and productive'** *(The Sea Kingdoms)* freshness to the world, qualities that have greatly complemented and assisted the endeavours of the English. There is an integrity about the Irish in particular that is admired throughout the world, to the extent that Ireland, or **'I-love-you-land'** as I saw a Chinese woman refer to it on television, has become a place of pilgrimage for people from all over the world irrespective of their racial origins. In the upcoming essay, *The Demystification Of Religion,* the section 'Australia's role in the world' emphasises the special contribution the Irish people have made to the world with their refusal to conform to the artificial, sophisticated and dishonest way of living. In terms of what was explained in the *Resignation* essay, it is almost as if the Irish as a nation refused to accept resignation, preferring to remain a somewhat mad 'ship at sea'. In comparison, the English applied their **'cool and practical'** temperament to humanity's fundamental task of accumulating knowledge using the relatively safe guise of eccentricity and ridiculous degrees of sophistication to valve-off the resultant corruption and angst. In their preference to stay ignorant and honest rather than become sophisticated and false, the Irish served to keep the English honest, preventing them from becoming too alienated from the soul's true world. Once when unloading my suitcase from a cab in an up-market part of London I was approached by a short man who smelt of alcohol. In a strong Irish accent he offered to carry my suitcase. When I said that I was happy to carry it myself he said, **'That's OK, you carry it but I'll come along and you can still give me the tip.'** The logic was confusing but it was very amusing and I found his off-centred way of thinking and his unconcerned, even disrespectful, attitude to all the sophistication around him immensely refreshing and relieving. We ended up sitting on the suitcase on the edge of the footpath talking about Ireland. Such is the effect the Irish have on people.

The British Isles certainly were an extraordinary refuge for relative innocence in the world. As will be explained in 'Australia's role in the world', the island continent of Australia has become the latest and undoubtedly last shelter for relative innocence in the world. 'Undoubtedly last' because there are no other large hide-outs left for innocence on Earth, and with communication and travel technology as sophisticated as it is, even its island isolation can no longer shield Australia from the march of upset and alienation.

The point is, when the history of the human race is taught in the future, it will be taught honestly, it will acknowledge the varying degrees of upset inherent within different races. As emphasised, with understanding of the human condition—understanding the reason why humans unavoidably and necessarily became corrupted—it at last becomes both safe and necessary to acknowledge the various states of upset/corruption/alienation.

Teachers will explain how 'This was a shy, naive, undisciplined, procrastinating teenager-equivalent race of people; while this race was a bold, adventurous, take on the world 20-year-old-equivalent race; this other race was an angry, embattled, but focused and disciplined 30-year-old-equivalent race; while this race was an overly embattled, defeated 40-year-old-equivalent race; and this race was a spent, bitter 50-year-old-equivalent race.'

Using this honest presentation, it will be explained that the relatively innocent northern Germanic races, who stayed isolated in the outreaches of Norway, Sweden and Denmark, along with the relatively innocent Celts who survived in the isolated British Isles, originally came from the Caucasian Steppes, what is now the Georgian region of southern Russia. The original stock were nomadic herders who had domesticated cattle and horses on the grasslands of the Steppes. Unlike hunter-gatherers, the herding lifestyle meant humans lived in permanently close quarters. This greater interaction with other humans living under the duress of the human condition meant that over time the herders progressed from the 'teenage-equivalent' psychological stage to the adventurous '20-year-old-equivalent' stage. It was these adventurous 20-year-old equivalents who left the Steppes and colonised neighbouring territories, eventually reaching Scandinavia and the British Isles.

Before elaborating on this migration from the Steppes, it needs to be more fully explained how the greater interaction with other humans led to the transition from the 'teenage-equivalent' stage to the '20-year-old-equivalent' stage.

In general, living with the human condition and the resulting upset and corruption was a toughening process; innocence was lost and people became adapted to a more angry, egocentric and alienated existence. The greater the exposure to upset and the unavoidable corruption of humans' instinctive self or soul, the greater the selection for 'tougher', more corruption-adapted people and the 'tougher' races became.

To examine the particular stage of advancing from the 'teenage-equivalent' stage to the '20-equivalent' stage, it is useful to take an example from society to illustrate this transition. It is a long-held tradition in western societies to hold a 'coming of age' party for offspring when they turn 21. The 21 year old is traditionally given a key symbolising that they are at last ready to leave home and face the world. While adolescents resigned to living a life of denial at around 15, it normally took them another six years of procrastination before they had made sufficient mental adjustments to embrace the extremely dishonest resigned way of living. It was not until they reached 21 that they finally managed to make the necessary mental adjustments. There were two main adjustments they had to make. Firstly they had to block out the negative that living so falsely had eventually to end in the disaster of becoming utterly corrupted. Secondly, they had to train their mind to focus on the one positive there was in the resigned life, which was that at least there was the adventure to look forward to of trying to avoid that disaster as much, and for as long as possible. They may be going to 'go under'—become defeated and corrupted—but at least they could hope to make a good fight of it. In fact, by the age of 21, young adult men could have so put out of their mind the fact they had resigned and its corrupting consequences that they could delude themselves that they might even be able to win the egocentric struggle to prove their worth. As was explained in the *Resignation* essay in the section 'The moment of resignation', the 'struggle to prove their worth' through winning power, fame, fortune and glory in a claimed battle of the survival-of-the-fittest, was the resigned mind's distorted form of participation in humanity's great fight against ignorance.

While humans have largely lived in denial of the various stages of self-corruption, a Japanese proverb does acknowledge this pattern: **'At 10 man is an animal, at 20 a lunatic, at 30 a failure, at 40 a fraud and at 50 a criminal.'** With understanding of the human condition it is possible

to explain the meaning of these stages all humans go through as they age. Ten-year-olds were **'animals'** in the sense that their instinctive selves were unrepressed. Twenty-year-old young men were **'lunatics'** in the sense that they were swashbuckling cavaliers who believed they could 'take on the world'. They deluded themselves that they could actually defeat the oppressive foe of ignorance (or, from the resigned mind's point of view, that they could actually achieve satisfaction through winning power, fame, fortune and glory). Thirty-year-old men were **'failures'** in the sense that, even though they were still determinedly trying to defy the inevitable, they were being forced to accept that the task of defeating ignorance was going to be beyond them (from the resigned mind's point of view, they were being forced to accept that the corrupting life of seeking power, fame, fortune and glory was not going to be a genuinely meaningful and thus satisfying way of living). Forty-year-old men were **'frauds'** in the sense that they had become so corrupted and disenchanted with their efforts to 'conquer the world' that they suffered a 'mid-life crisis'; a crisis of confidence in what they had been doing in life that precipitated a decision to take up support of some form of 'idealism' in order to make themselves feel better about their corrupted state. Having had enough of the critically important, yet horribly corrupting battle to champion the ego over soul, they changed sides and became 'born-again' supporters of the soul's 'idealistic' world. Their fraudulence was that they were deluding themselves that they were at last on the side of good when in truth they were working against good in the sense that good depended on defying and defeating—not supporting—the ignorant 'idealistic' world of the soul. (This dishonest, fraudulent, 'pseudo-idealistic' way of living is explained more fully in an essay titled *Death by Dogma* that will be published shortly in . . . .
. . . . Fifty-year-old men were **'criminals'** in the sense that they were beaten on every front and had become bitter and vengeful; their attempts to win power, fame, fortune and glory had not proved satisfying, nor had the fraudulent, immensely deluded life of being born-again to the soul's world of 'ideality'.

It needs to be understood that the maturation process of each race, and indeed of the human race as a whole, mirrors that of an individual's journey. Given the human race is composed of individual humans, it too, on a greater level, must reflect the stage of development that the individuals within it are in. Further, since the psychological stage of one individual affects the psychological stage of his or her neighbours, everyone within a group of humans gradually comes to share a common stage. While the psychological stages humans pass through have long been acknowledged as **'infancy'**, **'childhood'**, **'adolescence'**, **'adulthood'** and **'maturity'**, the stages have never been properly explained because humans have largely been unable to deal with the psychology of their circumstances. With understanding of the human condition, which means understanding of human psychosis, we can now explain that **'Infancy'** is when consciousness first appears and the individual becomes self-aware, able to recognise that they exist, that they are at the centre of the changing events around them. **'Childhood'** is when humans become sufficiently intelligent to experiment or 'play' with the power of conscious free will. **'Adolescence'** is when humans become sufficiently developed in their intelligence to seek to understand themselves, in particular to understand why they have not been ideally behaved—understand the by then well developed issue of the human condition. It is the time when they are insecure about who they are, the time when they search for their identity. **'Adulthood'** is when humans have learnt who they are, in particular, have understood the dilemma of the human condition, and are applying themselves in a secure way to the task of complying with the meaning of existence which is to develop the order of matter on Earth. **'Maturity'** is when they have completed the adulthood stage of securely applying themselves to the task of ordering their life and world. In summary, infancy is 'I am', childhood is 'I can', adolescence is 'but who am I?', adulthood is 'I know who I am', and maturity is 'I have fulfilled who I am'.

As emphasised, these psychological stages apply to both the maturation of humanity and each individual human. In *Beyond* I described our ape ancestor as 'Infantman'. It was with this ancestor that the nurturing, love-indoctrination process took place. In time Infantman gave rise to 'Childman'. In the fossil record Childman is the Australopithecines. I have described the earliest Childman, Australopithecus afarensis, as 'Early Prime of Innocence Childman'. Afarensis lived from 4 to 2.8 million years ago. As our Australopithecine ancestors' intelligence developed they matured psychologically to become Australopithecus africanus, 'Middle Demonstrative Childman'. They were 'demonstrative' in the sense that they began to demonstrate and even taunt the world around them with their power of free will. Africanus lived from 3 to 1.8 million years ago. Africanus developed to become Australopithecus boisei, who I have described as 'Late Naughty Childman', for the troubled, 'naughty' behaviour that arose from having encountered the frustration of the emerging dilemma of the human condition. Boisei lived from 2.2 to 1.3 million years ago. (Since *Beyond* was published in 1991 more varieties of the Australopithecines have been discovered but they still fall into these broad categories of Early, Middle and Late Childman.)

Some 2 million years ago our ancestors' intelligence had developed to such an extent that it began to challenge the instinctive self for management of our ancestors' lives. The resulting conflict between instinct and intellect gave rise to the dilemma of the human condition. With the human condition fully emerged humanity entered the stage of insecure adolescence, the stage where humans search for their identity, for understanding of why they became divisively behaved. Insecure 'Adolescentman' has progressed from the early sobered teenage stage during which the issue of the human condition is wrestled with, through the 15-year-old resignation stage, the procrastinating late teenage stage, the 20s 'adventurous' stage, the 'embattled' 30s stage, the 'born-again' 40s stage, to where humanity is now, entering the burnt-out, end-play, 50s 'criminal' stage. The only solution powerful enough to prevent self-destruction from over corruption—the potential end of this stage—was to find the dignifying understanding of the human condition, which has now been achieved. Only by bringing understanding to the human condition could humanity avoid eventual self-destruction and move from the insecure state of adolescence and enter the secure state of adulthood.

As is explained in *Beyond,* Homo habilis was 'Sobered, Resigning, Procrastinating Adolescentman', the variety of early human who became sobered from trying to confront and think about the human condition, only to resign to a state of procrastination about having to adopt a resigned lifestyle. Homo habilis lived from 2 to 1.3 million years ago before maturing into 'Adventurous Adolescentman'—Homo erectus—the variety of early humans who migrated all over the world from our species' ancestral home in Africa. They lived from 1.5 to 0.4 million years ago before maturing into 'Embattled, Angry, Disciplined Adolescentman', Homo sapiens. Homo sapiens lived from 0.5 million years ago to 50,000 years ago before maturing into 'Born-Again, Sophisticated Adolescentman'—Homo sapiens sapiens—who has lived from 60,000 years ago to the present. This 'born-again' 40-

year-old stage is where humanity is currently progressing toward the burnt-out criminal stage. The born-again 40-year-old-equivalent stage could also be termed the 'Civilisedman' stage. A civilised person is an overly corrupt person who, instead of living out their extreme anger and frustration, has learnt to restrain and conceal them. A civilised person is someone who, despite their corrupt, alienated, non-'ideal' reality, has decided to take up a 'born-again' attitude of *behaving* in a 'good', 'ideal' way. Emphasis is placed on the word 'behaving' because they were only pretending to be good or ideal; their real state of upset is concealed. Civility was humans' earliest way of being 'born-again' to 'ideality'. Today, all humans are adapted to behaving in a civilised way. For the most part we are instinctively restrained from living out and revealing our real angers and frustrations. We restrain and hide our true feelings. This 'dark side' of the human make-up that is rarely confronted and acknowledged is what those literary figures quoted in the *Introduction* did bravely acknowledge. Author Morris West confessed that with enough **'provocation I could commit any crime in the calendar'** (*A View from the Ridge,* 1996, p.78 of 143).

The question arises, if all humans who have lived since 60,000 years ago belong to the 40-year-old-equivalent, Born-Again, Civilised, Sophisticated Adolescentman stage, why were the hunter-gatherers of recent times described earlier as being in the procrastinating teenage-equivalent stage, and the herding Caucasians described as having entered the adventurous 20-year-old-equivalent stage? What is being described is another level of refinement of the already established stages. The first T-model Ford car had all the basic elements of a car in place but that did not mean the elements could not become much more refined. The relatively innocent hunter-gatherer Bushmen people of the Kalahari Desert have all the basic adjustments in place for managing extreme upset. They are civilised, instinctively restrained from living out all their upsets; they don't generally attack when they feel frustrated and angry. They have a form of marriage to artificially contain sexual adventurousness. They clothe their genitals to dampen lust. The women love to wear adornments such as jewellery; they are adapted to being sex objects. The men love hunting animals; they find relief from attacking innocence. Men and women don't relate to each other as well as their own gender; there is a lack of understanding between the sexes. They make jokes about their fraudulent state; they employ a sense of humour to lighten the load of the agony of being so corrupted and fraudulent.

They employ fatigue-inducing dance to access their repressed soul. In short they are members of 'Born-Again, Sophisticated Adolescent-man'. While they have these basic adjustments for managing extreme upset well in place, they are still a relatively innocent race compared to the adventurous, high-spirited Vikings, or the embattled, angry Mongols who raided Europe.

As explained, the maturation of an individual human through the stages of ever-increasing corruption follows the same pattern as the maturation of a race of people, and just as one human could be forced to adapt to new levels of corruption through greater exposure to corruption than another human, some races could be forced to adapt to new levels of corruption through greater exposure to corruption. As each new stage of corruption began to emerge in a society of people, those individuals who happened to have a genetic make-up that enabled them to tolerate and cope with that new level of corruption survived better and thus, over time, that race became genetically adapted to the new, more corrupt reality.

To clarify what is meant by the terms 'tolerating' and 'coping with' corruption, it is necessary to explain the two ways in which corruption destroyed innocence. The first was by the more corrupt deliberately oppressing and even killing innocence because of its implied criticism; the second by the more innocent finding themselves unable to tolerate and cope with corruption. Both methods will be elaborated upon in the following passages.

Throughout the 2 million years humans lived unable to explain and defend their corrupted state, innocence has unwittingly confronted, exposed and condemned the more corrupted. Innocence has been oblivious to its impact upon those more corrupted, as the following quote demonstrates: **'In the small fenced-in waiting area outside the departure hall an African woman sits with her wares spread out upon the grass. On the way to the plane I notice a framed piece of needle point she has hung on the fence that reads: "I love those who hate me for nothing"'** (*River of Second Chances* by Eric Ransdell, *Outside* mag. Dec. 1990). This reaction was evident also in the words of Christ when he quoted Psalm 35:19 and 69:4: **'They hated me without reason'** (John 15:25). We can explain now that Christ was crucified because the upset world hated his exceptional innocence and the exposing and confronting honesty of what he had to say. Tragically it has been the lot of innocence, be it innocent humans or nature, to suffer for its unjust condemnation of the upset, corrupted world. The biblical story of Cain and

Abel encapsulates the innocence-destroying, toughening process that has been occurring: **'Abel kept flocks,** [he lived the life of a shepherd, staying close to nature and innocence] **and Cain worked the soil** [he became settled and through greater interaction with other humans became corrupted]**...Cain was** [became] **very angry, and his face was downcast** [he became depressed about his corrupted state]**...**[and] **Cain attacked his** [relatively innocent and thus condemning] **brother Abel and killed him'** (Genesis 4:2,5,8).

While this replacement of the more innocent through oppression and even murder commonly occurred, the more corrupt have most often replaced the more innocent through the latter finding themselves unable to accept and cope with each new level of corruption, each new level of compromise of the soul's ideal world. Sir Laurens van der Post recognised the difficulty innocence has had coping with the more advanced levels of compromise of the cooperative, selfless, loving world that humans' original instinctive self or soul expects, when he wrote: **'Nor should we forget that there were races in the world which vanished not because of the wars we waged against them but simply because contact with us was more than their simple natural spirit could endure'** (*The Dark Eye in Africa*, 1955, p.101 of 159). In *The Lost World of the Kalahari* Sir Laurens related the story of how a member of the relatively innocent Bushmen race could not cope with having his natural spirit compromised: **'You know I once saw a little Bushman imprisoned in one of our gaols because he killed a giant bustard which according to the police, was a crime, since the bird was royal game and protected. He was dying because he couldn't bear being shut up and having his freedom of movement stopped. When asked why he was ill he could only say that he missed seeing the sun set over the Kalahari. Physically the doctor couldn't find anything wrong with him but he died none the less!'** (1958, p.236 of 253). Sir Laurens was more specific when he wrote about the Bushmen, that **'mere contact with twentieth-century life seemed lethal to the Bushman. He was essentially so innocent and natural a person that he had only to come near us for a sort of radioactive fall-out from our unnatural world to produce a fatal leukaemia in his spirit'** (*The Heart of the Hunter*, 1961, p.111 of 233). Sir Laurens recognised that innocence has been cruelly denied and neglected when he wrote, **'There was indeed a cruelly denied and neglected first child of life, a Bushman in each of us'** (ibid. p.126). The following extract from a poem by D.H. Lawrence also acknowledges how hardened the human race has become: **'In the dust, where we have buried/The silent races and their abominations** [their condemning and thus detestable innocence]/**We have buried so much of the delicate magic of life'** *(Son of Woman: The Story of D.H. Lawrence,* D.H. Lawrence, 1931, p.227 of 402).

Through the selection of those individuals genetically better suited to a corrupt world, all humans are now variously genetically adapted to this corruption and to life under the duress of the human condition. (Incidentally this means there is a great deal of genetic preconditioning in the make-up of a human. It is not simply a case of 'nature' [genes] or 'nurture' [environment] alone impacting upon our lives, but a case of both exerting their influence.) Humans and humanity's journey under the duress of the human condition has been a toughening process, a process where innocence was lost and corrupt reality adapted to. For a race of teenage-equivalents, such as the original Caucasians, to mature to 20-year-old-equivalents, selection pressure had to occur for individuals who could block out the negatives of the dishonest, resigned lifestyle and focus on the adventure of trying to champion the ego. In the case of Caucasians, that selection pressure would have been produced by their herding lifestyle that forced greater interaction between people.

The change in the 'Caucasians' to the adventurous 20-year-old stage occurred around 2000 BC, a time when they were able to colonise almost all the neighbouring regions before any other race emerged as 20-year-old equivalent adventurers. The only other race to reach this threshold in time to compete with the Caucasians were Semitic herders who colonised parts of the Middle East from their homeland somewhere on the Arabian Peninsular or North Africa. The descendants of the Semitic herders included the Phoenicians, Chaldeans, Aramaeans, Arabs and Hebrews. The original Caucasian herders called themselves Aryas, which meant 'noble of birth and race'. These strongly patriarchal, proud, tall, fair-skinned people, with their great herds of cattle and chestnut horses, their flocks of sheep and goats and packs of dogs migrated from their ancestral homeland on the Caucasian Steppes to colonise India, Europe and most of the Middle East. The Romans, Greeks, Slavs, Celts and the Germanic races all descend from this Aryan stock: **'Since *c.* 2000 waves of Aryan warriors have gushed from the Caucasian steppes. They have conquered the Indus Valley and set up the kingdoms of the Hittites and of Mitanni** [in the Middle East]**; by 1000 BC they will found five more kingdoms in the Middle East. Greek-speakers have built Mycenaean civilization. Descendant of other Aryans, the Hyksos** [**the "Shepherd Kings"**], **have ruled Egypt'** *(The Last Two Million Years,* 1974 edn, p.269 of 488). As evidence of their common ancestry, all races of Aryan stock share a similar 'Indo-European' language. For example, the English word 'mother' is 'mutter' in German, 'máthair' in Gaelic, 'mater' in Latin, 'meter' in Greek, 'mati' in Old Church Slavonic, 'matr' in Sanskrit and 'mater' in Indo-European. It is a measure of how assertive and pervasive the Aryans have been that these Indo-European tongues have been adopted by half the world. Aryans have in general been extremely influential in the world, but it is those relatively isolated Aryans from Scandinavia and the British Isles who have exerted the most influence in recent centuries.

Over time the original 20-year-old-equivalent Aryan stock from the Caucasian Steppes was bled dry of its relative innocence in endless wars in their new homelands. Because of their moral strength, the innocent are the first to stand against oppression and thus the first to be eliminated by an oppressor. The expression, **'the good die young'**, is an acknowledgment of this reality. As explained, it was a toughening process—an adaption—to the new level of corruption which the more innocent found most difficult to accept and adjust to. The result of this loss of innocence was that the original 20-year-old-equivalent Aryan stock advanced to more embattled and upset 30, 40 and 50-year-old-equivalents. Even the original 20-year-old-equivalent stock who remained on the Caucasian Steppes were eventually overrun by their even more upset-advanced neighbours, the extremely angry, utterly ruthless 35-year-old-equivalent Huns from Mongolia, who swept across Europe around 400 AD. Some idea of the amount of blood-letting that took place in Europe and the Middle East over the last four millennia can be gauged from this account of the outcome of just one war in central Europe: **'The flowing blood of these murdered men, ten million gallons steaming human blood could substitute for a whole day the gigantic water masses of the Niagara...Make a chain of these ten million murdered murderers, placing them head to head and foot to foot, and you will have an uninterrupted line measuring ten thousand miles, a grave ten thousand miles long'** (*Roumania Yesterday and Today*, Mrs Will Gordon, 1918, p.251 of 270). The loss of innocence has been immense. Races and their empires that were once great are now just shells of their former glory; an article in *Time* magazine described Italians now as, **'Creative and hardworking yet corrupt and ruthless'** (3 Aug. 1992). It can now be explained psychologically what was actually happening when history books talked of civilisations having 'peaked' and become 'decadent', such as Olive Schreiner described as having occurred in Rome.

It has to be emphasised that under the duress of the human condition all races eventually became overly corrupted, corruption of soul being the price of humanity's heroic search for knowledge. In this journey from innocence to exhaustion of soul the most creative period was the toughened and disciplined, but not yet overly corrupted, 30-year-old equivalent stage. As each race and its associated civilisation passed through this stage it made its particularly creative contribution to the human journey. This was when civilisations were at their 'peak', however, inevitably, they entered a more corrupted 'decadent' stage. The Mediterranean, Middle East and Indian civilisations all made extensive contributions to the human journey during their energetic and creative 30-year-old equivalent stage. The Egyptians and peoples from the fertile crescent of the Tigris and Euphrates delta in the Middle East began the civilisation of the 'known world', for example they invented the wheel, mathematics and writing. Greeks and Romans laid the foundations for 'western civilisation'. The great religions of the world, Hinduism, Buddhism, Judaism, Christianity and Islam, were developed in India and the Middle East. These parts of the world are now amongst the most exhausted but were once the most vital and creative.

With understanding of the human condition the various stages of soul corruption can be compassionately understood. To become corrupted was an unavoidable consequence of having to participate in humanity's heroic journey to defy ignorance and find understanding, ultimately self-understanding. To illustrate how races progress from innocence to corruption of soul I have used the history of the Aryan Anglo-Saxons and Celts. I have done this because they are currently in their 'peak' state of contributing to the human journey to enlightenment, and because that journey is in its crucial final stage where a great deal of honest explanation of the events that are taking place is needed. However, I could have chosen the history of the Aryan Greeks and Romans, the Aryan Indians, the Middle Eastern Semites, the Chinese and other races of Asia, or the Aztecs and Incas of Central and South America to illustrate the same journey. Each of their rich histories would have shown the same pattern of progressing from a state of innocence through an operational, exceptionally creative 30-year-old equivalent stage and on to a more corrupted, soul-burnt-out, 'decadent' state.

It also should be emphasised that even races at the more corrupted end of the alienation spectrum still contribute to the human journey. Every individual and every race always sought to contain and minimise the negative aspects of their particular condition and develop and maximise the positive aspects of that condition. The Italians for example, despite having progressed past their 'peak', still contribute to the human journey on many fronts. For example, their mature sophistication has made them masters in the **'creative'** world of design.

In the story of the Aryan Anglo-Saxons and Celts, evidence suggests that only in the northern and western hideaways of Scandinavia and the British Isles did the original Aryans remain in a state of relative innocence. Even amongst the relatively innocent Anglo-Saxons and Celts of the British Isles, the disciplined 'coolness' and focused 'practicalness' of the Anglo-Saxons indicates they progressed further in their loss of innocence than the still fiery, passionate 20-year-old equivalent Celts.

The southern edge of the Aryans' original range, the north African desert, offered no refuge nor did the eastern edge offer a hideout from all the wars that raged during the last 4,000 years. It was only in the north and the western edge of the range that there was any refuge from the march of corruption, and what is so remarkable is that this western region of England received a double-dose of the relative innocence, absorbing those from the northern edge of the original range.

The point is that a denial-free history of the human race will acknowledge not only the different levels of innocence amongst races, but how those levels have affected the human journey. The only difference of significance between individuals and races is their different degrees of corruption, their alienation, yet alienation is never acknowledged. This raises a whole debate that has never been allowed to take place; what happens when races at different levels of corruption attempt to coexist, and what ultimately is the solution to the problems that arise?

Much of the earlier part of this book has demonstrated the benefits of being able to be honest about the different states of corruption amongst individual humans. At last so much about humans can be compassionately explained, thereby subsiding the upset of soul, the psychosis in humans. Similar benefits can be derived from being honest about the different levels of corruption in races of humans and how these varying levels have impacted upon society.

In the late 1800s British colonists brought Indians to Fiji as indentured labour for farming sugar cane. As a result 44 percent of the Fijian population is now Indian and a serious conflict has arisen between the Indian and native Fijians. The issue is the Indian Fijians have been so industrious and successful that they now monopolise Fijian business to the extent that the native Fijians feel their country is being taken over by the Indian Fijians; for their part the Indian Fijians feel that they are being discriminated against. Indian sugar growers in particular feel this inequity for while they are only allowed to lease land from native Fijians (who control 90 percent of the land) they produce 90 percent of the country's sugar. Furthermore, since independence in 1970, native Fijians have ensured their domination of the political process.

The Indian Fijians come from a very ancient civilisation in India, one where innocence has long ago given way to more corruption-adapted humans. In comparison, native Fijians are still relatively innocent, yet to become embattled, hardened and upset-adapted. It is a case of a teenage-equivalent race having to coexist and compete with a 40-year-old equivalent race.

An 18-year-old won't wash the car—they do not yet know of the extent of the hardships that arise in a human condition-afflicted world and thus they do not yet accept the responsibilities that arise in that world. A 40-year-old on the other hand has had sufficient innocence knocked out of him and has encountered a sufficiently uncertain world to be able to apply himself with immense discipline. He readily washes the car, sweeps the drive and works diligently, for instance, to pay the mortgage on his home. The truth is 40-year-olds are so aware of the reality of their human condition-afflicted world where employment and income is never certain that in private moments they shake with fear, hardly able to believe they are fortunate enough to own a car and a home.

The frightening reality of the world of a 40-year-old arises from the fact that the more humans became corrupted and thus selfish, the more everyone else had to become selfish to successfully compete with them. Loss of innocence caused further loss of innocence; corruption was a self-fuelling process. At a certain point in this escalation of corruption life was no longer functional because everyone was too cynical and thus selfish for there to be any effective cooperative effort. That level of dysfunctionality, where everyone was only concerned for themselves equates to the level of corruption in the world of 40-year-olds.

Native Fijians are still sufficiently free of corruption, still sufficiently innocent and naive about the potential reality of life under the duress of the human condition to dislike applying themselves. Indians on the other hand had that naivety eliminated from their make-up long ago and are born to be cynical and realistic.

Native Fijians are renown for being a happy, smiling, generous race who like to sit around talking, playing music and singing. To them materialism is not so important. Native Africans share these same characteristics. Cecil Rhodes, the English pioneer developer of South Africa once famously complained, **'we have to teach the natives to want'**. To build the railway from Mombasa to Nairobi in Kenya, Indians had to be hired from India because the native Africans would not 'work'. In earlier times when the act of enslaving was tolerated, the native Africans were highly sought for their relative innocence and naivety because it meant they were generous and pliable rather than cynical and obstinate like people from more upset-adapted races. I have read that there are now economic maps of the world that do not have Africa on them. From a business perspective, Africa is apparently not considered to be an economically operational place. South Africa, with its high proportion of Europeans, is the only exception; as an illustration of this comparison, on a map of Africa, South Africa is distinguished by its crisscross patterns of roads and railways.

While it was not safe to acknowledge different levels of innocence, excuses had to be contrived for racial differences. For example I have often heard it said that, 'If Africa is dsyfunctional it is because the white missionaries and then the white colonialists so disrespected native culture and so ill-treated the native people that the native people were left economically and spiritually crippled by the experience.' It is true that innocence can be psychologically devastated by encounters with more corruption-adapted people, but that is not what is being argued. What is being asserted is that native Africans *could* compete if their culture had been respected and they had not been treated in a patronising way, and that is not true. What is true is that there *are* relatively innocent races who are naive about life under the duress of the human condition; and there *are* races who are relatively operational under the duress of the human condition; and there *are* races who are dsyfunctional because they are overly corrupted, cynical and selfish—just as there are individual humans in these various stages of corruption.

Yet again it has to be emphasised that these stages of corruption are a result of humanity's *necessary* battle to solve the human condition and while some races are more innocent than others, racist notions of some races being superior or inferior—such as Hitler's view of the superiority of the Aryan race, especially that of the Nordic branch—have no credibility. Equally lacking in credibility are those assertions that humans and races do not become more corrupt as they age. As stressed, without the reconciling understanding of the human condition everyone who was innocent eventually became corrupted. Would a person's 18-year-old self disown their 30-year-old self and would their 30-year-old self disown their 40-year-old self? Of course not. No one is better or worse than anyone else, merely differently corrupted as a result of the *necessary* battle to champion our intellect over the ignorance of our instinctive self or soul. What is being said in this essay explains everyone yet criticises no one.

The ability to be honest about the different levels of exhaustion or corruption of soul in different races, enables them to finally coexist and complement each other according to their different attributes. My second book *Beyond* is dedicated to the relatively innocent native Bushmen of the Kalahari because their innocence verified and sustained my innocence, allowing me to confront the issue of the human condition without being intimidated by the world of denial. As Sir Laurens van der Post hoped, the Bushmen's **'inner life might reveal a pattern which reconciled the spiritual opposites in the human being and made him whole...it might start the first movement towards a reconciliation'** (*The Heart of the Hunter,* 1961, pp.135–136 of 233). The more innocent have soundness to contribute while the more corrupt contribute necessary self-discipline and tough perseverance. Sir Laurens described the role of the more innocent when he wrote: **'We need primitive nature, the First Man in ourselves, it seems, as the lungs need air and the body food and water...I thought finally that of all the nostalgias that haunt the human heart the greatest of them all, for me, is an everlasting longing to bring what is youngest home to what is oldest, in us all'** (*The Lost World of the Kalahari,* 1958, p.151 of 253).

In his 1955 book *The Dark Eye in Africa,* an examination of the problem of blacks and whites in Africa, Sir Laurens used the word 'primitive' where I use the term 'teenage-equivalent race', and 'civilised' where I use the terms 'disciplined 30-year-old equivalent race' and 'born-again 40-year-old equivalent race'. With his terms, Sir Laurens made the same point that I have just made when he wrote: **'To my black and coloured countrymen who may read this book I would like to explain my use of the words "primitive" and "civilised" man. I use these words only because I know no others to denote the general difference of being which undeniably exists between indigenous and European man in Africa. I am, however, fully conscious of their limitations and relativity. I do not think of the European as a being superior to the black one. I think of both as being different and of the differences as honourable differences equal before God. The more I know of primitive man in Africa the more I respect him and the more I realise how much and how profoundly we must learn from him. I believe our need of him is as great as his is of us. I see us as two halves designed by life to make a whole. In fact, as I watch the darkening scene I see this need of the one for the other to be so great as to create fresh hope that this may yet save Africa from disaster'** (pp.19–20 of 159).

Relatively innocent, naive native races and over-exhausted, more cynical races, such as those in the Middle East, have both found it difficult competing with races from the middle of the spectrum, those that are toughened but not yet overly corrupted and can still effectively cooperate and function. What is occurring in the world today, in 2003, is that an operational 30-year-old equivalent race, namely the Anglo-Saxons, are proving so successful—for example they are most responsible for the so-called 'economic globalisation' of the world—that the relatively innocent teenage-equivalent races and the overly corrupted 40-year-old equivalent races are finding they are unable to compete and as a result their populace is being left impoverished. This ever-widening rift between the haves and have-nots is fuelling immense frustration amongst the have-nots. The break-out of terrorism in the world and the wars in the Middle East are the result of this frustration.

Obviously what is needed is a more equitable world where wealth is more evenly distributed. The denial-free interpretation of this is that the more operational races need to be prepared to share their good fortune with the less operational races. The fortunate have a responsibility to be merciful to those less fortunate.

The problem is that without understanding of the human condition no differentiation could be made between races according to how innocent or corrupt and thus how operational or non-operational they were. In order to avoid prejudice, everyone had to be allowed equal opportunity, yet obviously the more operational were always going to be ultimately more materially successful.

Inequality was an unavoidable price of living in denial.

The result of this free enterprise system of equal opportunity for everyone is a world that has divided into an impoverished majority and an obscenely wealthy minority. On the face of it nothing could appear more unjust and dangerous than that situation. There is however an accompanying development to this unfair pattern that—and this will come as a surprise to a world that has been living in denial—is even more dangerous.

The more the world became unjustly unequal the more people tried to counter this with the only means at their disposal—the imposition of freedom-denying dogmatic forms of restraint. What is so dangerous is that this development has the potential to shut down the freedom of expression needed to allow the reconciling, dignifying, peace-bringing, humanity-liberating understanding of the human condition to emerge.

In the situation where inequality was to a significant degree unavoidable the real danger was not of inequality eventually destroying the world, but of dogmatic forms of restraint eventually denying the freedom needed to search for understanding, ultimately self-understanding. In a world dependent on denial, the dangers associated with both the political left and the political right escalated as time went on. However the political view that has the potential to destroy the world is that of the freedom-restraining political left, *not* the free enterprise political right. An article by Geoffrey Wheatcroft in the *Australian Financial Review* recognised that **'the great twin political problems of the age are the brutality of the right, and the dishonesty of the left'** (29 Jan. 1999). The scientist philosopher Carl von Weizsäcker similarly stated that, **'The sin of modern capitalism is cynicism (about human nature), and the sin of socialism is lying'** (mentioned in a speech by Prof. Charles Birch that is reproduced in the Geelong Grammar School mag. *The Corian,* Sept-Oct. 1980). The dishonest pseudo-idealism of the left is by far more dangerous to humanity than the cynicism and brutality of the right.

Nietzsche was emphasising this surprising situation where fighting for freedom becomes more critical than the critical problem of inequality and its poverty when he said, **'You are not yet free, you still *search* for freedom. Your search has fatigued you...But, by my love and hope I entreat you: do not reject the hero in your soul! Keep holy your highest hope!...War and courage have done more great things than charity. Not your pity but your bravery has saved the unfortunate'** *(Thus Spoke Zarathustra: A Book for Everyone and No One,* 1892; tr. R.J. Hollingdale, 1961, pp.70,71,74 of 342).

Sir Laurens van der Post spoke equally strongly against pseudo-idealism: **'the so-called liberal socialist elements in modern society are profoundly decadent today because they are not honest with themselves ...They give people an ideological and not a real idea of what life should be about...They feel good by being highly moral...They have parted company with reality in the name of idealism...there is this enormous trend which accompanies industrialized societies, which is to produce a kind of collective man who becomes indifferent to the individual values: real societies depend for their renewal and creation on individuals...There is a very disturbing, pathological, non-rational element in the criticism of the Prime Minister** [Margaret Thatcher]. **It amazes me how no one recognizes how shrill, hysterical and out of control a phenomenon it is...I think socialism, which has a nineteenth-century inspiration and was valid really only in a nineteenth-century context when the working classes had no vote, has long since been out of date and been like a rotting corpse whose smell in our midst has tainted the political atmosphere far too long'** (*A Walk with a White Bushman,* 1986, pp.90–93).

The point is that without the freedom for individuals to question and to search for (and ultimately to deliver) understanding of the human condition, eventually alienation and its effects *will* destroy the human race, it will bring it into a state of abject physical and psychological poverty. Terminal poverty, the death of the human race, is what occurs if freedom is extinguished. The following extract from the aforementioned essay *Death by Dogma* explains the extreme danger posed by the left wing, pseudo-idealistic, freedom-denying, ideal-world-imposing dogma of the so-called 'politically correct' culture.

'We have reached the point in the human journey where the rapidly rising tide of the politically correct culture of deconstructionism, or post-modernism in affluent parts of the world, and its fundamentalist religious counterpart in poorer parts of the world, are threatening to stifle the freedom of expression upon which liberating enlightenment of the human condition depends.

In more affluent parts of the world the need to escape the escalating dilemmas and psychoses of life is becoming desperate and as a result the relief offered by idealistic, self-affirming, emotional causes is becoming increasingly irresistible. In poorer parts of the world there is an ever-growing need to counter the ravages of disorder and wretchedness through strict obedience to ideal principles. But the great danger is that if the strategy of imposing idealism or "correctness" becomes universal then the *freedom* to differ will be denied and the *responsibility* to address and resolve (rather than to escape and repress) the dilemmas of life will be abandoned—and *real* progress will be halted.

A crisis emerges for humanity when the seductive qualities of delusion become irresistible and/or the need for ultra-restraint becomes a necessity for the bulk of humanity. In the on-rush of the psychosis-escalating struggle of the human journey it was inevitable that a point would eventually be reached where the need for relief and restraint would become so great that the delusion of dogma—the artificiality of imposing ideality to solve reality—would become invisible to its practitioners. We are now in that situation where the advocates of dogma have become blind to its short-comings and consequently are now on an all-out mission to seduce and intimidate everyone with their culture.

It has been said that **"postmodernism has peaked, and will die with the century"** (*A Strange Outbreak of Rocks in the Head* by Damian Grace, *Sydney Morning Herald,* 21 Jan. 1998), but this is a psychosis-driven situation where increasingly people *have to* variously live in adherence to pseudo forms of idealism to cope with their circumstances. Therefore, if you live in an affluent part of the world, no amount of opposition will halt the rise of politically correct culture, or, if you live in an impoverished part of the world, no amount of opposition to fundamentalist expressions of religion will halt its growth.

Throughout history there have been warnings of this great danger of pseudo-idealism seducing the world. George Orwell's famous prediction in his 1949 book *Nineteen Eighty-Four* that **"If you want a picture of the future, imagine a boot stamping on a human face** [freedom] **forever"** is relatively recent; the *Bible* presents powerful warnings in both the Old and New Testaments.

In the Old Testament the Book of Daniel focuses upon the eventual rise of pseudo-idealism in the world, describing the end result where the two ultimate forms of pseudo-idealism, namely politically correct deconstructionism in wealthier, more educated parts of the world, and fundamentalist expressions of religions in more impoverished parts of the world, take over: **"two kings, with their hearts bent on evil, will sit at the same table and lie to each other"** (11:27). Daniel concludes with this summary: **"He** [delusion] **will invade the kingdom** [of honesty] **when its people feel secure** [when delusion becomes a sufficiently universal culture]**, and he will seize it** [seize the kingdom of honesty] **through intrigue** [through the seduction of pseudo-idealism's imitation of idealism]**...Then they will set up the abomination that causes desolation** [pseudo-idealists will establish their culture of delusion that leads to oblivion]**. With flattery** [the do-good, feel-good self-affirmation that pseudo-idealism feeds off] **he will corrupt those who have violated the covenant** [pseudo-idealism will seduce the more upset away from true religion's infinitely more honest way of coping with the human condition], **but the people who know their God will firmly resist him** [the more secure, less evasive will not be deceived and must strongly resist the seductive tide]" (11:21,31,32).

In the New Testament Christ reiterated Daniel's stance, advising people to head for the hills when pseudo-idealism threatens to take over: **"So when you see the 'abomination that causes desolation,' spoken of through the prophet Daniel, standing where it does not belong** [claiming to be presenting the way to the human-condition-free, good-versus-evil-deconstructed, post-human condition, new age]—**let the reader understand—then let those who are in Judea flee to the mountains. Let no-one on the roof of his house go down to take anything out of the house. Let no-one in the field go back to get his cloak. How dreadful it will be in those days for pregnant women and nursing mothers! Pray that your flight will not take place in winter because those will be days of great distress** [mindless dogma and its consequences] **unequalled from the beginning of the world until now—and never to be equalled again. If those days had not been cut short** [by the arrival of the truth about the human condition], **no-one would survive"** (from Matt. 24 & Mark 13).

The situation where extreme delusion threatens to seduce the world has now arrived. Its influence and hold has become so great that political opposition will soon no longer stop it. As the prophets above anticipated, humanity has finally reached the end-play situation where only the truth about the human condition can save the situation.'

That completes the extract from *Death by Dogma*. For great prophets throughout history, from Daniel to Christ to Nietzsche to van der Post, to make such powerful no-holds-barred warnings about the dangers of pseudo-idealism, emphasises just how serious a threat pseudo-idealism is. The level of intimidation that pseudo-idealists recently inflicted on the handful of countries willing to stand against the dangerous dictator and tyrant, Saddam Hussein, was of such intensity that they almost succeeded in preventing the stand altogether. If those who are so frustrated by the inequality in the world that they resort to reckless retaliation—such as President Robert Mugabe of Zimbabwe, Colonel Gaddafi of Libya, Saddam Hussein of Iraq and Osama bin Laden in Afghanistan—are not strongly countered then the all-important culture of freedom in the world would have failed a crucial test, and as a result be on the way out forever. In the recent case of Saddam Hussein there were so few willing to defend freedom that it seems likely that it will not be long before freedom will succumb to the global assault of pseudo-idealism. George Orwell's prediction that by 1984 the **'boot** [of pseudo-idealism would be] **stamping on a human face** [freedom] **forever'** is apparently not many years off the mark.

An article written by Ramesh Ponnuru titled **'Empire of Freedom'**, published in the United States magazine *National Review* (24 Mar. 2003), was truthful when it talked of **'The Three Anglos'**, the United States President George W. Bush, England's Prime Minister Tony Blair and Australia's Prime Minister John Howard leading **'The English-speaking alliance to save the world'** in the war against Saddam Hussein. The article reinforces many of the points I have made. It says that a US internet entrepreneur, James Bennett, uses the term **'Anglosphere'** to describe a coalition of English-speaking countries, the US, the UK, Australia, Canada, New Zealand, Ireland and South Africa that are **'characterised by a high degree of individualism and dynamism, and by a talent for assimilation.'** Bennett says it is **'no accident that it was in the Anglosphere that the industrial revolution and parliamentary democracy first emerged...Nor is it an accident that when French intellectuals and Malaysian prime ministers wish to denounce free markets, the phrase they use is "Anglo-Saxon capitalism".'**

What is happening in this situation where freedom is being almost annihilated and the inequality in the world is unbearably obscene is that humanity is entering end-play or end-game. If the reconciling understanding of the human condition had not been arrived at, the chasm between the rich and the poor and the resulting levels of frustration would only escalate, as would the difficulty of maintaining freedom in the world. The reconciling dignifying understanding of the human condition has arrived only just in time. As Christ said, **'If those days** [when **the abomination** of pseudo-idealism **that causes desolation**] **had not been cut short** [by the arrival of the truth about the human condition], **no-one would survive'**. The situation faced by the human race was that dangerous.

With the last minute arrival of understanding of the human condition, the innocent and upset states of the human condition can be finally reconciled and real and lasting peace can at last come to the human race. The battle to champion the human intellect over the ignorance of the instinctive self or soul of humans is finally resolved. The warrior's sword can be laid to rest and everyone can work side by side for the new world.

How the transition to the human condition-ameliorated, peaceful new world actually takes place was explained in detail in the *Resignation* essay in the section 'Renegotiating resignation'.

The Demystification Of Religion

In the earlier essays in this book, it was explained that humanity has, for good reason, been a species in denial and that as a consequence of that denial humans have, to varying degrees, been alienated from their original instinctive self or soul and its world. With acknowledgment of the overall extent of alienation within the human race (and therefore of the immense gulf between humanity's resigned, false, effectively dead world and the unresigned, true, all-sensitive world), and an appreciation of the immense differences in alienation from one individual to the next, it now becomes possible to explain and demystify all manner of mystery, superstition and abstract metaphysics—especially many previously impenetrable religious concepts.

This essay will explain, from a basis of first principle biology, such religious concepts as 'soul', 'conscience', humans' 'fall from grace' or 'original sin', 'evil', 'heaven and hell', the 'Garden of Eden', the 'golden age', the story of 'Adam and Eve', 'God', 'holy', 'the trinity', 'prophets', the 'messiah', Christ (**'the man that nobody knows'**), the 'saviour', 'the resurrection', Christ's miracles, 'the Virgin Mother', 'Noah's Ark', the story of 'Cain and Abel', 'David and Goliath', 'saints', 'false prophets', 'the abomination that causes desolation', 'the Antichrist', 'the four horsemen' of 'the apocalypse', 'the Battle of Armageddon', 'afterlife', religious texts, 'faith', religions themselves and the different roles of traditional and contemporary prophets and 'atheism'. Other previously unexplained concepts such as 'love', 'falling in love', 'sex' as humans practice it, 'humour', 'sexism', 'racism' and 'human personalities' will also be explained.

(All scripture citations in this work are taken from the New International Version of the *Bible.*)

Understanding the human condition makes it possible to acknowledge the immense differences in alienation between humans

In the preceding essays it has been explained that generation after generation of young adolescents have learnt they had no responsible choice but to deny and evade the issue of the human condition; and the degree to which they implemented this denial varied according to the amount of nurturing they received in childhood. This means that adults today are variously alienated from their true selves and the true world. In fact, there is a wide spectrum of alienation in the world that until now has largely been denied because acknowledgment of it would have hurtfully confronted and unjustly condemned those more alienated.

With the human condition explained and humans' necessarily corrupted state at last understood, the immense variation in the levels of alienation in the world can at last be safely and openly acknowledged. With that acknowledgment, many religious concepts can be demystified.

It is necessary to begin with a brief summary of how humans became corrupted and as a result variously alienated.

Early in the *Plato* essay it was explained that in our primate ancestry nurturing overcame genetic refinement's limitation to developing order, which was that it could not normally develop unconditional selflessness or 'love' in a species. Nurturing overcame this impasse and developed the utterly ordered, integrated, cooperative, selfless, loving, 'heavenly' state that humans once lived in, the instinctive memory of which is our 'soul', and the instinctive expectation within us of behaving as we did in that time, utterly cooperatively, is our 'conscience'. Nurturing was the main influence in the maturation of our species.

This utterly cooperative way of living became corrupted some 2 million years ago with the emergence of consciousness. Unlike the established *gene-based* learning system, which cannot develop insight and has only been capable of adapting or orientating species to situations, the emerging *nerve-based* learning system was insightful and had to *understand* the world. The already established, cooperatively behaving instinctive self, the product of the gene-based learning system, was in effect ignorant of the conscious self or intellect's need to master understanding and effectively criticised the *necessary* mistakes the intellect made as it attempted to gain understanding. The instinctive self left the intellect feeling it was bad or 'evil' for carrying out experiments in understanding. Lacking the biological understanding with which to refute this implicit and unjust criticism, all the intellect could do was retaliate against it (ie the intellect became angry), try to prove itself good and not bad (ie it became egocentric), and block the criticism out from its mind (ie it became alienated). From their 'heavenly', innocent, 'Garden of Eden', 'golden' state, humans' corrupted, 'fallen', 'sinful', 'hellish' state emerged. Angry, egocentric, alienated 'human nature' came into being.

Understanding that 'God' is the metaphysical religious term for the integrative, cooperative ideals of life (as was explained early in the *Plato* essay), we can understand the statement in Christian scripture that humans were once **'in the image of God** [cooperatively orientated]**'** (Genesis 1:27) as referring to this period when humans had become fully orientated by the genetic learning system to cooperative behaviour and were living perfectly cooperatively. Christ was referring to this cooperative period when he talked of a time when God **'loved me** [humans when they were like innocent Christ] **before the creation of the** [corrupt] **world'** (John 17:24), the time of **'the glory...before the** [corrupt] **world began'** (John 17:5). Humans were once innocent but with the emergence of consciousness a battle broke out between their newly emerged conscious self and their already established instinctive self. As it says in the *Bible,* **'God made mankind upright** [uncorrupted]**, but men have gone in search of many schemes** [understandings]**'** (Eccl. 7:29).

For 2 million years the task of conscious humans has been to stand up to, fight, and ultimately defeat with understanding, the inference that, because they were angry, egocentric and alienated (aggressively, competitively and selfishly behaved), humans were bad or 'evil' beings.

Tragically, this necessary battle against the ignorant inference that humans were fundamentally bad forced nurturing into the 'back seat'. The critically important search for knowledge by the conscious thinking self or ego (the definition of 'ego' is 'conscious thinking self') took precedence over nurturing. Since males were the group protectors and this threat of ignorance constituted a threat to the group, the responsibility of championing the ego over ignorance fell to men, and humanity changed from being a matriarchal, soulcentric, nurturing society to a patriarchal, egocentric, embattled society.

The effect of nurturing in our heritage was that every new generation coming into the world expected to receive an extraordinary amount of unconditional love. However, the reality for children entering this extremely embattled situation was that these instinctive expectations of encountering a loving, gentle world were brutally violated. Tragically, ever since the overwhelmingly difficult battle to champion the intellect over instinct imposed itself on human life, no mother has been able to nurture her offspring as much as all mothers were able to do before the battle emerged—and further, all fathers have been dysfunctionally egocentric and unable to reinforce their children with unconditional love. Further still, a society composed of such deficient parents could not collectively nurture its children as well as all groups did before the battle emerged.

As a result of this lack of love and reinforcement, children's pure (cooperation or integration orientated) instinctive selves were hurt and damaged in various ways and children were left variously mentally troubled. This was how the 'upset', the angers, egocentricities and alienations that resulted from the battle of the human condition were passed on to subsequent generations. As it says in the *Bible*, **'he** [God] **punishes the children and their children for the sin of the fathers to the third and fourth generation'** (see Exod. 34:7, 20:5; Deut. 5:9). The **'sin of the fathers'** originated with the battle of the human condition.

In the *Resignation* essay it was explained that when children reached early adolescence they tried to confront the issue of the human condition, tried to understand the corruption in the world around them and within them. Lacking the explanation of the human condition their efforts only led to suicidal depression, and at about 15 years of age they realised they had no choice but to resign themselves to a life of denying the whole depressing issue of the human condition.

While virtually all adults are resigned to living in denial of the issue of the human condition and the many truths that brings this issue into focus, the truth is there are vastly different degrees of childhood hurt and as a result there is a broad spectrum of alienation within humanity.

'Homes' at different distances from the incinerating bonfire of truth

Until now it has not been possible to openly acknowledge the immense differences in alienation amongst humans. Everyone has been living 'behind closed doors' in 'homes' at different distances from the incinerating bonfire of truth.

According to how much their instinctive expectations of being loved and reinforced had been met during their infancy and early childhood adults have varied in their degree of insecurity in the presence of the cooperative ideals and related truths. Their degree of insecurity was what governed the extent to which they had to live in denial of the issue of the human condition.

Individual humans each established an evasive way of living that suited their particular degree of insecurity. The more insecure an individual was, the more escapist, self-distracting, superficial and artificial their lifestyle had to be. (The stages that occurred in the development of self-corruption or psychological 'upset' under the duress of the human condition are described, and illustrated with cartoons, in *Beyond*.)

There were a number of factors, in addition to nurturing, that contributed to a person's level of corruption and therefore insecurity and therefore alienation. One of these was a person's age. While most corruption occurred during the sensitive, formative, impressionable years of infancy and early childhood, a certain amount of corruption occurred as life went on. Corruption has been a cumulative process, both in the individual and in our species (as was explained in the *Introduction,* under the heading 'The "deaf effect": the difficulty of reading about the human condition'). The older the person, the longer they had been living with and adapting to the heroic but corrupting battle to champion understanding over ignorance. Closely associated with age as a factor was the intensity and amount of corruption encountered throughout life. A person's level of IQ also contributed to corruption of the instinctive self or soul, for the more intelligent the person, the more likely they were to challenge their instinctive orientation to life and the faster they realised the need to adopt evasive denials to avoid depressing, unjust criticism.

A fourth factor that influenced the degree of corruption in a person's life—and as a result the degree of falseness and artificiality that until now they needed to employ to cope—was how genetically adapted their ancestors had become to living under the duress of the human condition. Because denial has been humans' only way of coping since the human condition emerged, alienation—with all its evasive, artificial, superficial, false, blind and cynical ways of thinking and living—has naturally become partially instinctive; there has been some selection for the ability to cope with the human condition. Those races that have been more exposed to the corrupting battle to champion the intellect over instinct have naturally become more adapted to cope with it. While it was not possible to explain the corrupt state of humanity, acknowledging differences in alienation between races and cultures—and indeed between sexes, ages and individuals—only led to unjust condemnation of the more corrupted; it led to false—so-called 'racist', 'sexist' and 'ageist'—inferences of superiority and inferiority.

As was explained in the previous essay, corruption or upset has been replacing innocence for some 2 million years, at times through the more corrupt feeling unjustly condemned by the more innocent and attacking them in retaliation for that unjust condemnation, but mostly by the more innocent finding themselves unable to accept and cope with the new more corrupt reality, the new level of compromise of the ideals. The biblical story of Cain and Abel acknowledges this toughening process, this genetic adaptation to the compromise and corruption of the ideals that resulted from the battle to champion the intellect: **'Abel kept flocks,** [stayed close to nature and innocence] **and Cain worked the soil** [became settled and began the discipline and drudge of the corrupting search for and accumulation of knowledge]**...Cain was** [became] **very angry, and his face was downcast** [depressed]... [and] **Cain attacked his** [relatively innocent and thus condemning] **brother Abel and killed him'** (Genesis 4:2,5,8).

Imagine the issue of the human condition and the related truth of the cooperative ideals of life as a bonfire, with each person, and indeed each race, constructing their 'homes' at various distances from that fire, wherever they found the heat bearable. A rare few could tolerate the full searing heat of the truth, metaphorically building right beside the bonfire, but most of humanity were much less able to cope with the truth and ranged themselves out from the hot flames of the bonfire accordingly. Each distance, each 'lifestyle', each person's 'home' had its own particular way of minimising the agony of life associated with its particular degree of insecurity. Each 'lifestyle' also had its own particular way of trying to contribute to the human journey as much as possible from that particular compromised, corrupted position. Life for almost all humans has been an extremely difficult juggling act of *having* to be evasive or false to some degree but also of *endeavouring* as much as possible to minimise and contain that level of false, escapist behaviour.

The different states of alienation or evasion or denial meant humans have had different needs, as these sayings attest, **'each to his own', 'one man's poison was another man's elixir'** and **'what was true for one person wasn't true for another'**. Further, humans could not admit, be honest about or even talk about what was really going on in their own and other people's lives. This disguise and denial is symbolised in the sketch below with each position represented by groups of houses rather than by groups of people. The deeper truth about the different states of alienation amongst humans has been hidden behind the 'closed doors' of their 'homes'. Humans have not been able to differentiate between these positions 'out from the bonfire', have not been able to admit what was really going on in those undifferentiated 'homes', not even to themselves, let alone others. What has been said here makes it possible to understand **'the Freudian notion that much of our inner life and motivation are unknown to us and that our desires, thoughts and actions are unconsciously motivated'** (Jason Cowley, *Australian Financial Review,* 31 Dec. 1998).

1998 JEREMY GRIFFITH

The cooperative ideal-fearing—'God-fearing'—denial-complying, evasive state that humans have had to live in is well described in the *Bible,* and similarly uses the analogy of fire, where it says, **'Let us not hear the voice of the Lord our God nor see this great fire any more, or we will die'** (Deut. 18:16). Plato also uses the powerful metaphor of fire in his cave allegory (refer to the *Plato* essay). So practiced did most humans become at evading the core question in human life of the human condition that today they allow virtually no confronting truth, no glaring light, to penetrate the safety of the darkness of their cave state of denial.

Resigned adults have had to employ various ways to cope with their particular encounter with the battle of the human condition. The immense tragedy of this situation was that without the ability to explain the human condition, it was not possible to explain and thus honestly defend the more evasive, dishonest ways or states of living against the criticism, express or implied, from the more innocent, less evasive and less dishonest states. That being the case, humanity was left with no choice but to accept the need to evade and deny the existence of different degrees of falseness or alienation in the world. Humans have even had difficulty acknowledging the existence of alienation—**'the dreaded A word'** as I once heard it referred to on television—let alone acknowledge that it differed immensely from person to person. A great silence, effectively denial, developed about the immense variations in the levels of alienation between people—and indeed about the extremely false state the whole human race was living in.

Now that we can explain, understand and honestly defend humanity's corrupted state, our alienation will subside and this historic denial of who is and isn't alienated can end. It is finally safe to confront the extent of humans' alienated condition and the immense differences in alienation between people. In fact now that we can acknowledge the truth it is critically important that we do so; the future of the human race depends on this honesty.

The real difference between humans has been the degree to which they were alienated. What has been termed 'personality' has, largely, been the expression of a person's particular state of alienation. This means that in the future when the human condition subsides and alienation doesn't occur, people will be very similar. While our world will lose some of its variety or 'colour' with this loss of different personalities, the incredible sensitivity and happiness that will come with being able to access the world of our soul again will lend our lives a depth of real beauty and a dimension that we have hardly dared to dream of. Even though people have not been able to acknowledge the issue of the human condition, they nevertheless occasionally do allow themselves to recognise its existence and at that moment to contemplate a world free of the human condition. I have encountered many comments to the effect that **'the future will be boring'** that have clearly been made in such moments. These comments are an understandable defensive reaction that people have in order to make themselves feel better about their agonising reality, yet to advocate staying in alienation, effectively remaining as dead people—with all the immense suffering that went with our estrangement from our true world and self—in order to preserve this 'colour' is absurd and a betrayal of all the human race has worked for. Once people understand what it will be like to be free of the human condition they will not seek to hold on to this paltry aspect of it.

Prophets and the concept of the 'Virgin Mother' demystified

In the process of elaborating on the truth of the great spectrum of alienation in the human race, a number of religious concepts have already been demystified. With a deeper appreciation of the great differences in alienation between people it is now possible to demystify the major religious concept of 'prophet'. (Some of this explanation was provided as a necessary part of deciphering Plato's cave allegory in the essay of that name, under the heading 'There were people who could live in the sun's light, but they were unbearably condemning for the cave prisoners.')

Considering the spectrum of alienation amongst the billions of people on Earth, there were always going to exist a few people who were sufficiently nurtured with unconditional love, and sheltered from alienation during their infancy and childhood, and who were thus uncorrupted. Such people were sufficiently free of divisive competitiveness, aggression and selfish egocentricity to not feel condemned by the integrative cooperative, loving, selfless, Godly ideals of life, and who thus did not have to resign themselves to a life of living in denial of those ideals and, as a result, become alienated from their soul and all the truth it knows. Not condemned by ideality they did not have to block out ideality to cope with their reality; if you are not at odds with the ideals you do not have to deny them. It was pointed out in the *Plato* essay that Plato hypothesised such innocence thus: **'But suppose...that such natures were cut loose, when they**

were still children, from the dead weight of worldliness, fastened on them by sensual indulgences like gluttony, which distorts their minds' vision to lower things, and suppose that when so freed they were turned towards the truth, then the same faculty in them would have as keen a vision of truth as it has of the objects on which it is at present turned' (*Plato The Republic,* tr. H.D.P. Lee, 1955, p.284 of 405).

It follows that the more a person's soul was hurt during infancy and childhood, and the more intelligent they were, the earlier resignation occurred. This occurs because the more intelligent were more likely to challenge their instinctive orientation to life and thus become corrupted, and also were faster to realise the need to adopt evasive denials to avoid any depressing, unjust criticism. These factors could contribute to resignation occurring as early as 12 or as late as 17, and in a few rare cases never occurring at all.

The people religions referred to as 'prophets' were these unresigned individuals who were capable of confronting and talking truthfully about the human condition.

Prophets have at times been thought of and referred to as superhuman, even supernatural figures, and, in some cases, as divine beings. The truth is, there has been nothing mystical or supernatural about prophets—as a group they were simply one of the various types of people that occurred along the immense spectrum of alienation that existed amongst all humans. They were not better or worse than other people, simply unresigned and thus able to think and talk truthfully. While everyone else was living in denial they remained 'truth sayers'. They were from the uncorrupted end of the alienation spectrum, sound in self, not separated from their true self. In fact the word 'holy' used to describe prophets literally means 'whole' or 'entire'; it has the same origins as the Saxon word 'whole', and thus confirms the prophet's wholeness or soundness or lack of alienation. (With regard to the earlier comment about intelligence contributing to alienation—this is not to say prophets are not intelligent, merely that if they were they had to have been especially nurtured to stay innocent.)

The renowned Australian poet, Les Murray, recognised that when the truth arrives it will simply be the truth that resigned minds have repressed, rather than something supernatural. In the poem, *First Essay on Interest,* he wrote, **'What we have received/is the ordinary mail of the otherworld, wholly common/not postmarked divine'** (*The People's Otherworld,* 1983, p.8).

The psychiatrist R.D. Laing once said, **'Each child is a new beginning, a potential prophet'** *(The Politics of Experience* and *The Bird of Paradise,* 1967, p.26 of 156). This strongly argues the point that if a person were sufficiently nurtured and sheltered during their upbringing for them to remain unaware or 'innocent' of hurt, that person could avoid resignation and thus remain an unevasive or denial-free thinker or prophet as an adult. (It should be explained that the 'ships at sea' category of people, mentioned in *Resignation,* were also people who did not resign; however they should have resigned because, unlike prophets, they were not sufficiently innocent to live safely, uncondemned in the presence of the cooperative ideals of life.)

Obviously with the state of the world as it is, encounters with the soul-corrupted, human condition-embattled state are almost unavoidable, and as a result exceptionally innocent adults, individuals who are capable of avoiding having to resign and who can thus grow up as unresigned prophets are going to be extremely rare. Playwright Samuel Beckett was only slightly exaggerating the brevity today of a truly soulful, happy, innocent life when he wrote, **'They give birth astride of a grave, the light gleams an instant, then it's night once more'** *(Waiting for Godot,* 1955). In fact, to nurture someone with sufficient love and trust in life for them to be innocent enough to not have to resign required an exceptionally innocent mother, in fact a mother who herself had not resigned and taken up a life of evasive lying and selfish preoccupation with seeking self-justification. Only a mother who was unresigned and thus had not become part of, and familiar with, the dishonest, corrupt world could raise a child oblivious to corruption, a child uninitiated into the corrupt world and thus unaccepting of corruption in their adult life. A child who could grow up as it were 'outside the square' of the corrupt world, someone who could see the 'wrongness' or non-ideality in people's behaviour. Once resigned, a woman knows about the world of resignation and no longer trusts in the existence of a world that is not dishonest and wayward. It is a trust in a world still 'as it should be' that allows a mother to reinforce in a child not only a similar belief but also an *expectation* of such a world.

Resigned mothers tried to hide their alienation from their children, but the reality is if a mother knew of the corrupt world, her children would invariably know of it also and psychologically adapt to it. Alienation was invisible to those alienated, but to the innocent—and children are born innocent—it was clearly visible.

Unable to acknowledge the importance a mother's innocence played in producing an unresigned prophet without exposing themselves to the criticising condemnation of their resigned, corrupted state, humans sought a metaphor that represented the truth but did not confront them too directly with it. That metaphor for an innocent mother was 'virginity'. As was explained in the first section of the previous essay, sex was at base an attack on innocence and so the symbol of innocence or purity in women *is* virginity. Of course to give birth to a child a mother could not be a virgin, but when seeking a metaphor for an innocent, unresigned mother, the concept of a 'virgin mother' was ideal. The real truth of the concept of the Virgin Mother of the Christ Child is that she was a virgin in terms of corruption, her soul was virginal. D.H. Lawrence recognised the essential innocence of Christ's mother when, in reference to her, he wrote **'Oh, oh, all the women in the world are dead, oh there's just one'.**

How much innocence and thus soundness humans retained into adulthood depended primarily on the quality of nurturing or mothering they received as infants. As the prophet Mohammed said, **'Paradise lies under the feet of the mother.'** There is also the famous saying, **'The hand that rocks the cradle rules the world'** and, as Olive Schreiner noted, **'There was never a great man who had not a great mother—it is hardly an exaggeration. The first six years of our life make us; all that is added later is veneer'** (*The Story of an African Farm,* 1883, p.193 of 300). The expletive that describes a corrupt man—**'son of a bitch'**—goes to the heart of the matter. In his 1973 television series and book, *The Ascent of Man,* historian Jacob Bronowski, alluded to the significance of nurturing and the purity that resulted from it when he said **'But, far more deeply,** [the brain] **depends on the long preparation of human childhood...The real vision of the human being is the child wonder, the Virgin and Child, the Holy Family'** (pp.424,425 of 448).

Assigning to prophets a supernatural and even divine status was the insecure world's way of acknowledging their difference without acknowledging their confronting and condemning innocence. By differentiating prophets in some mystical, supernatural way, people did not have to compare their own corrupted state with the innocent, uncorrupted purity of the prophet. Without the ability to understand and explain humans' corrupted state, the uncorrupted innocence and denial-free truthfulness of prophets was unbearably confronting and created the need to at least 'explain them away', if not remove them physically through persecution and even murder.

The consequence of almost universal corruption in the world is that there have been very few unresigned, unevasive, denial-free adults or prophets. Considering unresigned individuals also had to survive extreme oppression, persecution and ostracism during their early and middle adult years, because of their confronting denial-free truthfulness, it is remarkable that any survived sufficiently intact to become exceptionally penetrating thinkers in later life. The loneliness unresigned individuals experienced trying to maintain a denial-free way of thinking when continually subjected to immense coercion to abandon that way of thinking by virtually everyone else on Earth was described by Christ when he lamented, **'Foxes have holes and birds of the air have nests, but the Son of Man has no place to lay his head'** (Matt. 8:20). Amongst those who survived long enough to succeed in penetrating the human situation, a significant number were murdered for their honesty before their insights could gain a foothold of appreciation and support in the world. Anthony Barnett, Emeritus Professor of Zoology at the Australian National University in Canberra, highlighted the rarity of exceptional prophets, once saying, **'In all written history there are only two or three people who have been able to think on...** [a macro, all-confronting] **scale about the human condition'** (in a recorded interview, Jan. 1983).

The following is an account of one of the handful of truly sound and thus truthful thinkers or prophets to emerge in recorded history: **'sometime around or before 600 BC—perhaps as early as 1200 BC—there came forth from the windy steppes of northeastern Iran a prophet who utterly transformed the Persian faith. The prophet was Zarathustra—or Zoroaster, as the Greeks would style his name. Ahuramazda** [the supreme being or wise lord] **had appeared to Zoroaster in a vision, in which the god had revealed himself to be the one supreme deity, all seeing and all powerful. He represented both light and truth, and was creator of all things, fountainhead of all virtue. Ranged against him stood the powers of darkness, the angels of evil and keepers of the lie. The universe was seen as a battleground in which these opposing forces contended, both in the sphere of political conquest and in the depths of each man's soul. But in time the light would shine out, scattering the darkness, and truth would prevail. A day of reckoning would arrive in which the blessed would achieve a heavenly salvation, while all others would find themselves roasting in fiery purgatory. The concept of a single, all-powerful god was not entirely new. Egypt had flirted with the notion in the 14th century BC under the pharaoh Akhenaten, and the Jews had been tending towards it for centuries. But**

Zoroaster gave monotheism a powerful new impetus. And his view of moral struggle—light against darkness, truth versus falsehood—was a spiritual innovation of profound importance...promoting a standard of virtuous behaviour that illuminated the best years of Persian rule' (Time-Life History of the World, *A Soaring Spirit 600-400 BC,* 1988, p.37 of 176).

In the **'moral struggle'** of **'light against darkness, truth versus falsehood'**, where **'the keepers of the lie'**, the people maintaining the denial, were dragging humanity into ever greater states of darkness or falsehood or alienation, it was the handful of exceptionally sound unresigned individuals, the 'prophets', that have emerged during recorded history, who civilised humanity, **'promot**[ed] **a standard of virtuous behaviour'**. Prophets sufficiently resurrected the repressed truths to guide angry, upset, corrupted, resigned humans back onto a path that contained a semblance of cooperativeness, **'light and truth'**. As will be described shortly when religions are explained, this realignment brought about by prophets has been the basis of religions.

It required exceptional soundness to stand up to the **'powers of darkness, the angels of evil'**, the almost total denial or block-out of the resigned world. Typically, these exceptional individuals stood alone in the marketplaces or out in the desert and railed against all the lies, irrespective of the consequences to themselves, of persecution and even death. Free of denial, prophets were able to confront the truth of integrative meaning or God and thus think and talk truthfully all day long. For example, the *Bible* says, of one of the most unevasive thinkers in recorded history, the prophet Moses of the Old Testament, **'no prophet has risen in Israel like Moses, whom the Lord knew face to face'** (Deut. 34:10). Exceptional prophets such as Zarathustra or Moses, and the others around whom the great religions of the world were formed, could think completely honestly, uncondemned and thus free of denial of the integrative, cooperative ideals or God. As Jacob, another Old Testament prophet, marvelled, **'I have seen God face to face and yet I am still alive'** (Genesis 32:30). This ability was to the dismay of the rest of humanity whose survival from the threat of suicidal depression depended on maintaining denial. Only innocence has been able to face the cooperative ideals or God **'face to face'**, or, as the prophet Isaiah described it, **'delight in the fear of the Lord'** (Isa. 11:3).

Moses described how **'The Lord spoke to you** [the Israelite nation] **face to face out of the fire on the mountain.** [Only because] **At that time I stood between the Lord and you to declare to you the word of the Lord,**

because you were afraid of the fire' (Deut. 5:4,5). In 'Exodus' God warned Moses that **'no-one may see me and live'** (Exod. 33:20). As has been mentioned before in this book, 'Deuteronomy' provides a very clear description of resigned people's fear of the truth of cooperative meaning and the resultant issue of the human condition: **'Let us not hear the voice of the Lord our God nor see this great fire any more, or we will die'** (Deut. 18:16). Job pleaded for relief from confrontation with the human condition when he lamented, **'Why then did you bring me out of the womb?...Turn away from me so I can have a moment's joy before I go to the place of no return, to the land of gloom and deep shadow, to the land of deepest night** [depression]**'** (Job 10:18,20-22).

Again, as Christ described the situation, **'Everyone who does evil** [becomes divisively behaved] **hates the light** [hates the truth of cooperative meaning]**, and will not come into the light for fear that his deeds will be exposed...men** [who are divisively behaved] **loved darkness instead of light...everyone who sins is a slave to sin'** (John 3:20, 3:19, 8:34).

It should be explained that even amongst prophets there have been different classes of prophets. Earlier in the *Plato* essay I listed quite a number of contemporary denial-free thinkers or prophets. They were Jean-Jacques Rousseau (1712–1778), William Blake (1757–1827), William Wordsworth (1770–1850), Arthur Schopenhauer (1788–1833), Charles Darwin (1809–1882), Søren Kierkegaard (1813–1855), Friedrich Nietzsche (1844–1900), Olive Schreiner (1855–1920), Sigmund Freud (1856–1939), A. B. 'Banjo' Paterson (1864–1941), Eugène Marais (1872–1936), Nikolai Berdyaev (1874–1948), Carl Jung (1875-1961), Pierre Teilhard de Chardin (1881–1955), Kahlil Gibran (1883–1931), D.W. Winnicott (1896–1971), Sir James Darling (1899–1995), Antoine de Saint-Exupéry (1900–1944), Louis Leakey (1903–1972), Joseph Campbell (1904–1987), Erich Neumann (1905–1960), Arthur Koestler (1905–1983), Sir Laurens van der Post (1906–1996), Simone Weil (1909–1943), Albert Camus (1913–1960), Ilya Prigogine (1917–), Charles Birch (1918–), Robert A. Johnson (1921–), John Morton (1924–), R.D. Laing (1927–1989), Dian Fossey (1938–1985), Stuart Kauffman (1939–) and Paul Davies (1946–).

The question is how does this list of 33 contemporary denial-free thinkers or prophets reconcile with Anthony Barnett's assertion that, **'In all written history there are only two or three people who have been able to think on...** [a macro, all-confronting] **scale about the human condition'**? In the above material I have talked about *exceptional* prophets in recognition of the fact that some prophets were much more able than others to confront and think effectively about the issue of the human condition. The main distinction is between individuals who were sufficiently nurtured and thus secure in self to avoid having to resign to a life of denial—these are the exceptional prophets—and individuals who had resigned during adolescence but who were later able to find their way back to a sufficiently denial-free way of thinking to be considered a prophet.

It was described in the *Resignation* essay how some artists could so develop their capacity to bring out the beauty that exists on Earth, and by association bring into focus all the deeper human condition-related issues that such purity raises, that they could take themselves to the brink of suicidal depression. One of the greatest artists, Van Gogh, became so tormented he did suicide. The point is if a resigned person happened to take up a path that led back to the world of the soul and all the confronting truths that resided there, and had the courage, determination and gifts to pursue that extremely difficult path far enough, they could manage to reveal sufficient truth to be recognised as a denial-free thinker or prophet. Of course, as described in the *Resignation* essay, such heroic return-trips to the world of truth could lead to extreme depression and even suicide. If someone had become resigned during their adolescence then by definition they cannot have been sound enough to face the truth without being suicidally depressed by it. Resigned people fought their way back to the truth at their peril.

My view is that of the 33 denial-free thinkers or prophets listed above perhaps only two or three of them are exceptional prophets, unresigned individuals. To escape oppression from the effects of a world afflicted by the human condition in your infancy was extremely rare. For example Sir Laurens van der Post, whom I regard as an exceptional prophet, was the 13th of 15 children and yet his siblings, despite having the same mother, apparently could not escape becoming so corrupted during their upbringing as to avoid resignation. As was described in the *Resignation* essay, under the heading 'The extent of humans' fear of the issue of the human condition', Carl Jung had to struggle mightily to access some truth about the human situation. He was obviously not an unresigned, exceptional denial-free thinker or prophet. The difference in the ability to think freely about the human condition between someone who resigned to a life of denial and someone who did not is immense. While an unresigned mind can wander around in the world of the truth with ease, once someone has adopted denial of the world of truth it takes remarkable commitment and effort to reconnect with and attempt to again 'walk around' in that world.

The issue of women prophets should also be addressed. There is one woman prophet in the Old Testament of the *Bible,* Ruth, and in the list given above I have included Schreiner, Fossey and Weil as prophets.

As was explained in the first section of the previous essay, 'Bringing peace to the war between the sexes', men took up the task of championing the ego over the ignorant instincts of our soul. Our instincts were ignorant of our intellect's need to search for knowledge. It was emphasised that women, not responsible for the fight against ignorance, and so not participating in the battle itself, did not and could not be expected to understand the validity of the effects of the battle. Women had little empathy with and respect for the effects of the battle, in particular the egocentricity, alienation and anger it produced in men. As a result women tended to be soul-sympathetic, not ego-sympathetic. In *Jung and the Story of Our Time* Sir Laurens related a dream Carl Jung had about a blind woman named Salome. He wrote that **'Salome was young, beautiful *and* blind'**, adding that **'Salome was blind because the anima** [the feminine, soul part of humans] **is incapable of seeing'**. Women's 'blindness' has been their inability to empathise with, and thus sympathise with, the corruption of self that men's role of having to champion the ego over the ignorance of the soul produced in men. They are not as it were 'mainframed', as intuitively aware of the battle as men—in the same way as men have never been as 'mainframed' to the role of nurturing as women intuitively are.

Since women are not sympathetic to the conscious thinking self or ego's battle with the ignorant soul they are not in an effective position to mediate in the battle, and since a prophet's work was to reconcile the warring sides, a woman could not be a prophet in the truest sense. Women's lack of appreciation of the battle allowed them a greater ability than resigned men to reveal the truths about our soul's world; to breach the etiquette of denial. For example, it was, Deidre Macken who wrote the *Science Friction* article (referred to in the *Plato* essay) that was so exposing of science's deliberate denial of integrative meaning or teleology. However, women have not been in a strong position to grapple with the issue of the human condition. Women were relatively unaware of the struggle to exonerate humanity. Being as they were relatively unaware of the legitimacy and importance of the battle they tended to condemn rather than help and enlighten. To be an effective prophet you had to ultimately be able to bring love to the 'dark side', rather than simply criticise it.

As will be made clear when miracles are explained, Christ was able to heal people because he was able to bring uncorrupted love to their tormented, human condition-afflicted state. If Christ were an exceptionally sound person, but had no deep intuitive appreciation of the legitimacy of the battle going on on Earth, he would not have been able to offer people the strong, sound, honest, centring empathy that was so relieving and thus healing for them.

An exceptional prophet contained the qualities of both women and men in one being. A resigned man was virtually devoid of soul and as a result lacked any ability to communicate with the soul, but did intuitively appreciate the battle that was going on on Earth. A resigned woman was soul-empathetic but lacked a deep, intuitive appreciation of the battle. An exceptional prophet on the other hand was both imbued with soul *and* aware of the battle.

It was because women lacked empathy with the battle going on on Earth that they have had to work through men; women were man's **'helper'** (Genesis 2:20). Soul helped the intellect, it did not lead it. Christ did not choose 12 men as disciples because of the cultural conditioning of his day, as some have claimed. Christ was never influenced by arbitrary tastes and attitudes. He was only influenced by the truth emanating from his soul, and the truth is women have not been in the best position to mediate in the battle going on in the world. Humanity's recent 2-million-year adolescent search for its identity, search for understanding of itself, particularly for the reason for its divisive nature, has been a patriarchal journey, just as humanity's infancy and childhood was a matriarchal stage.

As was mentioned in the previous essay, the prophet Friedrich Nietzsche acknowledged women's greater ability to nurture, that women have had to work through men and that women haven't been aware of, or mainframed to, the deeper battle that has been waging on Earth. He wrote: **'Woman understands children better than a man...The man's happiness is: I will. The woman's happiness is: He will. "Behold, now the world has become perfect!"—thus thinks every woman when she obeys with all her love. And woman has to obey and find a depth for her surface. Woman's nature is surface, a changeable, stormy film upon shallow waters. But a man's nature is deep, its torrent roars in subterranean caves: woman**

senses its power but does not comprehend it' *(Thus Spoke Zarathustra: A Book for Everyone and No One,* 1892; tr. R.J. Hollingdale, 1961, p.92 of 342).

The reason Ruth has been included in the *Bible* as a prophet, and the reason I have recognised Schreiner, Fossey and Weil as prophets, is that they have each gone a long way towards meeting that definition of a prophet as being a denial-free thinker. Even though they were not able to grapple with the human condition directly they revealed so much honesty about the human situation that they greatly contributed to the destruction of the denial. While in some of her writing Olive Schreiner railed against men and in so doing revealed her blindness to the real nature of the battle going on in the world—in her famous 1883 book, *The Story of an African Farm,* she has her leading character say that men are **'fools'** (p.194 of 300) and that men's oppression of women is not part of God's plan, that **'He** [God] **knows nothing of'** it (p.189)—her extraordinary honesty about the corrupt state of the world, such as the lengthy passage quoted in the *Resignation* essay, was a revelation in a world drowned in denial. It is suggested that Schreiner, Fossey and Weil were able to reveal the truth because they had women's ability to expose the truth and possessed the exceptional courage, determination and gifts necessary to do so.

What hasn't been explained so far is what a false prophet is and how they differ from the true prophets described above. False prophets will be explained and the distinction between them and true prophets will be given shortly under the section 'The Apocalypse and the Battle of Armageddon explained.'

The story of Noah's Ark explained

Now that humans are at last able to acknowledge the broad spectrum of alienation and the occurrence of resignation, humanity is finally in a position to safely explain the true meaning of the biblical story, Noah's Ark. In light of this explanation, it is obvious that the great flood in this story represents a cataclysmic event that occurred in the human journey through consciousness. Clearly that event was when resignation became an almost universal phenomenon.

There was a time when all humans were sufficiently innocent, free of corruption, to be able to go through life without resigning themselves to blocking out ideality in order to cope with their reality; a time when all humans were prophets. However as the search for knowledge developed and corruption and alienation increased, more and more people needed to block out ideality to cope with their reality, needed to resign. Eventually there came a time when the average adolescent was growing up sufficiently corrupted or non-ideal to have to adopt the strategy of resignation to avoid the depression that their divisive condition would have otherwise caused—a time when resignation and the competitive and aggressive egocentric way of living became widespread, a normal part of human life.

The story of Noah's Ark metaphorically describes this time when resignation and its denial and oppression of the soul and all the associated truths became virtually universal amongst humans. It metaphorically describes the time when resignation 'flooded' the world and our soul and all its truths went under, 'drowned'. It describes the time when our soul was pushed into our subconscious, out of conscious awareness, and in its place the highly competitive egocentric way of living emerged. The only creatures to escape the horror of resignation, 'drowning', were the animals and the very few prophets, symbolised by the survival of the animals and Noah: **'Noah was a righteous man, blameless among the people of his time, and he walked with God** [he did not have to deny integrative meaning]**...God saw how corrupt the earth had become, for all the people on earth had corrupted their ways** [had become resigned and egocentric and living in denial of God]**. So God said to Noah, "...make yourself an ark...I am going to bring floodwaters on the earth...Everything on earth will perish** [the soul and all the denial-free truths will perish when people resign to a life of denial]**. But I will establish my covenant with you** [but from here on I will depend on prophets to preserve the truth of integrative meaning and all the other great truths that relate to it]**, and you will enter the ark...Go into the ark** [don't resign]**, you and your whole family, because I have found you righteous in this generation"'** (Genesis 6:9,12,14,17,18; 7:1).

The Bushmen people of southern Africa have a word for prophets that employs the analogy of a harvest. In Sir Laurens van der Post 1958 book *The Lost World of the Kalahari* he describes meeting a Bushman **'prophet and healer'** named **'Samutchoso'**, meaning **'He who was left after the reaping'** (pp.159&129 of 253). Christ too has been referred to as **'the firstborn from among the dead'** (Col. 1:18).

Demystifying the role of religions

With understanding of how people have lived in mortal fear of integrative meaning or God and thus have had to resign themselves to living a life of almost total denial of it—imprisoned in a deep, dark **'cave'** well away from the glare of the **'sun'** as Plato described it in his allegory—it is not difficult to understand the immensely valuable role 'religions' came to play in the human journey.

While the resigned state of denial saved humans from condemnation, it could eventually lead to a state of so much darkness, so much soul-death and meaninglessness, that the despair from living in that state was worse than the condemnation it provided an escape from. When this happened, life became so bereft of access to our species' original, cooperatively orientated, happy, healthy, God-imbued state that people began to feel a desperate need to find their way back to some connection with it. Humans were forever burying their condemning, ignorant, idealism-demanding instinctive soul or 'child within', only to find they wanted to unearth 'the child' again; they deliberately blocked out and lost sight of the cooperative meaning of life but as a result became so disorientated and lost that they had to try to find their way back to it again. As Fiona Miller said in her extraordinarily truthful poem (quoted in the *Resignation* essay), after resignation humans **'spend the rest of life trying to find the meaning of life and confused in its maze'**. Such was the oscillation, or Yin and Yang, or dialectic of life, under the duress of the human condition. On one hand humans were forever struggling with the consequences of living in denial, yet on the other were trying to face the unconfrontable truth about their corrupted, divisive condition.

The value of religions was that they provided an avenue for humans to be realigned or re-integrated with the true, cooperative, loving world of soul that they had previously devoted all their time denying, repressing and burying. In fact the word 'religion' comes from the Latin *re-ligare,* which means 'to bind' or integrate. By embracing and deferring to the true world and true words of the prophet around whom their religion was created, alienated people were able to live through the prophet's unresigned access to that true soulful world. They could be re-associated with or 'born-again' to the true world. (Incidentally, the common use of the term 'born-again', a term introduced by Christ [see John 3:3], represents a slip of resigned humans' evasive guard because it implicitly acknowledges that at some stage in life humans had killed off their true self, died in soul, an event which can now be seen as resignation. Similarly, humans' use of the word 'alienated' is an implicit admission that humans have become dissociated from their true self.)

Through religions people could relieve the pain they suffered as a result of their alienation and dishonesty by acknowledging the denial-free, honest world of the prophet their religion was built around. Religions saved people from themselves, from living so dishonestly, from the agony and guilt of being so false. Christ for example was called 'the saviour'. He saved people from having to live so falsely by providing people with a way to be 'born-again' to a state of *relative* honesty. By supporting the prophet, people were *indirectly* recognising the prophet's soundness and *indirectly* admitting to their own lack of soundness. The words 'relative' and 'indirectly' have been italicised for emphasis because the person was not totally or directly admitting to their corrupted, alienated state. While this admission of corruption and alienation was implicit in their actions, people were not having to openly admit to it. To do so was something they could not afford to do while they could not explain their corrupted, alienated state. To openly admit to their corruption while they could not explain it would have led to dangerous suicidal depression. Adopting a religion was a way for humans to be as honest as they could possibly be, short of being suicidally honest. As Carl Jung was fond of saying about the Christian religion, **'in Christianity the voice of God can still be heard'** (*The Undiscovered Self: Present and Future,* 1961).

Above all, to be religious was an expression of 'faith', faith that there was another world and state to which humans had lost access, and faith that one day that state would be regained.

In the book, *The Last Two Million Years,* under the heading **'Beginnings of Religion'**, it is stated that **'Man's religious urge through the ages has found expression in a bewildering variety of beliefs, ideas and practices. But one factor seems permanent and universal in religious experience: the sense of a supernatural "other world" which, though invisible, is believed to have power over men's lives. Even in the few religions which do not recognise a God, or are indifferent to the idea of one, such as early Buddhism or Jainism, this supernatural world is assumed to exist. Since the dawn of human consciousness men and women have regarded the supernatural world with a mixture of awe, fear and hope, and sought to bring**

their lives into harmony with it' (Reader's Digest, 1973, p 284 of 488). This **'other world'** that humans have regarded with the extremely contradictory mix of **'awe, fear and hope'** is in fact not **'supernatural'**, as in a realm beyond or above the natural physical world; rather it is 'ultranatural', the true, natural world that humans blocked out, became alienated from, when they resigned. In fact it was a reverse-of-the-truth lie to describe this 'other world' as supernatural, because it is humans' resigned, alienated state that is the unnatural or supernatural state.

In the *Plato* essay R.D. Laing talked about humans being separated from their true soulful state by **'a veil which is more like fifty feet of solid concrete'**. In part the quote was **'The condition of alienation, of being asleep, of being unconscious, of being out of one's mind, is the condition of the normal man** [p.24 of 156]**...between *us* and It** [our soul] **there is a veil which is more like fifty feet of solid concrete. *Deus absconditus*. Or we have absconded** [p.118]**'** (*The Politics of Experience* and *The Bird of Paradise,* 1967). We can now appreciate Laing's famous comment that **'Insanity is a perfectly natural adjustment to an insane world.'** The resigned evasive world was a state of such deep denial of the truths about our world and our condition that it *was* effectively 'an insane world', and escaping to a state of derangement or insanity *did* make a lot of sense. It is now also possible to understand the comment made by Diogenes, the Greek philosopher who was Plato's contemporary, that **'Most men are within a finger's breadth of being mad'**. (Incidentally, since holding onto such a false state has been extremely difficult and humans have been doing it for some 2 million years, natural selection must have excluded many people who could not adopt denial. This means that all humans today must have some instinctive propensity to block out the true world, a situation that makes the task of having people now confront the truth all the more difficult.)

While religions could not liberate humans from the agony of the human condition because they could not explain why humans are good and not bad (that enlightening explanation depended on science first finding sufficient understanding of the workings of our world), in providing people with an indirect way to reconnect with the soul's true world, they provided them with an invaluable way to withdraw from the brink of madness.

Religions could offer a degree of relief from the agony of the human condition but only with understanding of the human condition, which is at last found, could it end. It should be emphasised that the new culture that is coming has nothing to do with a new religion. As stated in my earlier two books, **'this work brings about the end of faith and belief and the beginning of knowing'**.

The Apocalypse and the Battle of Armageddon explained

With the dignifying understanding of the human condition available humans can now end their alienation and stop lying. In fact now that their corrupted condition is explained it is at last both safe and necessary for humans to be honest and acknowledge their alienation. It is also safe and necessary to acknowledge that levels of alienation vary immensely from person to person and race to race. Understandably it is an immense shock for people to suddenly confront the existence of their alienation, having lived virtually in total denial of it, and initially they will be inclined to resist acknowledging it. When a deep denial is challenged the immediate psychological response is to vigorously and urgently try to maintain it.

Since suicidal depression has historically been associated with trying to confront the previously inexplicable human condition, humans have come to fear the issue and the truth of their alienated state to an almost paralysing degree. This historic fear has to somehow be overcome, because it threatens to oppose and deny the arrival of the all-important, 2-million-years-searched-for, liberating truth about ourselves.

Alvin Toffler coined the term 'future shock', defining it as **'the shattering stress and disorientation that we induce in individuals by subjecting them to too much change in too short a time'** (*Future Shock,* 1970, p.4). The arrival of understanding of the human condition brings the real 'future shock', 'culture shock', 'paradigm shift', 'sea change' and 'brave new world' humans have long anticipated. For 2 million years humans have lived in deep denial of the truth about their corrupted condition; suddenly, they are now face-to-face with truth day, honesty day, exposure day, self-confrontation day, 'judgment day', 'revelation day', 'the day of reckoning', when all their denials or alienations are exposed. Humans cannot help feeling that the 'foundations' of their existence are being destroyed, despite the fact that these foundations—their old artificial defences—are actually being superseded by the real support structure, namely the actual *under-*

standing of their divisive state that they have always needed and sought. An anonymous Turkish poet spoke truthfully when he said that judgment day is **'Not the day of judgment but the day of understanding'** (*National Geographic,* Nov. 1987).

In using the biological understanding of the human condition, it is possible to decipher all our mythologies and clearly understand their meanings at last. Christian mythology anticipates the arrival of the liberating but all-exposing truth about ourselves as being like the arrival of a terrifying storm of thunder, lightning and hailstones. In Luke for example, it says the all-exposing truth will be **'like the lightning, which flashes and lights up the sky from one end to the other'**, and like a time when **'fire and sulphur rained down from heaven'** (Luke 17:24, 29; see also Matt. 24, Mark 13, Rev. 16).

While all the historic denials, disguises and evasions that humans have justifiably employed to cope are made obsolete when understanding of the human condition arrives, they are also suddenly exposed, made transparent. The truth destroys the lie. This exposure or, more to the point, the resistance to this exposure, signifies a very serious impasse to accepting the arrival of the liberating understanding of human nature.

It makes sense that the people who will especially fear the arrival of the truth about humans' alienated state and who will most resist exposure of their condition will be those who are most alienated. While the greater truth is that we can all welcome with relief the arrival of understanding of the human condition, those who suffered most from the effects of the human condition, those who were most hurt in childhood and as a result are the most insecure and alienated, will initially be the most fearful of this liberating but all-exposing understanding.

It is the arrival of this liberating but naked truth about humans, a truth that destroys the historic denial that resigned humans have hidden behind and which humans are variously in fear of, that is anticipated in the Book of Revelation in the *Bible.* The Book of Revelation is sometimes called the Apocalypse, with both titles derivative of the Greek word *apokalypsis* which means revelation. What is being uncovered or unveiled or revealed is the denied truth about humans.

Until understanding of the human condition arrived, humanity had to deny the immense differences in alienation between people. This was because without the explanation for humans' corrupted, alienated state, acknowledging any differences in that state (acknowledging who was more and who was less soul-destroyed during childhood) would have led to unjust condemnation of the more corrupted. It would have led to unjust, prejudiced, 'racist', 'sexist', and 'ageist' assertions of inferiority and superiority. To avoid unjust and condemning differentiation, no acknowledgment of differentiation was allowed; a great lie denying the existence of alienation had to be maintained. Everyone lived concealed from exposure 'behind closed doors', inside their 'homes'. The end result of this practice of denial of the existence and extent of alienation was that no one knew who was and was not alienated.

Now that the liberating, Earth-saving, but at the same time extremely exposing truth about humans has arrived, this 'concealment' is suddenly ineffective; the age-old denial of humans' state of alienation is exposed, the truth is revealed, the difference in alienation between people is suddenly apparent: **'On the day the Lord gives you relief from suffering and turmoil and cruel bondage...The desert and the parched land will be glad; the wilderness will rejoice and blossom..."...your God will come, he will come with vengeance; with divine retribution he will come to save you." Then will the eyes of the blind be opened and the ears of the deaf unstopped...Your nakedness will be exposed'** (Isa. 14:3; 35:1,4,5; 47:3).

With the arrival of the understanding of the human condition everyone who is resigned and living in denial will have their denial exposed and they will find it difficult to cope with that sudden exposure. The first reaction will be to try to maintain their denial by opposing this all-important, humanity-liberating breakthrough. However, those who are relatively secure, less alienated, with less to be exposed, will have enough strength to overcome this inclination. Those who are less secure, more alienated, with more to be exposed, will find it more difficult. Indeed many will be so afraid of the emergence of the truth about the human condition that they will feel they cannot contain their inclination to oppose it and will vigorously try to maintain the denial. A battle will break out between those who feel they cannot cope with the truth and want to stay in denial and those who can cope with the truth and want to end the denial. In fact the last great battle to be fought, described in the Book of Revelation as the 'Battle of Armageddon', is this battle, which has already commenced, between the more insecure and thus afraid and those who are relatively secure and relatively unafraid of the truth about humans.

The immense differences in alienation in the world that, until now, people have hidden from one another—even those they supposedly know intimately—will suddenly become apparent, with the less alienated being able to support and welcome the truth and the more alienated resisting it. This situation is perfectly described in the *Bible* where it says—immediately after describing the arrival of the truth as being **'like the lightning, which flashes and lights up the sky from one end to the other'** (Luke 17:24)—that **'two people will be in one bed; one will be taken and the other left. Two women will be grinding corn together; one will be taken and the other left'** (Luke 17:34,35; see also Matt. 24:40). With the arrival of understanding of the human condition some will immediately take up the truth and become part of the new world and others will initially feel they have to oppose the truth.

The word 'initially' needs to be emphasised because the greater truth is that confronting the human condition and ending our species' historic state of denial is manageable and can be negotiated by all humans. While some will initially resist the arrival of the truth and as a result will be temporarily left behind, it will not be long before everybody takes up support of the 2-million-year-searched-for liberating truth. The final 'Battle of Armageddon', between those who initially find it overwhelmingly difficult to cope with the truth about humans and those that don't, will be short.

As was explained in the *Resignation* essay in the section 'Renegotiating resignation', the reason why the understanding of the human condition can be coped with is that it is possible to support the understandings involved without overly confronting them. By doing so you avoid standing in the way of freeing humanity from the human condition. In fact the relief and enthusiasm derived from being able to adopt such an all-truthful, all-meaningful and all-liberating way of living will swiftly carry all before it. The entrenched denial will be swept away and the battle will conclude shortly after. A song by the band Hunters and Collectors, included in the *Introduction,* contains a description of the emergence of an immense army sweeping all before it in support of these liberating understandings of the human condition. Here are the relevant lyrics: **'Woke up this morning from the strangest dream/I was in the biggest army the world had ever seen/We were marching as one on the road to the Holy Grail//Started out seeking fortune and glory/It's a short song but it's a hell of a story...Well have you heard about the Great Crusade?/We ran into millions but nobody got paid/Yeah we razed four corners of the globe for the Holy Grail//All the locals scattered...they were hiding in the snow'**. In the *Bible* the prophets Joel and Isaiah described the same event using similar imagery. Joel said: **'Blow the trumpet in Zion; sound the alarm on my holy hill. Let all who live in the land tremble, for the day of the Lord** [the denial-free understanding of the human condition] **is coming...Like dawn spreading across the mountains a large and mighty army comes, such as never was of old nor ever will be in ages to come. Before them fire devours, behind them a flame blazes. Before them the land is like the garden of Eden, behind them, a desert waste—nothing escapes them. They have the appearance of horses; they gallop along like cavalry. With a noise like that of chariots...like a mighty army drawn up for battle. At the sight of them, nations are in anguish; every face turns pale. They charge like warriors; they scale walls like soldiers. They all march in line, not swerving from their course. They do not jostle each other...For the day of the Lord is near in the valley of decision. The sun and moon will be darkened, and the stars no longer shine. The Lord will roar from Zion and thunder from Jerusalem; the earth and the sky will tremble..."In that day the mountains will drip new wine, and the hills will flow with milk; all the ravines of Judah will run with water...Their bloodguilt, which I have not pardoned, I will pardon** [dignifying understanding of humans' corrupted state is finally found]**"'** (Joel 2,3). Isaiah similarly said: **'He lifts up a banner for the distant nations, he whistles for those at the ends of the earth. Here they come, swiftly and speedily! Not one of them grows tired or stumbles, not one slumbers or sleeps; not a belt is loosened at the waist, not a sandal thong is broken. Their arrows are sharp, all their bows are strung; their horses' hoofs seem like flint, their chariot wheels like a whirlwind. Their roar is like that of the lion, they roar like young lions; they growl as they seize their prey and carry it off with no-one to rescue. In that day they will roar over it like the roaring of the sea. And if one looks at the land, he will see darkness and distress; even the light will be darkened by the clouds'** (Isa. 5:26–30).

As was emphasised in the conclusion of the *Plato* essay, the most exciting and challenging adjustment humanity has ever had to make lies directly ahead. As difficult as it will be, adjusting to the truth about ourselves can be managed. Humans would never have had the strength to pursue the immense 2-million-year-long journey to find knowledge, ultimately self-knowledge, if they had not always believed that when they finally found the truth about themselves they would be able to adjust to it.

The most difficult aspect of this journey to understanding is its beginning, when there exists only a small group of people supporting these unevasive, immensely confronting understandings. It is in this early stage, where the public at large has still to appreciate the new way of living that the arrival of self-understanding makes possible, that the battle is at its most ferocious. It is this foundation stage of humanity's homeward journey that we at the Foundation for Humanity's Adulthood are negotiating and pioneering. This is the time when immense, unwavering courage is required. So deeply entrenched has our species' denial been that the task of reversing the practice is analogous to turning the Amazon River around in its bed. Humanity has been living in a state of denial for so long it is almost as if it cannot change its ways, but it can.

With regard to the fear of being 'not chosen' on 'judgment day', it might be mentioned that with alienation in the world escalating to terminally dangerous levels in the wealthier parts of the world (manifesting in cynical opportunism breaking out in the boardrooms of the world's biggest companies), and with the gulf between the rich and poor becoming so deep that it is as if there are two unconnected continents on Earth (manifesting in the 2001 September 11th terrorist attack in New York), there has been an upsurge in apocalyptic sentiment. A series of books by Tim LaHaye and Jerry B. Jenkins containing fictional apocalyptic stories selectively based on the scriptures, especially Revelation, have sold an astonishing 50 million copies (*Time* mag. 1 June 2002). The biggest seller of these books was *Left Behind: A Novel of the Earth's Last Days* (1995), the first of the *'Left Behind'* series. This focus on those 'left behind' plays on the anxiety that people intuitively feel of being unable to cope with the truth about the human condition when it arrives, of being left behind, unable to face the truth. The parable of two in the bed and only one being chosen taps into people's intuitive fear of knowing themselves to be extremely alienated and thus at risk of not being 'chosen'. It has to be emphasised that it is not an individual that does the 'choosing', it is the confronting truth that separates those who can accept the truth and those who find they cannot. Again it has to be emphasised that while some will be slow to adjust to the new way of living in support of these compassionate understandings, no one will be permanently 'left behind'.

Obviously, if the truth about humans is to emerge, the extremely entrenched denial of the issue of the human condition in the world today *must* be defied—at least until wider support emerges—by the few who are both young enough to take up the new ideas and secure enough to cope with the confronting nature of the ideas.

One of the problems to be overcome in introducing these denial-free understandings is that there have been so many 'false prophets' promoting artificial, pseudo forms of ideality that the whole business of bringing ideality to the world has been extremely discredited. These false forms of ideality, such as the New Age Movement, Environmentalism, Feminism, the Politically Correct Deconstructionist Movement, the Peace Movement, are such superficially satisfying forms of idealism to live through that when the real ideality arrives, namely the reconciling understanding of the human condition, people actually prefer the non-confronting, false forms of ideality. Most disconcerting of all is that those who have seen through, and been offended by, these false forms of ideality sceptically assume the real ideality is merely another of the false forms of ideality. Real ideality is discredited by the false forms of ideality that have failed before it.

These false or pseudo forms of ideality are extremely seductive because they give people relief from the horror of their corrupted state by allowing them to feel good about themselves without having to confront their corrupted state. In a world where people are rapidly becoming more corrupted and in need of relief from their condition, pseudo-idealism has become a plague. In fact it has gained such a foothold that it now threatens to control the world and lead it to a totally non-confronting, truthless state of oblivion. Pseudo-idealism *is* the 'Antichrist' because it is at base anti-truth, opposed to the truth which Christ so wholly represented. The following is a brief explanation of the extreme danger of pseudo-idealism.

Three of the Christian gospels (see particularly Matthew 24, Mark 13 and Luke 17) accurately predict that before the frightening storm of the truth about humans arrives the human journey would come under very real threat of being stopped dead by the emergence of what is described as the **'abomination that causes desolation'**. This **'abomination'** that halts progress and leads nowhere is the great lie that asserts there is no difference in alienation between humans. The gospels predict the emergence of the *artificially achieved* undifferentiated, criticism-free world; a world where no fundamental difference between the **'two people...in one bed'** is acknowledged. The emergence of **'abomination that causes desolation'** is described as the **'sign...of the end of the age'** (Matt. 24:3). It is a dangerous time that we now live in, where the more insecure have established a dishonest, pseudo-idealistic, 'politically correct', 'postmodern', artificially 'deconstructed', artificially undifferentiated, artificially criticism-free world. Our task as humans has been to *understand* the dilemma of our condition and in so doing render it obsolete and allow it to subside, *not* to abandon this search by transcending the whole issue and attempting to artificially impose an ideal state. That was a false way of living, promoted by false prophets; it halted progress towards self-knowledge and could only end in untenable levels of alienation, total, eternal 'darkness'.

False prophets claimed that the way to achieve a reconciled, 'new age' for humans was for everyone to simply dogmatically adopt ideality, transcend their reality, deny their own and others' alienation, 'think positively', 'embrace their human potential and be at one with the cosmos', and so on, when the real path to a reconciled new world for humans depended on going in the opposite direction, confronting the reality of humans' corrupted state, penetrating the almost universal denial and solving the underlying issue of the human condition. True prophets attacked the denial while false prophets sought ways to embellish it. An essay titled *Death by Dogma,* that is to be published shortly in another book I have written titled . . . . will explain in detail this immense danger of pseudo-idealism.

This time when pseudo-idealistic dogma would try to control the world was also anticipated by the prophet Daniel who, after warning of the **'abominations'** that lead to **'desolations'**, accurately concluded, **'but the people who know their God** [those who are less alienated] **will** [must] **firmly resist** [the **abomination** of the dishonest world of pseudo-idealism]**'** (Dan. 11:32).

The rise to dominance of pseudo-idealism in wealthier parts of the world and its counterpart of fundamentalism in the poorer parts of the world is the outward **'sign of the end of the age'**. With alienation at last able to be acknowledged and its extent revealed, it can be understood that the 'four horsemen'—'war, death, famine and pestilence'—that herald the apocalypse, as described in Revelation in the *Bible,* are the closest the alienated mind could come to acknowledging the fact of a terminally alienated world.

The demystification of God

The explanation of 'God' that is presented in *Beyond* was expressed in summary form in the *Plato* essay. In this, God was explained as being the metaphysical, religious acknowledgment of the fundamental truth of integrative, order-developing, cooperative meaning or purpose to existence, a direction to life that results from the law of physics called negative entropy. In support of this explanation the pre-eminent physicist, Stephen Hawking, said **'I would use the term God as the embodiment of the laws of physics'** (in interview, *Master of the Universe,* BBC, 1989). It was also mentioned that the leading physicist, Paul Davies, said that **'these laws of physics are the correct place to look for God or meaning or purpose'** (in interview, *God Only Knows, Compass,* ABC-TV, 23 Mar. 1997), and that **'humans came about as a result of the underlying laws of physics'** (in interview, *Paul Davies—More Big Questions: Are We Alone in the Universe?,* SBS-TV, 1999). This recent quote from Stephen Hawking was also mentioned: **'The overwhelming impression is of order** [in the universe]**. The more we discover about the universe, the more we find that it is governed by rational laws. If one liked, one could say that this order was the work of God. Einstein thought so...We could call order by the name of God, but it would be an impersonal God. There's not much personal about the laws of physics'** *(Sydney Morning Herald,* 27–28 Apr. 2002).

This demystification of God as being the law of negative entropy has given rise to three particular concerns:

Firstly, there is the difficulty of the exposure that occurs with the demystification of God.

Secondly, there is the perception that Hawking alludes to above of the demystification seeming to make God **'impersonal'**, of seeming to present an interpretation of God that does not account for the spiritual dimensions that people have experienced and come to associate with God.

Thirdly, there is the concern that if God is negative entropy and negative entropy ends when the universe ends, either in the 'heat death' or the 'big crunch', then God is not infinite as religions have maintained. Each of these concerns will be addressed in turn.

The difficulty of exposure that the demystification of God brings

The *Plato* essay dealt at length with the problem of the confronting nature of the truth of integrative meaning. Integrative meaning implies that humans should behave cooperatively, selflessly and lovingly, but human behaviour has been quite the opposite; competitive, selfish and aggressive. If humans accepted the truth of integrative meaning without understanding of their divisive condition they would be left feeling suicidally depressed. While the human condition was not able to be explained humans had no choice but to live in denial of the truth of integrative meaning. In Plato's cave allegory they had no option but to hide in a deep, dark **'cave'** away from the exposing glare of the **'sun'** and the **'fire'**.

As has been explained, religions were a way for humans to be honest about their corrupted, alienated condition without having to openly admit and therefore confront it. While the admission of being corrupted and alienated was implicit when they became 'born-again' to supporting the prophet's sound words and life, they were not having to openly admit to it; it was not explicit.

Similarly, being able to use the word 'God' in an abstract sense, instead of directly acknowledging the cooperative, loving ideals and meaning of life—the **'absolute good'** as Plato described it—saved humans from suicidal depression. The term 'God' was sufficiently abstract and thus remote to not confront humans directly with the truth of their corrupted, alienated state. This undefined gap of abstraction has protected people from exposure.

The difficulty with demystifying the concept of 'God' is that it destroys the comfort zone that the gap of abstraction provided. Having God demystified as negative entropy makes its integrative, cooperative, loving and selfless meaning inescapable.

It should be emphasised that prior to the demystification of God many people felt that even the abstract description of God was too confronting. While humans have not been able to explain the human condition, explain why they weren't 'bad' or 'evil' for being corrupted, they have always intrinsically believed they weren't fundamentally bad or evil. If humans had believed that they were fundamentally evil they wouldn't have been able to stand against this implication the way that they have for 2 million years. As a result, those humans who became extremely corrupted frequently took refuge in this greater truth and refused to accept any implied condemnation of themselves. While the concept of God was relatively abstract and thus not overly confronting, for some people it was sufficiently condemning to cause them to become atheists, asserting they did not believe in God. For similar reasons, with corruption, and its product block-out or denial or alienation increasing to extremely high levels in recent times, many people have turned to religions that do not emphasise God, such as Buddhism. It is worth quoting again the words from Deuteronomy in the *Bible* that articulate how hurtful confronting God can be: **'Let us not hear the voice of the Lord our God nor see this great fire any more, or we will die'** (18:16).

Another means for alleviating the condemnation that people could feel from acknowledging God was to broaden the gap of abstraction, step further away from tolerating any interpretation of God, or of seeing meaning in any religious concept for that matter. People could deliberately avoid and resist any interpretation of what 'prophets' are, or what 'the resurrection' or the 'Virgin Mother' or the 'miracles' of Christ represent, or of the integratively orientated biological nature of our soul or psyche (soul is derived from the Greek word 'psyche'), and so on. The more insecure people became the more fundamentalist or literalist they needed to be. Mindlessness saved people from hurtful mindfulness. Humans have been retreating deeper and deeper into the dark cave of denial for 2 million years. In *Jung and The Story of Our Time,* Sir Laurens van der Post railed against this increasing mindlessness of religion: **'Yet the churches continue to exhort man without any knowledge of what is the soul of modern man and how starved and empty it has become...They behave as if a repetition of the message of the Cross and a reiteration of the miracles and parables of Christ is enough. Yet, if they took Christ's message seriously, they would not ignore the empiric material and testimony of the nature of the soul and its experience of God that** [Carl] **Jung has presented to the world. He did his utmost to make us understand the reality of man's psyche and its relationship to God. But they ignore the call'** (1976, p.232 of 275).

Some people have rejected all religions because even those religions that do not acknowledge God have not been sufficiently free of condemning implications. Secularism, the rejection of all forms of religious faith and worship, is on the rise throughout the world. The soon to be published *Death by Dogma* essay will document the development and support of 'guilt-free' forms of idealism such as Communism which, unlike its predecessors, does not contain any recognition of the world of soundness. Instead, Communism dogmatically demanded an idealistic social or communal world and denied the depressing notion of God and associated guilt. The limitation of communism is that while there is no confronting innocent prophet present, there is an obvious acknowledgment of the condemning cooperative ideals. In time, as levels of insecurity rose, the need developed for an even more guilt-free form of idealism to live through. This was supplied by the New Age Movement in which all the negatives of humans' corrupted condition were transcended in favour of taking up a completely escapist, 'human-potential' stressing, 'self-affirming', 'motivational', 'feel-good' approach. The limitation of the New Age Movement was that while it did not remind humans of the cooperative ideals, the focus was still on the issue of humans' variously troubled, alienated state. The next level of delusion dispensed with the problem of alienation by simply denying its existence.

The Feminist Movement maintains that there is no difference between people, especially not between men and women. In particular it denies the legitimacy of the egocentric male dimension to life. The limitation of Feminism is that while it dispensed with the problem of humans' divisive reality, humans are still the focus of attention and that is confronting. The solution that emerged was the Environmental or Green Movement where there is no need to confront and think about the human state since all focus is away from self onto the environment. While it has been said that **'The environment became the last best cause, the ultimate guilt-free issue'** (*Time* mag. 31 Dec. 1990), there was still a limitation. In the Environment Movement there is still a condemning moral component. If we are not responsible with the environment, 'good', we are behaving immorally, 'bad'. In addition, the purity or innocence of nature contrasts with humans' lack of it. At this stage a form of pure 'idealism' had to be developed where any confrontation with the, by now, extremely confronting and depressing moral dilemma of the human condition was totally avoided. This need for a totally guiltless form of 'idealism' was met by the development of the Politically Correct Movement and its intellectual equivalent, the Deconstructionist or Post-modern Movement. These are pure forms of dogma that fabricate, demand and impose equality in complete denial of the reality of the underlying issue of the reasons for the different levels of alienation between individuals, sexes, ages, generations, races and cultures.

While humans lacked scientific explanation, metaphors and abstract descriptions were useful tools for explaining the world; however, in their bid to avoid exposure of their alienated condition some people took some metaphors and abstractions literally. People have actually gone in search of the wreck of Noah's Ark, while many think that the story of David and Goliath is simply the description of a child slaying a giant man, and that the fabled poem, *The Man From Snowy River* is merely a chase for a horse. (The truthful meaning behind the stories of David and Goliath and *The Man From Snowy River* will be explained shortly.) Many believe Christ literally rose from the dead, literally walked on water, and that his mother was literally a virgin. Many believe God *is* a supernatural, beyond-this-world being, an actual person up in the clouds, seated upon a throne surrounded by people with wings. As physicist Paul Davies has said, **'A lot of people are hostile to science because it demystifies nature. They prefer the mystery. They would rather live in ignorance of the way the world works and our place in it...many religious people still cling to an image of a God-of-the-gaps, a cosmic magician'** (from Davies' 1995 Templeton Prize acceptance speech).

Alienation is alienation; it is the need to avoid exposure, to maintain separation from the truth, to stay hidden in the cave of denial. For those sufficiently insecure/alienated, the 'abstraction gap' had to be preserved. The more insecure people were, the more religious they tended to be—the bigger the church, the larger the gap between the adherents and their soul—until they became so insecure that God became too confronting to acknowledge, at which point the many less confronting forms of idealism mentioned above *had to be* developed.

This situation suddenly changes when the dignifying understanding of the human condition arrives, as it now has. It is now, at last, both safe and necessary to demystify God, however for the more alienated it may initially seem too much truth to bear. Bringing God down to Earth, so to speak, destroys the up-in-the-clouds perception of God, and some people will prefer to stay with this transcendent, beyond-this-world, abstracted view. . . . .

. . . .

. . . .

. . . .

. . . .

. . . .

. . . .

Apology: As explained in Notes to the Reader, the dotted lines indicate text that has temporarily been withdrawn because it deals with issues before the courts. Once the legal restrictions end the fully restored pages will be available on the FHA's website, or from the FHA.

. . . .

As was explained in the *Plato* essay, science has been mechanistic not holistic; it has avoided acknowledging integrative meaning because science, like humanity, had to live in denial of integrative meaning while humans' divisive nature could not be explained. While some scientists have dared to acknowledge integrative meaning the majority have complied with this need to live in denial of it. In the *Plato* essay, an article titled *Science Friction* was quoted saying that the few scientists who have **'dared to take a holistic approach'** have been seen by the scientific orthodoxy as committing **'scientific heresy'**. The article said that scientists taking the **'holistic approach'**, such as the Australians **'physicist Paul Davies and biologist Charles Birch'**, are trying **'to cross the great divide between science and religion'**, and are **'not afraid of terms such as "purpose" and "meaning"'**, adding that **'Quite a number of biologists got upset** [about this new development] **because they don't want to open the gates to teleology—the idea that there is goal-directed change is an anaethema'** (*Sydney Morning Herald, Good Weekend* mag. 16 Nov. 1991). The late, highly acclaimed biologist, Stephen Jay Gould, was one notable scientist who strongly opposed any attempt at reconciling science or mechanism with religion or holism. To quote a review of one of his last books, *Rocks Of Ages: Science And Religion In The Fullness Of Life* (1999), **'Science and religion, he** [Gould] **argues, are separate do-**

mains of knowledge: the former deals with facts, the latter with meaning' (*Sun Herald, Sunday Life* mag. 5 May 2002). Gould is well known for being an ardent mechanist, for strenuously opposing any acceptance of the existence of meaning in existence.

. . . .

As was explained in the *Plato* essay, while science and humanity had a responsibility to be mechanistic and not holistic, to live in denial of the issue of the human condition, ... The opportunity for understanding of the human condition to emerge was always to be preserved.

Since science and religion provide two different perspectives on the one subject, namely the human situation, they must ultimately be able to be reconciled. In fact the reconciliation of religion and science, God and humans, holism and mechanism, ideality and reality, good and evil was the goal of the whole human journey of the last 2 million years! The whole objective of all the efforts of conscious humans has been to solve the dilemma of the human condition, find understanding of human nature. The 1964 Nobel Prize-winning physicist Charles H. Townes emphasised this objective of reconciling science and religion when he said, **'For they** [religion and science] **both represent man's efforts to understand his universe and must ultimately be dealing with the same substance. As we understand more in each realm, the two must grow together...converge they must'** (*The Convergence of Science and Religion, Zygon,* Vol.1 No.3, 1966). Sir James Darling, one of Australia's greatest educators, and former Chairman of the Australian Broadcasting Corporation . . . . has said: **'The scientist can no more deny or devaluate the truths of spiritual experience than the theologian can neglect the truths of science: and the two truths must be reconcilable, and it must be of importance to each of us that they should be reconciled'** (from a 1954 address pub. in Darling's 1962 book, *The*

Education of a Civilized Man, p.68 of 223).

In his 1991 commendation for my book *Beyond The Human Condition,* the Emeritus Professor of Zoology at the University of Auckland and fellow of St John's Theological College, Auckland, John Morton emphasised the importance of reconciling science and religion when he wrote: **'*Beyond The Human Condition* is a book about anthropology and the human future. So it is necessarily about Christianity and importantly relates it—as Christianity must ultimately be related—to biology. It is a forward view of humanity's moral progress and destiny'**.

Australian biologist Charles Birch has also said that **'Those who say that science and religion do not mix understand neither'** (from Birch's 1990 Templeton Prize acceptance speech). Ultimately science had to become holistic and religion tolerant of having its abstract concepts interpreted scientifically. In his 1976 book, *Jung and The Story of Our Time,* Sir Laurens van der Post wrote that **'There was, for instance, this progressive rift between religion and the scientific spirit. At their best, both seemed to be bound on the same quest. But I was...dismayed at their mutual failure to understand each other'** (p.38 of 275).

In religious texts and in the words of the prophets around which the great religions were founded there is acknowledgment that religions themselves looked forward to a time when their role of supporting humanity would be brought to an end by the arrival of understanding of the human condition. In Genesis in the *Bible* it says that one day **'you will be like God, knowing'** (3:5). Christ was anticipating the time when humans would become all-knowing when he said, **'another Counsellor to be with you forever—the Spirit of truth...** [this] **Counsellor, the Holy Spirit, whom the Father will send in my name, will teach you all things** [in particular it will make it possible to explain the riddle of life, the dilemma of the human condition] **and will remind you of everything I have said to you** [make clear what he was only able to talk about in abstract terms]**'** (John 14:16,17,26). As will be pointed out when the Trinity is explained, the **'Holy Ghost'** or **'Spirit'** is the nerve-based learning system, our conscious mind, which had to overcome the impasse of the dilemma of the human condition that occurred in the development of conscious thought. So the **'Spirit of truth'** is the human intellect, of which science is the ultimate expression. It is science that elicits sufficient understanding of the mechanisms of the workings of our world to make clarification of the human condition possible.

Christ was also looking forward to the time when knowledge could finally make dogma and metaphysics obsolete when he said, **'Though**

I have been speaking figuratively, a time is coming when I will no longer use this kind of language but will tell you plainly about my Father' (John 16:25). Again, as will be clarified when the Trinity is explained, the **'I'** Christ refers to is the instinctive self or soul, of which Christ himself was an uncorrupted, perfect personification. As was explained in the *Plato* essay—when it was explained that science was the liberator and soul the synthesiser—while mechanistic science had to find all the details about the workings of our world that make clarification of the human condition possible, a denial-free, unevasive thinker or prophet was required to assemble the truth about the human condition from science's hard-won but evasively presented insights. Science found all the pieces of the jigsaw of explanation but presented those pieces evasively, 'picture-side down', and as such it was impossible to assemble the 'jigsaw'. The 'jigsaw's' assemblage depended on somebody being able to flip over those pieces, expose all the hurtful truths; it depended on the involvement of a denial-free thinker or prophet. When Christ talked about the instinctive self, **'I'**, telling humanity **'plainly'** about the **'Father'**, about God, he was recognising this involvement of an unevasive thinker or prophet in assembling the biological explanation of the human condition from science's hard-won insights, and in the process necessarily demystifying and obsoleting religion.

It should be pointed out that in John 14, 15 and 16 where Christ talks about the **'Counsellor, the Holy Spirit, whom the Father will send in my name...teach**[ing] **you all things'** and, **'a time** [that] **is coming when I will no longer use this kind of language but will tell you plainly about my Father'**, he is talking about the *combined* efforts of the intellect and soul, in the form of science and a denial-free thinker, being involved in the liberation of humanity. Both were necessary. The denial-free thinker had to draw upon what science has been able to impart about the workings of our world in order to synthesise the liberating truth about the human condition. For example in John 16:7,8,13 it says **'When he** [the Counsellor] **comes, he will convict the world of guilt in regard to sin and righteousness and judgment...but when he, the Spirit of truth comes, he will guide you into all truth. He will not speak on his own; he will speak only what he hears** [from the Father/God]**...All that belongs to the Father is mine** [I am free of denial]**. That is why I said the Spirit** [the intellect/science] **will take from what is mine** [will depend on the denial-free soul to be able to synthesise the truth about the human condition] **and make it known to you.'** Clearly, to be speaking only from God, the **'he'** being referred to is the soul-guided prophet. But what is being referred to by **'Counsellor'**, **'the Spirit of truth'**, coming to **'guide you into all truth'** and by so doing bring about **'judgment'** day or exposure day or truth day, is the science-guided intellect because it is only with the insights into the workings of our world that science has found can the truth about the human condition be synthesised.

It needs to be emphasised that the cooperatively orientated instinctive self contributed a key but minor concluding role in the liberation of humanity from ignorance. The difficult work was done by science. It had to patiently investigate reality, find first principle-based explanation of how our world works, in particular how the nerve-based and gene-based learning systems operate (the understanding that makes clarification of the human condition possible), all the while carefully avoiding any condemning truths about humans. Humanity's real liberator or so-called 'messiah' was science. What was missing from the beginning of the human journey to enlightenment was not soundness, that has always been there to varying degrees, but sufficient knowledge to make clarification of humans' fundamental goodness possible. The concepts presented in these pages are a synthesis (and reconciliation) of biology, physics, chemistry, philosophy, psychology and indeed all the scientific disciplines. These explanations of the human condition are synthesised from the wealth of detail won at great personal sacrifice by the warriors of all the scientific traditions. However, science was only the institution created by humanity to investigate the mechanisms of our world and depended on the supportive structure of civilisation for its existence, and as such humanity as a whole is responsible for liberating itself from the human condition. In truth every human who has ever lived has contributed to this breakthrough.

In the Buddhist faith there is an anticipation similar to Christ's of a time when humans would be liberated from the human condition. Buddha said, **'In the future they will every one be Buddhas/And will reach Perfect Enlightenment/In domains in all directions/Each will have the same title/Simultaneously on wisdom-thrones/They will prove the Supreme Wisdom'** (Buddha [Siddartha Gautama] 560–480 BC, *The Lotus Sutra,* ch. 9; tr. W.E. Soothill, 1987, p.148 of 275).

It is clear that both Christ and Buddha anticipated a time when understanding would replace dogma, a time when, through the development of science, denial-free thinkers would be able to tell humanity **'plainly'** about God and therefore about the nature of humans' relationship with the cooperative ideals or God. They looked forward to a time when understanding would replace **'figurative'** speech or dogma, and when self-understanding or **'Perfect Enlightenment'** would ameliorate the insecurity of the human condition thereby making **'every one'** secure in self like denial-free prophets or **'Buddhas'**—thus ending the need for faith and religion.

Christ and Buddha anticipated an end to the need for religion, a time when understanding would replace dogma. The words of the prophets confirm that religions are not being threatened or destroyed by the arrival of understanding of the human condition, as some people fear, rather they are being fulfilled.

What *is* being obsoleted is the gap created by the use of abstract, metaphysical, mystical terms, the gap in which people were able to hide from excessive exposure to the truth of their corrupted condition, but, as stressed, the whole point of the human journey was to achieve that demystification. Humans are an understanding variety of animal, which means understanding is our responsibility and destiny. The purpose of science is to be a winnower of mystery, metaphysics and superstition, ultimately to explain the human condition, dignify humans and in the process demystify God.

Where is the spirituality in negative entropy?

It was mentioned that there are three particular concerns that occur in people's minds when it is suggested that God is the laws of physics, in particular, the law of negative entropy. The first which has just been dealt with was the difficulty of the exposure that the demystification of God brings.

The second concern is that the explanation of God as the law of negative entropy seems utterly incapable of accounting for the spiritual dimension that people have come to associate with God. Hawking raised this concern when he said, **'We could call order by the name of God, but it would be an impersonal God. There's not much personal about the laws of physics'** (*Sydney Morning Herald,* 27–28 Apr. 2002). Interpreting God as the laws of physics, in particular as the law of negative entropy, seems to destroy the **'personal'**, comforting, spiritual dimension that humans have come to associate with God.

This problem has also been raised by Templeton Prize-winning biologist, Charles Birch. (The prestigious and financially rewarding

Templeton Prize is awarded for **'increasing man's understanding of God'** [*The Templeton Prize,* Vol.3 1988–1992, p.108 of 153].) He and another Templeton Prize-winner, physicist Paul Davies, were the two scientists mentioned in the *Science Friction* article as committing **'scientific heresy'** by **'dar**[ing] **to take a holistic approach'**, where they are **'not afraid of terms such as "purpose" and "meaning"'**, and by so doing are **'cross**[ing] **the great divide between science and religion'**. Birch and Davies, like Hawking, are scientists who have dared to recognise the integrative purpose and meaning to existence—as the titles of some of their books intimate: Birch wrote *Nature and God* (1965), *On Purpose* (1990), and *Biology and The Riddle of Life* (1999); Davies wrote *God and the New Physics* (1983), *The Cosmic Blueprint* (1987) and *The Mind of God: Science and the Search for Ultimate Meaning* (1992). In a 1996 television program titled *Talking Heads,* in a panel discussion about the concept of God in which both Birch and Davies participated, Birch emphasised that **'For most people when you use the word "God" you conjure up a picture of a supernatural, magical magician...** [and] **the importance of science is that it has been a winnower of...** [such] **ideas about God that are wrong'** (SBS-TV, 28 Jan. 1996). He then went on to raise the issue Hawking alluded to, that even in the holistic, integrative meaning-accepting, order-developing, purposeful, negative entropy-driven interpretation of God, something seems to be missing, that the strictly materialistic, physics-based interpretation appears to be deficient, saying **'I have had experiences, and still have experiences, that I don't think materialism can explain.'**

While Birch accepts that the idea of a supernatural, magical God is wrong, he is still concerned that there are dimensions to the concept of God that a law of physics alone does not appear to account for.

What needs to be factored in to clarify this apparent deficiency is the extent of human alienation. Once it is appreciated just how alienated humans have become, as a result of having to live with the horror of the human condition, the dimensions to God, the **'experiences'** that Birch has had that he does not **'think materialism can explain'**, in fact the whole spiritual aspect of life, suddenly becomes accounted for in the concept of God as the law of negative entropy.

Negative entropy dictates that matter develops larger and more stable wholes. It integrates or orders matter. As has been explained, for a larger whole to develop it is vital that the parts of the whole act unconditionally selflessly. Simply stated, selfishness is disintegrative while selflessness is integrative. As has been explained 'unconditional selflessness' is the definition of 'love'. 'God' is the personification of the integrative cooperative, loving, selfless ideals of life. The old Christian word for love was **'caritas'**, which means charity or giving or selflessness (see the *Bible*, Col. 3:14, 1 Cor. 13:1-13, 10:24; John 15:13)—so 'God is love', or unconditional selflessness, or commitment to integration. Mechanistic science has practised denying this truth of the integrative meaning or purpose or direction to life. While the existence of negative entropy, or the Second Path of the Second Law of Thermodynamics as it is sometimes referred, has been acknowledged by mechanistic science, it's real significance, that of giving rise to an integrative purpose to existence, has been staunchly avoided. Once this avoidance or denial is penetrated and the real significance of negative entropy (as being 'love' itself) is admitted then negative entropy is not at all devoid of spirituality.

With regard to mechanistic science avoiding the real significance of negative entropy, Arthur Koestler said: **'I referred to "the active striving of living matter towards the optimal realization of the planet's evolutionary potential". In a similar vein, the veteran biologist and Nobel prize winner Albert Szent-Györgyi proposed to replace "negentropy", and its negative connotations, by the positive term *syntropy*, which he defines as an "innate drive in the living matter to perfect itself". He also called attention to its equivalent on the psychological level as "a drive towards synthesis, towards growth, towards wholeness and self-perfection"'** (*Janus: A Summing Up*, 1978, 223 of 354). It is precisely negative entropy's **'drive towards' 'wholeness and self-perfection'** that so unjustly condemned humans and necessitated their reluctance to recognise the real significance of negative entropy in science.

Sir James Darling was quoted earlier saying, **'The scientist can no more deny or devaluate the truths of spiritual experience than the theologian can neglect the truths of science: and the two truths must be reconcilable, and it must be of importance to each of us that they should be reconciled.'** As Darling points out, it has been in religions that the spiritual dimension to life has been maintained, albeit represented abstractly. For example, the abstract concept of God contained within it a recognition of an all-loving and all-beautiful dimension to life, a dimension that mechanistic science determinedly ignored. 'Love' for example is one of the most used words in human life, yet mechanistic science hasn't even had a definition for it.

Mechanistic science is just one expression of the alienation in the world, the estrangement from all the truth and beauty that really exists. It was explained in the *Resignation* essay that if people were not resigned and alienated they would have the kind of capabilities and sensitivities that savants have. This comparison gives the resigned, denial-maintaining, alienated world an accurate measure of just how spiritually bereft its world really is. The *Macquarie Dictionary* (3rd edn, 1998) defines 'spirit' as **'the vital principle in humans, animating the body or mediating between body and soul'**, so our 'spirit' is really our aliveness, our sensitivity to all of existence, and it is this capacity to be super-aware that largely died with resignation. Humans have had to live in a dark cave of denial; they have had to block out all the sunlight in the world, all the truth and beauty that our soul knows because of the soul's unjust criticism of the intellect. Mechanistic science had to comply with that strategy of denial.

I have known Charles Birch since I was a student in biology at Sydney University and I know of his great love of animals. I know that Charles has extraordinary empathy for his cats for example, and that that sensitivity extends beyond what conventional science is able to account for. Mechanistic science is alienated science, it is science that complies with the tragic and costly resigned strategy of denial of all things related to the all-sensitive world of our soul.

When humans resigned there were so many truths and aspects of the world that they had to block out from their mind, live in denial of. However, in order that their world would not be completely devoid of truth, meaning and beauty, what resigned humans did was gradually allow the deeper confronting truths to be acknowledged in relatively non-confronting abstract, mystical, metaphysical descriptions—in such terms as 'God', 'soul', 'spirit', 'prophet', 'sin', 'heaven' and 'hell'. Comfortable with all the deeper truths—in fact, with the whole 'spirituality' of life—described only in non-confronting abstract, mystical terms, it is a shock to have the arrangement breached. The gulf that now exists between what we have been able to acknowledge in scientific terms and the deeper truths that we have allowed through our defences using abstract description in religion, is immense. The so-called 'spiritual world' seems like another universe to resigned, alienated humans, but it is not.

For example humans are now able to actually understand, in first principle scientific terms, that God is love. This rational insight does not destroy the spirituality that humans have come to associate with God, rather it brings immense confirmation and reinforcement of that truth. Similarly monotheism, the existence of one great overriding, all-pervading God, can now be clearly and rationally understood as the law of negative entropy. In particular, understanding the human condition, namely how the development of order of matter led to humans' corrupted, divisive condition, explains how and why 'God' created the suffering in the world. It also explains how that suffering can at last be brought to an end. The relieving and reinforcing demystifications go on and on. It can be seen that the world becomes far more meaningful than it was before explanation. While humans could not understand and confront God, their world was for the most part a meaningless empty place—a dark cave-like existence as Plato described it. That hollow, empty, dark existence has gone.

During my 30s and 40s, when I became increasingly immersed in the task of explaining the human condition, my mother often commented that this world of explanation that I was becoming so preoccupied with seemed to her to have lost all the enthralment, beauty and magic that my world had been full of in my youth. In truth the ability to understand the world of humans liberates our mind and allows it to at last truly access all the beauty and magic in the world.

The demystification of our world doesn't mean the specialness of our world is removed, rather it means the specialness is made more tangible. Mechanistic science, like all forms of denial, has largely been a process of stripping the beauty and truth from the world whereas the denial-free explanation of the world restores all that beauty and truth. It allows humans to live with a full awareness of, and ability to savour, the beauty of integration. It is alienation that is devoid of spirituality, not the truth.

Humans' deep reverence for God as a deity is a direct measure of, and counter response to, the frailty of human life under the duress of the human condition. With understanding of our world all the beauty and truth that really exists out there at last becomes fully accessible. As Tim Macartney-Snape described it in his Foreword to *Beyond*, **'It is like having mist lift from country you've never seen in clear weather'** (p.18 of 203).

If negative entropy ends with the 'heat death' or 'big crunch' end of the universe, where does that leave God?

The third concern that enters people's minds when it is suggested that God is the law of negative entropy is what happens to God when the universe comes to an end, as physicists say it will, either through expanding until eventually the stars burn out—the so-called 'heat death' of the universe—or alternatively, the eventual contraction of the universe until the so-called 'big crunch' occurs in which everything is obliterated, including time. Does God come to an end when the universe comes to an end? Does this mean that God's omnipotent (all-powerful), omnipresent (all-present), omniscient (all-knowing) qualities are not eternal?

While Charles Birch has acknowledged that **'a deeper religion no longer envisions God as omnipotent creator outside a mechanical universe'** (Templeton Prize acceptance speech, 1990), he is troubled by the possibility of God ending with the end of the universe. In the aforementioned *Talking Heads* program, physicist Paul Davies pointed to the eventual destruction of our universe many trillions of years in the future, either through 'heat death' or 'big crunch', after which Birch asked Davies: **'But does *everything* come to an end? I mean, the physical universe will come to an end in what you are saying** [but] **is anything saved in the system?...The notion of God really collapses if God also collapses with the physical universe. Is there anything which is saved in the notion of God after the big crunch or whatever is going to happen at the end?...You talk of there being meaning and purpose and if you then think that at some stage the whole physical thing is going to collapse, where is the meaning and purpose then?...It only has meaning so long as it physically lasts and I think religions are very concerned about the possibility of there being something which goes beyond the physical state of the universe.'**

Davies' response emphasised the limitation of viewing our universe from within it, saying: **'We come back to the question of time. The physicists are very comfortable thinking of the whole of time as laid out all at once—the timescape. There may be a bottom edge—the beginning, the big bang—and a top edge—the end, the big crunch—but the whole thing is just there. There is a directionality to it, but you can stand outside of this spacetimescape as a whole and that can be something with meaning. There**

is nothing to say that it has to go on forever, that you can have a bottom edge but no top edge.'

As has been mentioned, physicist Stephen Hawking said **'I would use the term God as the embodiment of the laws of physics.'** The point is God is not only the law of negative entropy, but *all* of the laws of physics. In fact as Davies says, **'you can stand outside of the spacetimescape as a whole and that can be something with meaning.'** The whole phenomenon of the existence of the universe, and even its potential demise, is meaningful.

Humans made the negative entropy-driven integrative meaning of existence 'God' because, of all the laws of physics, that was the one that deeply troubled humans, the one humans have been so at odds with.

Integrative meaning was the one truth that humans lived in mortal fear of. It stood over humans as an awesome presence. It *was* as a God in the immensity of its significance. Humans 'wrestled' with 'God', with the issue of their apparent inconsistency with the cooperative, integrative meaning of life. 'God' condemned humans but in their deeper selves humans have always known that while they appeared to be at odds with 'God' there was a greater truth—that they one day hoped to find, and now have—that would explain that they were not in fact unGodly beings, that they were part of God's plan after all, that there was a biological purpose to their divisive behaviour. Humans were able to give expression to this belief that they weren't unGodly beings by recognising that while 'God' was something fearful, 'He' was also something loving and compassionate. 'God' was a frightening entity in people's lives, but 'He' was also immensely comforting, forgiving and redeeming. Because of the insecurity that resulted from the dilemma of the human condition humans have derived immense comfort from being able to associate with the idea of an all-powerful, all-present, all-knowing, overseeing, loving God. It can be seen that the whole dilemma of the human condition was intimately tied to the concept of 'God'.

The point is that the issue of 'God' for humans has really been the issue of the human condition; clarify and resolve the human condition and the whole issue of 'God' becomes redundant. The overseeing, fearsome, awesome, and at the same time, comforting, loving and forgiving concept of God, ends. The whole situation for humans changes.

Once free of the agony of the human condition humans will view themselves as part of a universe that is influenced by various forces and subject to various laws, each of which has to be lived with, however humans won't be condemned and oppressed by any particular force or law, as they have been for 2 million years by negative entropy.

Afterlife explained

Another concern the demystification of God as negative entropy raises in people's minds is whether this demystification destroys the possibility of there being an afterlife.

To answer such a question the problem of the human condition, humans' immense insecurity and resulting denial and alienation, again needs to be taken into consideration.

Now that it is possible to understand the human condition, understand that humans were actually good and not bad, were indeed part of 'God's' great plan—a profound part of the development of order of matter on Earth—it can be understood that all human effort since time immemorial has been meaningful. Humans can now understand that each and every human life is extraordinarily significant and that their efforts on Earth, the real essence of their being, *do* carry on and endure. The *spirit* of humans, the enormous courage that they have exhibited on the journey to enlightenment through the incredible darkness, loneliness and hardship of having to live in denial, lives on in each of us and is carried on in all subsequent generations.

In fact all of life is part of this great mission to develop order on Earth, and so all of life has passed on its spirit, its courage, its will, its determination to all subsequent life. In the *Plato* essay, a passage from Sir Laurens van der Post about the **'great procession'** and **'victory parade'** of life beautifully illustrated this point.

The truth is no one or no thing in essence dies! We and all things are eternal and everlasting in the sense that their meaningfulness is eternal and everlasting, irrespective of what happens to our universe. Sir Laurens van der Post described how humans carry on in an even greater way after their physical death when he wrote: **'We make a great mistake when we think that people whose lives have been intimately woven into our own, cease to influence us when they die...The dead become part of the dynamics of our spirit, of the basic symbolism of our minds. They**

join the infinite ranks of the past, as vast as the hosts of the future, and so much greater than our own little huddle of people in the present' (*The Face Beside the Fire,* 1953, p.63 of 311).

The problem for humans has been that they have been insecure as a result of the human condition. They have doubted that they were part of 'God's plan'. They sometimes feared that they might be evil beings, a blight on the planet, something even worse than meaningless. Living in this horrible state of insecurity for some 2 million years meant continually struggling to believe in themselves, in their worth and ultimately in their relevance to the future. They worried that there may not be an afterlife for them, in the sense of their positive contribution to the future. There were, however, times when humans were not overcome with depressing thoughts about their possible worthlessness and managed to realign themselves to the greater truth of their true worth or meaningfulness—a truth that they were fighting to one day be able to establish irrefutably. In these rare moments of clarity and security humans knew there was an afterlife for them, that they did contribute meaningfully to the future. Unable to explain it rationally, this greater truth, that humans were occasionally able to access, became expressed in literal terms, and these literal interpretations became central in religious beliefs. Over time, as humans became more alienated, the literalness became more simplistic, to the point where some people actually believe they are reincarnated as new humans or even as animals after death.

The other dimension to the concept of afterlife involves the concepts of heaven and hell. The insecurity of humans of doubting their worthiness or meaningfulness became expressed in the duality of a hell-versus-heaven afterlife. The more corrupted humans were the more they feared they were not meaningful and the idea of going to hell became the expression of that fear. Similarly, the more humans managed to avoid becoming corrupted, the more they felt they might be meaningful and the idea of going to heaven became an expression of that hope. From there, the idea of heaven and hell became developed as a form of reward or punishment for humans. The threat of going to hell deterred humans from living out their corrupted self's angers and frustrations. In reality, as has been explained, 'heaven' was the utterly integrated state that humans' once lived in, and will now return to, and 'hell' was the corrupted state that developed as a result of the battle of the human condition, a state which humanity is about to leave. Heaven and hell have been states here on Earth.

By analysing these various concerns people have with the demystification of God as being negative entropy, it can be seen that the common underlying problem is that people have been limited by their resigned state of alienation and cannot view their situation objectively. It is precisely this ability that prophets have of not being part of the resigned, alienated state that allows them to look truthfully in upon the alienated world. Davies pointed out that if **'you can stand outside of the spacetimescape** [and view it] **as a whole'** then **'that can be something with meaning.'** This is all very well but humans find it difficult to extract themselves from their position within the paradigm; to use a popular saying, they have trouble 'standing outside the square'. The same situation applies to the alienated paradigm—if you can stand outside it, then what's going on inside can be easily understood.

Another way of describing this situation is that people can't help projecting their alienated way of viewing the world onto every situation. Alienated humans are extremely limited in their ability to consider questions about 'God' and an 'afterlife' by the extreme selfishness of their alienated state. What essentially happened at resignation is that humans became selfish, they became *preoccupied* with trying to justify themselves and avoid any implication that they are bad. Resigned humans are preoccupied with their insecure state, always on the look out for something that will reinforce their worth. This preoccupation with self means they are selfish and this selfishness clouds their view of the world. They can't see things as they really are. In particular it is impossible for resigned, alienated humans to imagine a world free of resigned, selfish alienation.

Free of alienation, people in the future will be dramatically different to people today. In particular they will be full of love and generosity, devoid of selfishness. They will be secure in self and won't project their insecure view of the world onto everything they think about. If you ask someone who is in love with absolutely everything, is full of trust in life and generosity towards all things, including life and its limitations, and to whom everything is sparkling with beauty and the sheer magic of life, whether it matters that the universe is going to end one day, they may say that it doesn't.

The point is, it is virtually impossible for humans to see beyond their own generation's way of viewing the world. There are questions that cannot adequately be addressed until the new situation emerges and is adjusted to. To attempt to know, understand and accept the answers to some questions depends on first reaching a new paradigm or way of viewing the world. There are questions that should be left to the future. When Sir Laurens van der Post wrote that **'We live not only our own lives but, whether we know it or not, also the life of our time'** (title page, *Jung and the Story of Our Time,* 1976), he was recognising that humans are obliged to accept that they can only hope to live life within the confines of their generation—that they can be aware that there are going to be wonderful worlds for future generations, but that their responsibility is to fulfil the role allocated to their generation, and by so doing, ensure that that future potential will one day be realised. When I listen to the rhythmic chanting and singing of the Bushmen of the Kalahari, or the Australian Aborigines, I can sense the deep acceptance that they have for the immense distance there is still to travel from where they are to freedom from the human condition. While it conveys a feeling of being a long way from home, it is not lonely music, just deeply intuitive. Similarly when I listen to music from the 1930s and 1940s I feel that I can hear that era's acceptance that freedom from the human condition is not quite going to occur for their generation. They are toughing it out and bravely partying on into the night. The music of the 1960s is full of incredible optimism, while music of the late 20th century is full of manic, repetitive, mind-numbing, head-banging frustration. No more work towards enlightenment can be done and yet the answers still haven't emerged, and humanity is piling up on top of itself in an awful, utterly corrupted, alienated mess. Every race and every generation has intuitively known its exact position in the great human journey to enlightenment and freedom.

The Trinity demystified

As is fully explained in *Beyond,* the story of the development of matter on Earth involves a 'trinity' of 'characters'. They are the theme or purpose of existence, which is the integration or the development of order of matter, and the two great tools that have emerged for achieving it, namely the gene-based learning or information processing system (historically referred to as 'natural selection'), and the nerve-based learning or information processing system (the intellect).

Humans have long been aware of the existence of a trinity of fundamental 'characters' or forces at work in their world. Most of the great religions recognise a trinity.

Hinduism for example, recognises a trinity or 'Trimurti' of three powers or gods at work in the world. They are Brahma the Creator, Vishnu the Preserver and Shiva the Destroyer. These equate with negative entropy, the creator of order in the universe; the gene-based learning system which produced humans' perfect instinctive orientation to that order, and persists in us as its preserver; and the nerve-based learning system which gave rise to humans' corrupted divisive state.

In Christian doctrine the trinity is described as comprising God the Father, God the Son and God the Holy Ghost or Spirit. Again God the Father is integrative meaning, and God the Son and God the Holy Ghost or Spirit are, respectively, the two great tools for developing integration or order, the gene-based and nerve-based learning systems.

Christ is associated with God the Son because he was a pure, uncorrupted expression of the gene-based or genetic learning system's achievement of the development of order in the human species, humans' instinctive orientation to cooperative or integrative meaning. He was a human in which humans' original cooperatively orientated instinctive self or soul was uncorrupted. The common dictionary definition of a 'prophet' is **'someone who speaks for God'**. Since in biological terms God is the cooperative ideals of life, then, being free of alienation, a prophet *does* speak for God. Christ was an instinctive expression of the image of God. As Bruce Chatwin said in a moment of remarkable honesty in his 1989 book *What Am I Doing Here,* **'There is no contradiction between the Theory of Evolution and belief in God and His Son on earth. If Christ were the perfect instinctual specimen—and we have every reason to believe He was—He must be the Son of God. By the same token the First man was also Christ'** (p.65 of 367). As Chatwin points out, and as has been fully explained, there was a time when all humans were innocent, were prophets, uncorrupted expressions of our species' cooperatively orientated original instinctive self or soul.

The Holy Ghost or Spirit is the nerve-based learning system, our conscious mind, which had to overcome the impasse of the dilemma of the human condition that occurred in the development of conscious thought. As was explained in 'The demystification of God' section, the **'Spirit of truth'** is the human intellect of which science is the ultimate expression. It is science that finds sufficient understanding of the mechanisms of the workings of our world to make clarification of the human condition possible. Science was the messiah of the human journey.

Jesus Christ demystified

Christ has been described as **'the man no-one knows'** (*Great Lives Great Deeds*, 1966, p.448 of 448). As explained, the one type of person that humans have been unable to 'know', to confront the nature of, is unresigned prophets. The soundness of unresigned prophets was so confronting that humans deified them as a means of recognising their exceptional soundness without having to confront their own lack of soundness. By assigning to unresigned prophets a divine status as supernatural beings, unrelated to typical humans, people did not have to compare themselves with them. With reconciling understanding of the human condition found, it is at last safe and necessary to demystify unresigned prophets and admit and confront the fact that they were simply humans like everyone else, albeit ones who were exceptionally nurtured in infancy and exceptionally sheltered from corruption during childhood and thus, as adolescents, did not have to resign to a life of living in denial. The *Resignation* essay used the example of savants to illustrate the extraordinary sensitivities and capabilities of unresigned minds. While savants come from the opposite end of the soundness spectrum to unresigned prophets, their complete mental dissociation from the alienated world allowed them access to the extraordinary capabilities all humans would have if they were not resigned. Christ was clearly an unresigned prophet.

Being unresigned Christ could look into, and talk openly about, the human condition. Christ was recognising the power of the unresigned mind to think truthfully and effectively when he compared clever but resigned adult thinking to the thinking capabilities of the minds of unresigned children, when he said, **'you have hidden these things from the wise and learned, and revealed them to little children'** (Matt. 11:25). He also recognised that, unlike him, most people were living in a state of extreme denial of the truth of integrative meaning when he said, **'why is my language not clear to you? Because**

you are unable to hear what I say...The reason you do not hear is that you do not belong to God' (John 8:43-47).

If Christ had been resigned he would have been party to the 'lie', the great denial, and thus been unable to expose the lie. While resigned prophets variously succeeded in their attempt to expose this 'lie', the truth is humans cannot very easily undermine themselves, self-betray, 'rat on' their condition; a human's thinking cannot be alienation-free when they are alienated. A person cannot be a **'keeper of the lie'**, as Zarathustra described it, and also be capable of telling the truth. Christ made this logical point to those who were accusing him of being a dangerous, deluded charlatan when he said, **'Satan can't drive out Satan'** (Mark 3:23), and **'a bad tree cannot bear good fruit'** (Matt. 7:18).

Part of the process of deifying Christ has been to rob him of his human vigour and righteous strength. Christ has usually been portrayed as a sorrowful, soft, even weak, saintly, pacifist-like individual, when, like all true prophets, he was strong, defiant and even warlike. Prophets took on the world of lies. A prophet was not like a 'saint'. The 1971 *Encyclopedic World Dictionary* defines a saint as **'one of exceptional holiness of life'** and a prophet as **'one who speaks for God'**. A saint *lived* a holy life while a prophet defied denial, acknowledged integrative meaning and all the truths related to it. A saint was passive while a prophet was active. If Saint Francis of Assisi, loving the animals as much as he did, was strong enough, he would have taken that love into battle against the denial of the issue of the human condition, sought to address it and by so doing solve the cause of the corruption and destruction of Earth and its animals. Saints were people who had become 'born-again' to acknowledgment of, and adherence to, the soul's ideal world in a morally exemplary way, while unresigned prophets were people who had never left the state of innocence, had never resigned, and who sought to end the suffering in the world and bring permanent order to the divisive chaos in human life by addressing the underlying problem. In his 1966 book *Beautiful Losers,* the poet and musician Leonard Cohen clearly identified this difference between saints and prophets when he wrote that, **'A saint is someone who has achieved a remote human possibility. It is impossible to say what that possibility is. I think it has something to do with the energy of love. Contact with this energy results in the exercise of a kind of balance in the chaos of existence. A saint does not dissolve the chaos; if he did the world would have changed long ago. I do not think that a saint**

dissolves the chaos even for himself, for there is something arrogant and warlike in the notion of a man setting the universe in order.'

As will be explained later when the authority of prophets is raised, while prophets were 'warlike' they were the opposite of 'arrogant'. It was the extraordinary authoritativeness of unresigned, truthful thinkers that made them *appear* arrogant to insecure, resigned minds.

It should also be explained that the class of prophets who were resigned but who managed to work their way back to confronting the truths associated with the soul's true world, were not of the same category as saints who had become born-again to acknowledging and adhering to the ideal world of the soul. Saints did not try to penetrate the denial and dig up the truth about humans, as resigned prophets did, rather they simply transcended the whole issue of the human condition and adopted an ideal life, a life free of corrupt behaviour. One is a thinking state and the other is a non-thinking state. Saints adopted ideality to an exceptional degree, but avoided wrestling with the issue of why humans were not ideal.

While saints chose to influence the world by living an exemplary life, prophets sought to change the world by tackling the lies that people were practicing, and the issue of the human condition that lay behind those lies. Unresigned prophets, like Christ, were secure enough in self to confront 'God', the truth of integrative meaning and the associated issue of the human condition, **'face to face'**, as Moses described it. They were exceptionally capable of defying and penetrating the lies on Earth. They were immensely courageous in their defiance of lies or denial and they were immensely strong in the amount of soul they had guiding and supporting them. The unresigned prophet Kahlil Gibran had this to say on the subject: **'Humanity looks upon Jesus the Nazarene as a poor-born who suffered misery and humiliation with all of the weak. And He is pitied, for Humanity believes He was crucified painfully...And all that Humanity offers to Him is crying and wailing and lamentation. For centuries Humanity has been worshipping weakness in the person of the saviour. The Nazarene was not weak! He was strong and is strong! But the people refuse to heed the true meaning of strength. Jesus never lived a life of fear, nor did He die suffering or complaining...He lived as a leader; He was crucified as a crusader; He died with a heroism that frightened His killers and tormentors. Jesus was not a bird with broken wings; He was a raging tempest who broke all crooked wings. He feared not His persecutors nor His enemies. He suffered not before His killers. Free and brave and daring He was. He defied**

all despots and oppressors. He saw the contagious pustules and amputated them...He muted evil and He crushed Falsehood and He choked Treachery' *(The Treasured Writings of Kahlil Gibran,* 1951, pp.231–232 of 902).

In a public speech given in 1960 the incomparable Australian educator and resigned prophet Sir James Darling emphasised the need in life to be both sensitive *and* tough—precisely the capacities that unresigned prophets have, being sensitive enough to access the true world of the soul and tough enough to defy the all-dominating world of denial. Darling wrote that **'the future,** [Canon Raven] **has said, lies not with the predatory** [selfish] **and the immune** [alienated] **but with the sensitive** [innocent] **who live dangerously** [defy the world of denial]**. There is a threefold choice for the free man...He may grasp for himself what he can get and trample the needs and feelings of others beneath his feet: or he may try to withdraw from the world to a monastery...: or he may "take arms against a sea of troubles, and by opposing end them"...** [and so] **There remains the sensitive, on one proviso: he must be sensitive *and* tough. He must combine tenderness and awareness with fortitude, perseverance, and courage. The sensitivity is necessary because without it there is no life of the mind, no growing consciousness, no living conscience; nor is there any real communication one with another. It is necessary also if we accept Father Teilhard's extension of the idea of evolution as illuminating the end of life. Only by a growth of sensitivity can man progress from the alpha of original chaos to the omega of God's purpose for him...Sensitivity is not enough. Without toughness it may be only a thin skin...**[only from] **an inner core of strength are** [you] **enabled to fight back...Can such men be? Of course they can: and they are the leaders whom others will follow. In the world of books there are, for me, Antoine de Saint-Exupéry, or Laurens van der Post'** *(The Education of a Civilized Man,* 1962, pp.33–34 of 223).

The following essay, *One Solitary Life,* describes the value of the contribution Christ's soul-strength, toughness, and capacity to defy the world of denial has made in the world: **'Here is a man who was born in an obscure village, the child of a peasant woman. He grew up in another village. He worked in a carpenter shop until He was thirty. Then for three years He was an itinerant preacher. He never owned a home. He never wrote a book. He never held an office. He never had a family. He never went to college. He never set foot inside a big city. He never travelled two hundred miles from the place He was born. He never did one of the things that usually accompany greatness. He had no credentials but Himself. While still a young man, the tide of public opinion turned against Him. His friends ran away. One of them denied Him; another betrayed Him. He was turned**

over to His enemies. He went through the mockery of a trial. He was nailed to a cross between two thieves. While He was dying His executioners gambled for the only piece of property He had on earth—His coat. When He was dead, He was placed in a borrowed grave through the pity of a friend. Nineteen long centuries have come and gone, and today He is the central figure of the human race and the leader of the column of progress. I am far within my mark when I say that all the armies that ever marched, and all the navies that ever sailed, and all the parliaments that ever sat, and all the kings that ever reigned, put together, have not affected the life of Man upon this earth as powerfully as has the One Solitary Life!' (this essay is the common rendering of an original sermon by Dr James Allan Francis, *The Real Jesus and Other Sermons,* ch. 'Arise Sir Knight', 1926, p.123). The fact that much of humanity dates its history around the life of Christ—as either AD or BC—is another measure of the impact Christ has had on the human journey to enlightenment.

Incidentally the above passage says that Christ's three years of ministry before he was crucified took place in his early thirties. While many accounts state that Christ was in his early thirties when he died, in the *Bible* it says **'You** [Christ] **are not yet fifty years old'** (John 8:57). This strongly suggests that Christ was in his late forties when he undertook his ministry. The unresigned prophet, Sir Laurens van der Post's formative writing period—the time when he first articulated his exceptional insights into the human condition—occurred in his late forties and early fifties, with his centre-piece book, *The Lost World of the Kalahari,* published in 1958 when he was 51. The unresigned prophet Plato, who died when he was 80, wrote his seminal works, *Phaedo, Symposium* and *The Republic* in his 'middle period', when he was around 50 years old. The resigned prophet Sir James Darling's most powerful enunciation of his extraordinary vision for education was made in the annual Speech Day Address he gave at Geelong Grammar School in 1950 when he was 51 years old.

I have read somewhere that of all professions, the age when prophets 'hang out their shingles', establish their business, is the oldest. The journey of the maturation of a prophet is a long one. This journey is outlined in *Free* from page 180 onwards. In it I wrote that, 'While our basic alienation/personality was established in our infancy and early childhood we did not realise the full extent and consequences of it until we reached the middle and latter half of our life' (p.205 of 228). For most people the full flowering of their personality in mid-life resulted in a crisis because it meant having to confront the shortcomings of their corrupted, alienated reality. For the very few who were sound, this was when their soundness became manifest, when they learnt just how free of shortcomings they were.

Christ's miracles and resurrection demystified

Christ would not have been able to expose falseness and delusion, as he was able to do, had he been false and deluded. Honesty is therapy, it is the ultimate healing force—as Christ himself said, **'the truth will set you free'** (John 8:32)—and Christ's unresigned unevasiveness and resulting ability to expose falseness and delusion, while it could be extremely confronting, was also what centred people and made them well. It was what released them from their psychosis, from their crippling state of soul-denial. His honesty defeated their lies that were in effect killing them. His soundness healed people, brought them home from where they were living 'out in the cold', brought them back from their disconnected, estranged, separated, alienated state. Since the word 'psychiatry' comes from *psyche* meaning soul and *iatreia* meaning healing (literally 'soul healing'), Christ's completely denial-free mind meant he was the ultimate psychotherapist.

Immensely confronting as Christ's soundness was—it was the reason he was eventually murdered, crucified on a wooden cross—it was also immensely healing. His unevasiveness allowed Christ to see through people's denials and think truthfully and thus effectively about their situation. He could see where and why people were 'lost' or alienated. He could 'understand' them, not in terms of first principle explanation of their condition (in Christ's time there was no first principle scientific knowledge with which to explain the human condition), but in terms of being able to see into their situation, and this 'understanding' or appreciation represented true or pure love or compassion. To quote a reference to the work of psychoanalyst Carl Jung: **'Jung's statement that the schizophrenic ceases to be schizophrenic when he meets someone by whom he feels understood. When this happens most of the bizarrerie which is taken as the "signs" of the "disease" simply evaporates'** (R.D. Laing, *The Divided Self*, 1960, p.165 of 218).

The evasive explanation for Christ's ability to heal people was that he was able to perform supernatural feats called 'miracles'. The concept of miracles protected humans from having to admit to Christ's soundness or innocence and, by inference, their lack of soundness or innocence. Describing events in his life as miracles was a device for avoiding the unbearably confronting issue of the human condition.

The comedian Spike Milligan had this insecurity-defying comment to make about Christ's miracles: **'They made him do miracles..."Loaves and fishes, loaves and fishes, just like that!" This isn't indicative of the man. What he said and preached was enough. Why did he have to raise the dead? Did that make him holier? These are post-Jesus Christ PR stunts, raising the dead, walking on water. I find it an insult to the dignity of the man. I've written to *The Catholic Herald* about this. The outraged letters I've got! I said the Turin Shroud was a load of shit, I've said it for 15 years. Jesus Christ didn't need to do tricks'** (*Bulletin* mag. 26 Dec. 1989).

To understand the loaves and fishes 'miracle' that took place when Christ spoke to a gathering of people, it is only necessary to understand how astonishingly honest and penetrating the unresigned mind is compared with resigned minds. Unresigned prophets are extremely rare and what they can do is utterly astonishing to resigned minds. They can talk hour after hour about so many things that resigned minds cannot go anywhere near, and because they are thinking truthfully they can clear up issue after issue, make sense of all manner of mystery. What they can do *is* 'miraculous' to resigned minds. They venture into what the resigned mind knows of as a terrifyingly dangerous minefield and do somersaults out there, lie on their backs, run around, skip, go to sleep out there—they grapple with the human condition with impunity. It is all incredible to the resigned mind but really it is just that unresigned prophets have never had to adopt lying. For humans to sit down and listen to an extremely unevasive, truthful description and analysis of themselves and their world, *is* mesmerising—even though the resigned mind soon afterwards starts blocking out all the truths that were brought to the surface as it realises their confronting implications. The point is while they were listening to Christ talk, the listeners would have been astonished and enthralled, and so forgotten their hunger, satisfied with the distribution of what little pooled food there was available. Later on after the event, unable to acknowledge the astonishing truth about what really took place, the resigned mind had to find an evasive way of recognising the specialness of the occasion, and that was achieved by saying 'I remember a miracle occurred where a few fish and loaves of bread fed everybody'. Similarly, so overwhelmed would the audience have been by having so much truth emanate from someone, that, years later, some would evasively describe the impact of what happened by saying that when Christ finally departed at dusk and walked out through the shallows of the lake, to the disciples waiting in a boat to take him back across the lake he had 'walked on water'.

The concept of miracles is the resigned mind's way of acknowledging that an event is special without having to acknowledge its real significance. In fact the mystical interpretation the resigned world had to give to all the events in Christ's life hid very significant truths about those events. In the case of 'the resurrection' Christ's cause of offering himself as a retreat for humans from their corrupted state was martyred and thus did rise up after his death, but again, resigned humans had to find a way to mystify this important truth. Christ was murdered because resigned humans found his soundness and honesty unbearably confronting, but then resigned humans came to need and revere his soundness to such an extent that they deified him, brought him back into their lives in a very substantial way; metaphorically they did raise him from the dead (ie denied) state they had assigned him. To admit this however was to admit to being extremely corrupted which was unbearably confronting and so the best corrupted humans could do in terms of giving recognition to the marvel of Christ's—their souls'—death and resurrection in their lives, was to say that Christ had literally risen from the dead, actually come back to life. Being able to use the description of Christ having actually risen from the dead was far less confronting than having to acknowledge that they were corrupted and needing to be born-again through the recognition they were giving Christ's soundness.

One reason Christ had to accept martyrdom—he could have stayed away from Jerusalem where he knew there was extreme animosity towards him—was to set the standard for the resolve needed in those days to defy the denial of the soul's true world. The other reason was that while he was present he was extremely intimidating of the egos of resigned people. As was explained in the *Resignation* essay in the section 'Renegotiating resignation', one of the problems this denial-free information faces is that it produces a Mexican stand-off, a situation where the resigned mind finds it impossible to derive an egocentric win from the denial-free understandings—the truth destroys all the lies of people's delusions. The same egocentric stand-off was produced by the unevasive, denial-free soundness of Christ. It wasn't until Christ had died that people were easily able to acknowledge and take up the cause of his soundness. People could then derive an egocentric win from being someone prepared to acknowledge and live by his soundness—this being the psychological basis of the feel-good righteousness of being a Christian.

Christ clearly knew the effect of martyrdom because he explained that **'unless an ear of wheat falls to the ground and dies, it remains only a single seed. But if it dies, it produces many seeds'** (John 12:24). Throughout history we have seen the effect of martyrdom. For example Martin Luther King's cause, his **'dream'** of being able to **'speed up that day when all God's children, black men and white men, Jews and gentiles, Protestants and Catholics, will be able to join hands'** (Washington, 28 Aug. 1963), was given impetus when he was martyred. It is only after a gifted person dies that resigned humans have found themselves easily able to acknowledge their talent and contribution and this especially applies to the most threatening of all attributes namely innocence. There is no one more threatening to the resigned, evasive mind than a prophet. Christ knew his 'resurrection'—the public acknowledgment of his soundness—would follow his death because he said he would **'rise again'** after he died (see Matt. 20:19, 27:63; Mark 8:31, 9:31, 10:34; Luke 18:33, 24:7,46; John 11:24, 20:9).

What needs to be emphasised is that, unlike the situation now, Christ was introducing a religion. He was the embodiment of the ideals that resigned, corrupted humans could defer to and live through. He was the focus of attention. He was what was confronting. What is being introduced now is the denial-free *explanations* of the human condition and it is these explanations or understandings that people live in support of. An individual isn't the focus of attention. There is no religion involved, no faith or worship of a deity. It is the information that is doing the confronting and liberating. It is the information that resigned humans are initially at odds with, not an individual. As was described above, the Mexican stand-off is caused by the information confronting people and undermining their established practice of denial. This is significant because it means that it is not martyrdom but the logic of the explanations that ensures the denial-free state is resurrected. Recognition of the denial-free understandings of the human condition is not dependent on martyrdom but on the ability of the first-principle based, scientific explanations to make the truth undeniable. It is the accountability of the explanations, their capacity to explain the human condition and demystify our world that has, and will continue to, overcome human egocentricity and make the soundness of the understandings visible.

In fact martyrdom works against the objective of bringing first-principle based explanation of the human condition to humanity. With the blinds drawn on the world of denial it is now critical that I, the mind that was sufficiently unevasive to draw the blinds, present as much denial-free understanding as possible to avoid misunderstanding and confusion. To quote *Free,* **'a complete banquet—a feast of explanation'** (p.200 of 228) has to be laid out. The more unevasive understandings available the less dangerous procrastination there will be. The priority now is to avoid procrastination with lots of explanation of the nature of the new paradigm and how it works. The need now is not to leave the door ajar so that some might be tempted to try and shut it again, but to open it completely—put an end to any doubt, insecurity and fear about the new paradigm.

The main point being made is that resigned humans had to represent Christ's life in a non-confronting, abstract, mystical form. In particular, the way they acknowledged the soundness of Christ's life was in the reverence they gave his life, but again this had to be done in a non-confronting, abstract, mystical way—by making him divine, from a place far away up in the clouds called 'heaven', sitting beside a bearded person called 'God'. In truth, as has been explained, Christ was simply another human, with the only difference being that he had been sufficiently nurtured with love in his upbringing to not have to resign to a life of lying like virtually everyone else had to. Sir Laurens van der Post made this point about humans' religious images reflecting their degree of alienation, when he wrote: **'It seemed a self-evident truth that somehow the sheer geographical distance between a man and his "religious" images reflected the extent of his own inner nearness or separation from his sense of his own greatest meaning. If so this made the conventional Christian location of God in a remote blue Heaven just as alarming as, conversely, the descent of his Son to earth was reassuring'** (*Jung and the Story of Our Time,* 1976, p.31 of 275).

Why the psychologically desperate do not suffer from the 'deaf effect'

While talking about 'soundness centering people', it can now be explained why the psychologically desperate don't suffer from the 'deaf effect', the difficulty of reading about the human condition (as explained in the *Introduction).*

It is not difficult to understand that in living in such an extreme state of denial or alienation, resigned humans could reach a crisis point where they had become excessively alienated from their true self and as a result become what is labelled 'psychotic'. While it can now be understood that virtually all people have been psychotic to a significant degree, people with extreme psychosis—the ones society evasively refer to as being 'people with psychological problems'—can often 'hear' discussion of the human condition. There are two reasons for this. One is that they have become so desperate for relief from their exhausted, alienated state that they are ready and wanting to hear the realigning truth. The other is that people who have become extremely alienated may want to maintain their denials but under the stress of their circumstances, the rigidly maintained structure of denial in their mind breaks down, and the truth, as it were, is able to slip through these shattered defences. As was mentioned in *Resignation,* R.D. Laing was describing the 'shattered defence' access to the truth when he said that **'the cracked mind of the schizophrenic may *let in* light which does not enter the intact mind of many sane people whose minds are closed'** *(The Divided Self,* 1960, p.27 of 218). Laing went on to say that the German existentialist, Karl Jaspers, considered the biblical prophet Ezekiel **'a schizophrenic.'** While some biblical prophets may have accessed the soul's true world using shattered defence, those who had full and natural access to the soul were exceptionally sound, unresigned men, rather than exceptionally exhausted, alienated, separated from their true self, or schizophrenic people.

As mentioned, Christ was aware that his unevasive, human condition-confronting **'language was not clear'** to most people, that they couldn't 'hear' what he had to say. The very sick, however, were able to hear his unevasive words and were thus able to have their psychosis or soul-denial healed, along with all the outward, physical expressions of this psychosis. The truth is, the agony of the human condition has been the real agony in human life and the real cause of most physical sicknesses.

In summary, there are four categories of people who can readily 'hear' discussion about the human condition. They are those who have not yet reached the age of resignation; those who are exceptionally innocent and didn't have to resign to a life of denial of the issue of the human condition; those who are 'ships at sea', who should have resigned but didn't; and those who are exceptionally alienated and can hear through their shattered defence. All those in between, which is the great bulk of humanity, have great difficulty reading about the human condition.

Why prophets were 'without honour' in their 'own home' and 'own country'

It is timely to explain why it is that **'Only in his home town, among his relatives and in his own house is a prophet without honour'** (Mark 6:1-6).

The fundamental situation is that denial-free thinking, which is the essence of a prophet, has been an anathema—**'without honour'**—to the resigned world. The more corrupted humanity became the more confronting uncorrupted individuals were, and the more they were denied honour or recognition or acknowledgment. What alienation was separating itself from was the truthful, sound world of the soul. In a world almost devoid of soul, an emissary from the soul's world, which is what a prophet is, was actively denied and outcast by persecution, and even eliminated by murder. The basic activity of resigned humans was not to honour the sound, truthful, soulful existence that prophets represented.

However, while this is the fundamental situation, it has also been explained that while resigned humans needed to deny and repress the soul's truthful world, when they became overly corrupted they needed to find their way back to some truth in order to repair and heal their overly corrupted state. While the presence of truthful prophets was something humans tried to ignore and even destroy, over time they also came to need and appreciate—even to the point of revering—that presence.

While there emerged a need to recognise or honour prophets, it was still difficult to acknowledge their immediate presence because the resigned humans' ego came into play. As briefly mentioned earlier, it was difficult for resigned, egocentric humans to acknowledge the gifts of any individual when that individual was in their presence, nearby, or even still alive. The greater space and time between the presence of the especially gifted person and the average person, the easier it became for the average person to acknowledge their gifts without being made to feel inferior or worthless in comparison. Such was the level of insecurity in humans under the duress of the human condition. A great sportsperson often only received due credit for their achievements after they died, while many gifted individuals died in extreme poverty and anonymity, only to be resurrected and glorified by subsequent generations. Van Gogh managed to sell only one painting in his lifetime, yet his paintings now sell for millions of dollars.

While all talented and gifted people encountered this problem, there was no talent or gift as threatening to the ego of resigned humans as the gift of soundness. Egocentricity in resigned humans was all about trying to establish worthiness at the exclusion of the truth of their corrupted state. The presence of a sound prophet made that all-important exclusion almost impossible to maintain, and it therefore made the business of artificially deriving reinforcement from the world impossible.

A prophet's uncompromising truthfulness was both utterly confronting and utterly ego-deflating. It follows that the closer in both time and space resigned humans were to an unresigned prophet the more difficult it was for those humans to acknowledge the prophet's essential difference.

Christ suffered from this problem. To quote from the *Bible,* **'Now Jesus himself had pointed out that a prophet has no honour in his own country'** (John 4:44). Christ also said **'no prophet is accepted in his home town'** (Luke 4:16-30), and **'Only in his home town and in his own house is a prophet without honour'** (Matt. 13:54-57), and **'Jesus left there and went to his home town...When the Sabbath came, he began to teach in the synagogue, and many who heard him were amazed. "Where did this man get these things?" they asked...Isn't this Mary's son and the brother of James...Jesus said to them, "Only in his home town, among his relatives and in his own house is a prophet without honour."...he was amazed at their lack of faith'** (Mark 6:1-6).

It makes sense that the first place a prophet would go for support for his truthful way of thinking would be his own family, but 'a prophet is not recognised in his own home'. Tragically, his family is in fact the very last place he can expect to find support. Having grown up with the prophet, his family are the closest people to him in both time and space and so suffer most from the problems of being confronted by his truthfulness and having the artificiality of their world of reinforcement made transparent.

A prophet's mother is the closest of all people to him, but the reason for her inability to recognise the importance of his work is different to the other members of the family, who are resigned individuals. As has been explained, an unresigned prophet's mother is also necessarily unresigned. Being unresigned and thus having not taken up an egocentric attitude to life, the prophet's mother was not confronted by his honesty, or faced with having an artificial world of reinforcement made transparent. Being unresigned herself, the problem the mother of a prophet has is that she cannot see anything unusual about her unresigned son's way of thinking and behaving. To her he is just an extremely enthusiastic, energetic, soulful person, and as he grows up and begins to fight the world of denial, she, being a woman and relatively unaware of the nature of the battle, can be persuaded by her other children that what he is doing has no meaning, is unnecessarily uncompromising, destructive and even mad.

The potential trap for the prophet, of hopelessly trying to have his family appreciate and benefit from his work, has a dangerous capacity to exhaust and destroy him. A sound, unresigned person will naturally try extremely hard to have his family appreciate him and his work, however he simply has to be strong enough to at some stage realise the futility of trying to 'reach' his own family and be prepared to get on with his work without their support. In taking this step he can draw some comfort from the knowledge that all people constitute family and that his love for humanity simply has to take precedence over that for his own family. An unresigned prophet intuitively knows how precious his honest way of thinking is in a world that finds itself totally unable to think truthfully. He knows therefore that he must not fail to contribute his enlightening honesty to that world of terrible darkness and suffering, and so no matter how much he loves his family—and being free of corruption his love for them is without blemish and thus total—he must not cave in to their coercion to abandon his work. In terms of the value of the contribution the unresigned prophet Christ was able to make, the Australian educator, Sir James Darling, wrote that Christ's life **'was incalculably the most important event in human history, as we understand it, up to the present'** *(The Education of a Civilized Man,* 1962, p.206 of 223).

When Christ began his ministry and his family heard about it, they accused him of having gone mad, and acting on that belief, tried to take charge of him as if he did not know what he was doing. To quote the *Bible,* **'When his family heard about this, they went to take charge of him, for they said, "He is out of his mind"'** (Mark 3:21); **'For even his own brothers did not believe in him'** (John 7:5); **'but his own did not receive him'** (John 1:11); and **'Then Jesus' mother and brothers arrived. Standing outside, they sent someone in to call him. A crowd was sitting around him, and they told him, "Your mother and brothers are outside looking for you." "Who are my mother and my brothers?" he asked. Then he looked at those seated in a circle around him and said "Here are my mother and my brothers! Whoever does God's will is my brother and sister and mother"'** (Mark 3:31-35).

The persecution that unresigned prophets have had to endure for revealing the truth and the inevitable estrangement from their family is also described in the *Bible* in Psalm 69 of David: **'Those who hate me without reason outnumber the hairs of my head; many are my enemies without cause, those who seek to destroy me...O God of Israel. For I endure scorn for your sake, and shame covers my face. I am a stranger to my brothers, an alien to my own mother's sons; for zeal for your house consumes me** [I stand resolutely against the world of denial]**, and the insults of those who insult you fall on me'** (4,6,7,8,9).

Moses taught his people to stand by the truth against all odds, and, when they did, he said to them, **'You have been set apart to the Lord today, for you were against your own sons and brothers, and he has blessed you this day** [because you did not give in to the world of denial]' (Exod. 32:29).

The divisiveness of their work greatly impacted upon the personal lives of prophets. To stand against the world of denial was an extremely lonely occupation and it left many isolated from society and without honour in their own family. Christ, possibly more than any other prophet, knew this, stating, **'Do not suppose that I have come to bring peace to the earth. I did not come to bring peace, but a sword. For I have come to turn "a man against his father, a daughter against her mother, a daughter-in-law against her mother-in-law—a man's enemies will be the members of his own household." Anyone who loves his father or mother more than me is not worthy of me'** (Matt. 10:34-37), and, **'You will be betrayed by parents, brothers, relatives and friends, and they will put some of**

you to death' (Luke 21:16). It should be emphasised that Christ was introducing a religion to the world, he was establishing a place of soundness that people could defer to and live through when they became overly corrupted. When sound *explanation* arrives in the world, as it now has, while it will also be confronting of the world of denial, and thus divisive it shouldn't be *as* divisive as the situation faced by Christ because explanation can be understood. Religions were about supporting the embodiment of the ideals in the form of the prophet they were founded around. What happens with the arrival of understanding of the human condition is people live in support of those understandings. There is no faith involved: **'This is the end of faith and belief and the beginning of knowing'** (*Beyond*, p.166 of 203). With tolerance and patience, and a preparedness to accept logic, the explanations being presented can be evaluated as true or not. Faith can't be argued but logic can. If people are prepared to consider and accept reasoned argument there doesn't have to be conflict and division. The whole purpose of the human journey was to find understanding specifically because it ends the need for misunderstanding, both in ourselves and in others.

It should be explained here why some expressions of Christianity place so much emphasis on the Virgin Mother. The more corrupted humans became, the more important was nurturing, because nurturing is what was needed to bring about a less corrupted generation. Thus, the more a society became corrupted the more emphasis it placed on the Virgin Mother, the symbol of nurturing. Also, for an overly corrupted human it was much easier to defer to the Virgin Mary and the world of gentle nurturing, than to relate to the strong, confronting truthfulness of Christ.

Christ's acknowledgment that his mother and family sided against him and his ministry, and even questioned his sanity, puts those expressions of Christianity that emphasise the Virgin Mother in an extremely difficult and compromised position. The reader can imagine how difficult a truth this would be to accept for those that worship the Virgin Mary, and it follows that the Biblical passage that mentions Christ's family questioning his sanity, varies with different versions. The passage is Mark 3:21 and only the *New International Version* (NIV) gives what I believe is the real account. I am confident that the other versions have translated this passage in a way that does not imply that Christ's mother failed in her support of him. The underlinings have been added for emphasis.

The *New International Version* says 'When his family heard about this, they went to take charge of him for they said "He is out of his mind."'

The *King James Version* says 'And when his friends heard of it, they went out to lay hold on him: for they said, He is beside himself.'

The *New King James Version* says 'But when his own people heard about this, they went out to lay hold of him, for they said, "He is out of his mind."'

The *New American Standard Bible* says 'And when His own people heard of this, they went out to take custody of Him; for they were saying, "He has lost His senses."'

The New Revised Standard Version says 'When his family heard it, they went out to restrain him, for people were saying, "He has gone out of his mind."'

The Good News Bible says 'When his family heard about it, they set out to take charge of him, because people were saying, "He's gone mad!"'

I trust the *New International Version* because of the effort made in it to be accurate and because of the relative soundness or innocence of those involved in the translation. The preface to the NIV translation says that the project began in 1965 (a very idealistic and sound period) with more than 100 scholars working **'directly from the best available Hebrew, Aramaic and Greek texts'**. These scholars came from the United States, Great Britain, Canada, Australia and New Zealand (all relatively young, uncorrupted, innocent, sound cultures) and from Anglican, Presbyterian, Methodist and other denominations, none of which are regarded as particularly fundamentalist. I think it is significant that the Catholic Church, which is so orientated to worshipping the Virgin Mother, was not included in this process.

I might mention here, that within the various NIV editions of the *Bible,* I prefer the first edition over later ones. The 1960s, when work on the first edition commenced, was such an idealistic, innocent time. Subsequent decades were less idealistic and I think the revisions of the NIV *Bible* they made in those subsequent periods reflects this. In the first 1978 NIV edition, Matthew 24-27 reads, **'For as the lightning comes from the east and flashes to the west, so will be the coming of the Son of Man.'** In the second, 1983 NIV edition, this text has been changed to, **'For as lightning that comes from the east is visible even in the west, so will be the coming of the Son of Man.'** Christ was clearly using the most powerful metaphor available to express how exposing and thus confronting the truth about human nature would be when it arrives—that it will be like the onset of a great clashing and flashing thunderstorm. The revised version has corrupted this all-important point and changed it to using the brightness of lightning to illustrate the visibility of the second coming. It is the confrontational nature of the truth, not its visibility, that warrants the use of the metaphor of a thunderstorm.

For reasons explained, any unresigned, denial-free thinking prophet will necessarily experience great difficulty in having his family appreciate his work—as Christ said **'no prophet is accepted in his home town'**. The reason I have sought to clearly understand why unresigned prophets are not recognised in their own home is because I encountered this problem. As will be explained shortly, to be able to wander around in the realm of the human condition as easily as I have I must necessarily be an unresigned, denial-free thinker or prophet, albeit a contemporary one. In my family I am the second of four sons. My older brother saw my extreme idealism as **'dangerous'**, and the third brother described me as **'mad'**. This attitude led to a serious rift in 1991. My mother, to whom I attribute all my strength to be able to defy the world of denial, and whom I love more than anybody in the world, sided with these two brothers, even repeating to me that one of them thought I was mad. Incidentally, my father had died in a farming accident on our sheep property when I was 26 years of age, which was many years before this rift occurred. I had done all I possibly could over the years, short of self-destroying, to try and have my family understand the importance of my work, and once the rift occurred I had to decide whether or not to condone their effective dismissiveness of my work. Obviously I could not condone it and had to persevere with my work without them. Thankfully, my youngest brother, Simon, stood by me throughout this terrible experience and has continued to support me in this undertaking to bring understanding to the human condition. My other two brothers are relatively close to me in age while Simon is eight years younger than me, and it is the gap in both time and space that this age difference produced that has undoubtedly contributed to Simon being able to appreciate the real nature of my work.

Of the Members of the Foundation for Humanity's Adulthood, whose support of this project has been more precious than I am able to describe—this book is dedicated to them—the three who have been my closest supporters in this struggle to bring this denial-free information to the world have been my brother Simon, my partner Annie Williams, and the eminent Australian mountaineer, Tim Macartney-Snape.

Whenever an extremely difficult impasse for the resigned mind loomed in this undertaking, my brother Simon has always been there to tackle it. In particular Simon and two other FHA Members, Richard Biggs and Anthony Landahl, developed the techniques for dismantling the denial that resigned humans adopted at resignation, paving the way for others in the Foundation, and ultimately all resigned humans, to renegotiate resignation. The support Simon has given me personally, and given my life's work of bringing understanding to the human condition, all through the years we grew up together, through the trauma of our family rift, through the years of persecution I have been subjected to for daring to grapple with the human condition, and throughout the development of the FHA, has truly been saint-like. On a large card that all of us at the FHA gave Simon on his 50th birthday recently, I wrote: 'To the best brother a brother could ever have. You have given every last ounce of energy in your body to help me and I love you like there is no tomorrow, no yesterday, no anything else in the universe. Dear God thank you for Simon.'

Annie has worked by my side for 23 years now, helping with research, doing all the typing and computer work and looking after me practically. We began with one of the first word processors and the writing and research has been going on almost daily ever since. The contribution Annie has made to my work is incalculable and she is a model to all in the FHA of the selfless potential that these understandings make possible. I love her like the summer sun that envelopes everyone with its generous warmth.

My close friend, Tim Macartney-Snape, is a twice honoured Order of Australia recipient. He is a world renowned mountaineer, being the first Australian to climb Mt Everest. He is also a biologist and former student of Geelong Grammar School, the school that Simon and I and others in the FHA also attended. . . . .
. . . .
. . . .
. . . .

For the final essay in this book, Tim Macartney-Snape, in his role as Vice-President of the Foundation for Humanity's Adulthood, has written a summary of the work of the FHA and its Membership.

Denial-free books

All but an armful of the books written in human history comply with humans' strategy of evading the issue of the human condition. The few books that are not party to the 'noble lie' are, in the main, the great religious texts. Fortunately for the resigned world, the denial-free, unevasive words and truths in these texts are not overly confronting because the truths are expressed in pre-scientific, abstract, often metaphorical language that leaves them sufficiently obscure not to confront humans directly with their meaning and implications. Even so, in humans' unreconciled state, the way they have coped with the degree of confrontation these religious texts cause has been to dishonestly assign to them a divine, separate from humans, origin. (The honest ultra-natural—as opposed to the dishonest supernatural—origin of religious texts was explained earlier when prophets were explained.)

Martin Luther accurately compared books written from the blind, denial-complying, evasive position with those written from the unevasive, honest position, saying **'The Superiority of the Homer, Virgil, and other noble, fine, and profitable writers, have left us books of great antiquity; but they are nought to the Bible'** (*Great Thoughts on Great Truths,* gathered by Rev. E Davies, book undated but appears to be late 20th century, p.56 of 707).

Apart from the religious texts there are a few other books that have been written from the denial-free position. The FHA has a small but comprehensive library containing a collection of denial-free works. In particular, the collection contains works by Jean-Jacques Rousseau, William Blake, William Wordsworth, Arthur Schopenhauer, Charles Darwin, Søren Kierkegaard, Friedrich Nietzsche, Olive Schreiner, Sigmund Freud, A. B. 'Banjo' Paterson, Eugène Marais, Nikolai Berdyaev, Carl Jung, Pierre Teilhard de Chardin, Kahlil Gibran, D.W. Winnicott, Sir James Darling, Antoine de Saint-Exupéry, Louis and Richard Leakey, Joseph Campbell, Erich Neumann, Arthur Koestler, Sir Laurens van der Post, Simone Weil, Albert Camus, Ilya Prigogine, Charles Birch, Robert A. Johnson, John Morton, R.D. Laing, Dian Fossey, Stuart Kauffman and Paul Davies.

Virtually everything written and said by humans has been dedicated to maintaining the great lie by which humanity has lived. In fact everything about humans is now almost saturated with the lie.

We are almost completely fraudulent beings.

It is interesting that, in contrast with the situation today where society does not recognise its prophets and instead evasive intellectualism holds sway everywhere, the ancient Hebrews collected only the words of their prophets. Humanity does not have any records of the great authors or poets or playwrights or composers or artists or singers or astronomers or legal minds from the 5,000 year-history of the Israelites. All we have is the collection of the words of the few prophets that lived amongst the Israelites during those 5,000 years. That collection is the *Bible*.

The more corrupted and alienated people became as the search for knowledge continued, the more insecure they became and thus the more evasive they became of any condemning idealism. While prophets have been persecuted throughout history for being unevasive and for being so defiant of evasion, in earlier more innocent times, people were sufficiently secure to be able to acknowledge their prophets.

These early more innocent and thus less evasive civilisations even sought out their prophets to lead their societies. The Old Testament of the *Bible* is the documentation of the search for prophets to lead the Israelite nation. Moses, who was an exceptionally denial-free thinking unresigned prophet, lead the Israelites out of Egypt to the foothills of 'the promised land' of Palestine. The ancient Athenian society elected only uncorrupted, innocent shepherds to run their society. To quote Sir Laurens van der Post, **'He** [Pericles] **urged the Athenians therefore to go back to their ancient rule of choosing men who lived on and off the land and were reluctant to spend their lives in towns, and prepared to serve them purely out of sense of public duty and not like their present rulers who did so uniquely for personal power and advancement.'** Sir Laurens continued, **'Significantly in *The Bacchae*, the harbinger of the great catastrophe to come is "a city slicker with a smooth tongue"'** (in his foreword to Theodor Abt's book *Progress Without Loss of Soul*, 1983, p.xii of 389). In the biblical Book of Exodus, Moses took the advice of his father-in-law to **'select capable men from all the people—men who fear God, trustworthy men who hate dishonest gain—and appoint them as officials over thousands'** (Exod. 18:21). (Note the 'fear' of God mentioned here refers to respecting God, not the fear of integrative meaning that Isaiah for example was referring to when he said prophets **'delight in the fear of the Lord'** [Isa. 11:3].)

In another example of how humanity has become increasingly

insecure and thus dishonest I have noticed that often the earliest investigators of a scientific discipline revealed the most truth, and that as the discipline became more developed it became more sophisticated, which, if you look up the definition of 'sophisticated', means 'false'. Probably the best example of this phenomenon is Plato. What the *Plato* essay reveals is that philosophy has retreated from the clarity of the thinking Plato demonstrated over 300 years before Christ. As the 20th century philosopher Alfred North Whitehead said, the history of philosophy is nothing but **'a series of footnotes to Plato'**. In the study of anthropology, the renowned anthropologist Richard Leakey, who didn't attend university, was exceptionally honest in some of his early writings about the original innocent, cooperative purity of the human race. In his 1977 book *Origins,* which he wrote with Roger Lewin, he said: **'We emphatically reject this conventional wisdom** [that war and violence are in our genes]**...the clues that do impinge on the basic elements of human nature argue much more persuasively that we are a cooperative rather than an aggressive animal.'** I haven't been able to find such a clear admission from Leakey since *Origins.*

With humanity becoming increasingly evasive, superficial and alienated—retreating further and further into Plato's cave of denial—there is now such an emptiness in the world, such a lack of truth, such a jadedness from all the New Age quackery and other trite, superficial nonsense, that some 3 billion dollars has recently been committed in China to typing up all the Buddhist scriptures from wooden blocks onto computer, even going to the trouble of deciphering some of the ancient characters used in those scriptures whose meaning has been lost in antiquity. To go digging for truth in such ancient, pre-scientific texts is a tragedy, but it is a measure of how bereft of truth the world has become. The arrival now of the denial-free *scientific* description and explanation of the human condition will truly be like the summer rains that come to water the parched earth.

Humour, swearing and sex demystified

So complete has their denial been that resigned humans don't have an everyday word for the denial or evasion that they practice day in and day out—apart from the <u>swear</u> word, 'bullshit'.

Resigned adults are always talking about training people to think when the truth is that resigned adult minds are primarily concerned with training their minds to *avoid* deep, meaningful thought. They are full of denial, 'full of bullshit'.

Swearing is a powerful tool in that it tears through all the 'bullshit' or 'crap' that resigned adult life is full of. The truth is resigned adults are 'up to here in it'. They are so 'full of it', it *is* a 'fucking joke'.

The origins of humour have never been able to be properly explained, but once it is understood how false humans are, the source of humour becomes overtly apparent. For the most part, adult humans maintain a carefully constructed facade of denial but every now and again they make a mistake, 'slip-up', and the truth of their real situation is revealed, providing the basis for humour. When someone falls over, for instance, it's humorous because suddenly their carefully constructed image of togetherness disintegrates.

From time to time situations occur where the extreme denial, self-deception, delusion, artificiality, alienation becomes apparent and transparent, and in these moments the truth of the extreme falseness is revealed and seen for what it really is; so farcical it *is* funny, in fact a joke. On a more serious scale, when the extreme hypocrisy of human life is revealed, it *is* a 'fucking joke', ludicrous and appalling. The words of Tracy Chapman's 1986 song *Why?* acknowledges the extent of the hypocrisy in human life: **'Why do the babies starve/When there's enough food to feed the world/Why when there're so many of us/Are there people still alone/Why are the missiles called peace keepers/When they're aimed to kill/Why is a woman still not safe/When she's in her home/Love is hate, War is peace/No is yes, And we're all free/But somebody's gonna have to answer/The time is coming soon/Amidst all these questions and contradictions/There're some who seek the truth/But somebody's gonna have to answer/The time is coming soon/When the blind remove their blinders/And the speechless speak the truth.'**

To understand why 'fuck' is such a powerful swear word we only have to acknowledge the truth of what sex really is. As has been explained in the first section of the previous essay, 'Bringing peace to the war between the sexes', while sex at its noblest level was something that marvellously complemented the human journey, and as such has truly been an act of love, it has nevertheless, at base, been about attacking innocence (which women represent) for innocence's unjust condemnation of humans' (especially men's) lack of innocence. 'Fuck' means destroy or ruin, and what is being destroyed or ruined or sullied is innocence or purity. Such was the horror of the human condition; while humans were unable to explain their lack of innocence they had no choice but to use denial, retaliation and oppression to hold at bay the unjust criticism they were having to endure because of their divisive state. Sex has been such a preoccupation of humans and yet everyone lives in denial of the truth that it is at base an attack on innocence. This makes sex one of the biggest lies and thus jokes of all, which is why using the word 'fuck' is such a powerful attack on the world of lies, and thus such a powerful swear word. Swearing is one way to be honest, one way to tear through all the denial, a way of admitting humans are living in an ocean of dishonesty.

In one sense the civilised state, which requires that humans avoid swearing, was marvellous because it made life bearable by concealing the ugliness of humans' extremely false condition, but in another sense it made life *un*bearable because it hid the truth of humanity's extremely false condition. The importance of civility depends on your position; do you want to contribute to the maintenance of the lie or do you want to relieve the world of its lying?

The following comment about George Gurdjieff, a Russian philosopher who took on the world of denial, supports the civilised position: **'In his writings he used both humour and vulgarity to stimulate man's awareness of his own unworthiness and these "weapons" often detract from Gurdjieff's reputation as a serious philosopher'** *(Gurdjieff: An Approach to His Ideas,* Michel Waldberg, 1989).

Prophets adopt the position of relieving the world of its lying, and in doing so they are typically raw, defiant and irreverent—uncivilised. As it says about the defiant personality of prophets in the *Bible,* **'zeal for your** [denial-free] **house consumes me'** (Psalm 69:9 & John 2:17). As was mentioned earlier, prophets were **'warlike'** rather than saintly, capable of being both sensitive *and* tough. They have to be sensitive enough to access the true world of the soul and tough enough to defy the all-dominating world of denial. Kahlil Gibran was quoted earlier as saying, **'Jesus was not a bird with broken wings; He was a raging tempest who broke all crooked wings. He feared not His persecutors nor His enemies. He suffered not before His killers. Free and brave and daring He was. He defied all despots and oppressors. He saw the contagious pustules and amputated them...He muted evil and He crushed Falsehood and He choked Treachery.'** In a tirade against the intellectuals of his day, Christ said, **'You snakes! You brood of vipers!'**, repeatedly adding, **'Woe to you, teachers of the law and Pharisees, you hypocrites...you, blind guides!...You blind fools!'** (see Matt. 23).

. . . .

Apology: As explained in Notes to the Reader, the dotted lines indicate text that has temporarily been withdrawn because it deals with issues before the courts. Once the legal restrictions end the fully restored pages will be available on the FHA's website, or from the FHA.

. . . .

For me, swearing signals to those listening that I am not party to the hypocritical world of denial. Many young adults in their late teens and early 20s find relief from having the false world torn apart, for having just left the pre-resigned state where they could see the world was utterly false, they are still resistant to adopting the artificial, monstrously egocentric resigned adult world of lies. In opening up the issue of the human condition my fundamental task is to defy and expose the world of denial. I have to shatter the delusions and artificialities of that world.

R.D. Laing was a prophet who regularly used swearing to **'shock...the established order, the fool living in ignorance of his own ignorance'** (*R.D. Laing A Biography,* Adrian Laing, 1994, p.133 of 248). The following is one sample of Laing's use of swearing, and he is using it to emphasise just how corrupted, and dishonest about being corrupted, humanity has become: **'How do you plug a void plugging a void? How to inject nothing into fuck-all? How to come into a gone world? No piss, shit, smegma, come...** [etc], **will plug up the Hole. It's gone past all that, that, all that last desperate clutch. Come into gone. I do assure you. The dreadful has already happened'** (*The Politics of Experience* and *The Bird of Paradise,* 1967, p.153 of 156).

Carl Jung was another prophet whose **'language, which could be just as earthy as it was poetic, when he was roused in this profound regard was**

worthy of an inspired peasant, and words like "shitbags" and "pisspots" would roll from his lips in sentences of crushing correction' (Sir Laurens van der Post, *Jung and the Story of Our Time*, 1976, p.220 of 275).

Even the great Australian educator and prophet Sir James Darling was not averse to using strong language as this quote from a journalist who interviewed him records: **'Indeed; Darling's lengthy conversations during my visit revolved from intense thoughts on God, Socrates, Jung and the unconscious, to a joke with the copulating adjective rousingly pronounced'** (Janet Hawley, *Sydney Morning Herald, Good Weekend* mag. 19 Nov. 1988).

Recognising myself as a contemporary unresigned denial-free thinker or prophet

I have been described as a prophet—for instance in 1987, biologist, Dr Ronald Strahan of the Australian Museum, commented about my book *Free: The End Of The Human Condition:* **'I consider the book to be the work of a prophet'** (documented in the *Reviews* section of the FHA website). In an interview with Australian journalist Andrew Olle in April 1995, Professor John Morton, the Emeritus Professor of Zoology at the University of Auckland and fellow of St John's Theological College, Auckland, said of my book *Beyond The Human Condition,* **'This is prophesy I believe—a prophetic utterance'** (ABC Radio 2BL, 25 Apr. 1995). Earlier in the same month he had also told the FHA that **'Griffith is not the first prophet to be persecuted'** (personal communication Apr. 1995). I also recognise myself as an unresigned, denial-free, unevasive thinker or prophet. The reason I do is simply because one of the important tasks of bringing understanding to the human condition is to demystify abstractly expressed concepts and in the course of doing that I need to recognise that 'prophet' is the term that has historically been given to someone who is able to confront and look into the human condition. To be able to wander around in the realm of the human condition as I have, I must necessarily be an unresigned, denial-free thinker or prophet, albeit a contemporary one. By 'contemporary' I mean a prophet whose one concern is to bring understanding to the human condition, unlike pre-science prophets who could only offer their soundness as a basis for people to associate with if they wanted to be 'born-again' to the soul's world of soundness. Contemporary prophets have sought to bring rational understanding to the human condition, and by so doing obsolete the need for religion; traditional prophets could only create religions.

Acknowledging that I am a denial-free, unevasive thinker or prophet has . . . . but to effectively look into the human condition, as I have done, a person must be secure in self, the opposite of deluded, egocentric and arrogant. An individual cannot be secure and sound enough to be an honest thinker and be an insecure, unsound, deluded, arrogant, egocentric person hungry for reinforcement.

Since unresigned prophets were not insecure from a lack of reinforcement during their upbringing, their self-worth or self-esteem was intact. This means they were the *least* egocentric of people. In fact they tended to be childlike in their lack of sophistication and invisible to resigned egocentric people with their lack of imposing, pretentious presence—as Samuel was advised when he was trying to identify the prophet David: **'Do not consider his appearance...The Lord does not look at the things man looks at'** (Sam. 16:7). To be able to avoid becoming resigned an individual had to have had a secure upbringing.

To establish if someone is an unresigned prophet it was only necessary to establish their capacity to confront and look into the human condition. Being able to think truthfully, unresigned prophets were capable of making sense of all manner of mystery and in the process exposing all manner of delusion. They could not begin to do this if they were alienated, insecure, deluded, arrogant or egocentric. One state precludes the other. As Christ succinctly put it, **'Satan can't drive out Satan'** (Mark 3:23), and **'a bad tree cannot bear good fruit'** (Matt. 7:18).

The authority of an unresigned prophet

A person cannot be secure enough to look into the human condition and be an insecure, self-opinionated, arrogant, deluded megalomaniacal seeker of fame, fortune, power and glory. The question remains however, if you are secure then why do you need to claim to be a prophet and make such dogmatic statements as 'the human condition has been solved', and such and such an explanation 'is the truth', especially when some people, having no experience with secure unevasive thinkers or prophets, might think such statements are symptoms of insecure behaviour?

What needs to be explained is the difference between the authority of a denial-free, unresigned mind and the arrogance of a deluded mind. While resigned, evasive thinking is blind, uncertain and egocentric, unevasive thinking is not. Evading such fundamental truths as integrative meaning, the significance of nurturing in human life and the existence of humans' alienated state means evasive thinking progresses from a false basis and as a result is not in a strong position to know if ideas are true or not. Operating in a false framework, resigned thinking is insecure and uncertain. Unresigned thinking on the other hand, progressing from a truthful basis, has an infinitely greater capacity to know if an idea is right or wrong.

Living with the truth an unresigned mind can be certain in its thinking, it can know when its thinking is right and when it is wrong. If we are standing in a well-lit room and someone asks where the chair is in the room we can say exactly where it is without being considered arrogant. If we are metaphorically living in a dark room—in Plato's dark cave of denial—and someone asks where the chair is, all we can say is 'I think it is possibly over there somewhere'. Christ described this reality of the alienated state: **'The man who walks in the dark does not know where he is going'** (John 12:35). The reason people 'walk in the dark' was also explained by Christ using this light analogy when he said, **'everyone who does evil hates the light** [the truth], **and will not come into the light for fear that his deeds will be exposed'** (John 3:20).

Unresigned prophets could think truthfully. They were not uncertain in their thinking like resigned minds. They were not making an arrogant statement when they said 'the chair is there', rather they were simply making a truthful, authoritative statement. As was said in the *Bible* about Christ, he taught **'as one who had authority, and not as their teachers of the law'** (Matt. 7:29).

Denial-free, holistic thinkers have an authoritativeness that evasive, mechanistic thinkers aren't capable of and find difficult to comprehend. In fact resigned, evasive, mechanistic thinkers tend to project their insecure view of the world and assume the authoritativeness is arrogance. They forget that having adopted denial they have forfeited the ability to think truthfully and thus effectively. Victims of incest, after finding they cannot comprehend such abuse often decide they have no choice other than to block out any memory of it. 'Repressed memories', living in denial of an issue, is a common enough coping mechanism, but there is a down side or penalty to such practice. In the case of the incest victim, having blocked the issue from their minds they are in no position to think truthfully and thus effectively about their psychological state. Humans who are resigned to living in denial of the crux issue in all human affairs of the human condition—and along with it denial of all the truths that bring the human condition into focus, such as of integrative meaning and the extent of human alienation—have basically forfeited the ability to think truthfully. So fundamentally false is the paradigm they are living in that subjectivity for them cannot be trusted. Christ explained the comparative integrity of unresigned thought when he said, **'if I do judge, my decisions are right, because I am not alone. I stand with the Father who sent me** [I don't live in denial of and thus am truthfully guided by integrative meaning]' (John 8: 16).

Christ also forthrightly acknowledged his innocence, his uncorrupted, unresigned, alienation-free, sound state when he said, **'I and the Father are one'** (John 10:30), and **'The Father is in me, and I in him'** (John 14:10, 10:38).

Prophets have always been accused of being arrogant and deluded. Christ claimed to be the son of God (see Matt. 26:64, Mark 14:62, John 17:1) and he said such things as **'I am the way and the truth and the life'** (John 14:6). To the resigned mind such statements could appear as arrogant, and the person making them as deluded. In fact **'The Jews insisted "...he** [Christ] **must die, because he claimed to be the Son of God"'** (John 19:7) and Christ was subsequently mocked with a crown of thorns and put to death. In truth it was Christ's innocence and unresigned honesty that exposed and threatened the resigned mind and caused the resigned mind to murder him. His claimed arrogance in saying he was the son of God was merely the excuse used to eliminate his confronting presence. Ironically Christ's persecutors only ended up martyring his soundness and honesty. As mentioned, the martyrdom of Christ's soundness and truth is the real meaning of Christ's 'resurrection from the dead'.

While it was claimed that Christ was deluded and arrogant for saying he was the son of God, the truth is he was neither deluded nor arrogant. He was simply being honest, as history has verified. The soundness of Christ has stood the test of experience. He *was* an uncorrupted expression of humans' integratively orientated soul, he *was* the 'son of God', he was a pure expression of integrativeness. He could not have confronted the human condition as he was so clearly able to do if he was deluded and arrogant. History of course wasn't needed to verify Christ's soundness. Listening to his words, humans' repressed, unevasive, subconscious self instantly knew the depth of his soundness, while humans' surface, evasive, resigned self instantly knew how confronting that soundness was. Persecuting Christ for his innocent, denial-free honesty was acknowledgment by his persecutors that they knew full well he was the exact opposite of the deluded and arrogant person their evasive, resigned self was accusing him of being.

It is not difficult to find examples of the authority of unevasive truthful thinkers. In P.H. Butter's Introduction to his 1982 book, *William Blake Selected Poems,* he talked of William Blake being a prophet: **'The prophet is also a spokesman for God...Blake claimed in a letter in 1803 to have completed "the Grandest Poem that this World Contains...I may praise it, since I dare not pretend to be any other than the Secretary; the Authors are in Eternity." His belief in inspiration contributed to that "terrifying" honesty which T.S. Eliot saw in him'** (p.xiii of 267). Unlike resigned adults, prophets were not evasive of the truth of integrative or cooperative meaning or God. As it says in the *Bible* prophets were able to **'delight in the fear of the Lord'** (Isa. 11:3). This unevasiveness is what allowed prophets to think truthfully and thus effectively—to be, as the aforementioned definition of a prophet states, **'someone who speaks for God'**.

Like Blake, when he stated, **'I may praise it, since I dare not pretend to be any other than the Secretary; the Authors are in Eternity'**, Christ explained his denial-free, unevasive, integrative-meaning-accessing, sound, secure, non-egocentric state when he said, **'By myself I can do nothing; I judge only as I hear, and my judgment is just, for I seek not to please myself but him who sent me'** (John 5:30) and **'He who speaks on his own does so to gain honour for himself, but he who works for the honour of the one who sent him is a man of truth; there is nothing false about him'** (John 7:18). Prophets were the very essence of humility. They were the least egocentric of people. To quote Moses they **'served the Lord'** (see Deut. 34:5) not their ego. They stood by the truth of integrative meaning but when they did they were misrepresented as being arrogant.

The unresigned prophet Sir Laurens van der Post was misunderstood as being arrogant, egotistical and dogmatic, writing in one of his books that he was **'accused of always knowing better, and being dog-**

matic and domineering in my ways and in the advice I gave. As far as I was concerned, this was not in the least due to any hidden, egotistical agenda in my spirit' (*The Admiral's Baby*, 1996, p.142 of 340). In the *Plato* essay it was described how J.D.F. Jones attempted to 'crucify' Sir Laurens van der Post as a liar and a charlatan in his 2001 'biography'. I pointed out that if Sir Laurens were guilty of exaggeration, it wasn't because he was arrogant and egocentric, rather it was because of the need to counter the immense loneliness of being a denial-free thinker. The real motivation for J.D.F. Jones' book was that Sir Laurens had an unresigned prophet's authoritativeness and capability to reveal the truth about the human condition—attributes that J.D.F. Jones' resigned world found unbearable.

One of the greatest Indian chiefs to emerge during the invasion of the American west and the associated virtual genocide of the plains Indians by the Europeans, was the American Indian chief Tashunkewitko or Crazy Horse. The largest sculpture in the world is a sculpture of Crazy Horse that is slowly being carved out of a mountain in the Black Hills of South Dakota. Crazy Horse was undoubtedly an unresigned thinker or prophet. He was instrumental in the defeat of General Custer at the battle of Little Bighorn and was never defeated in battle. He was one of the few plains Indian chiefs who never sold out to the invading whites and in the end, like Christ, died a martyr's death with his own people accomplices to his murder. His complete incorruptibility was due to his ability to stay in the denial-free true world. As is explained in the *Resignation* essay, once you have resigned you have in a sense already sold out, your paradigm is not a clean one, and so it is much easier to sell out again. On the other hand if you haven't resigned, sold out, you are holding onto something pure, true and precious and that gives you immense strength. The following quote about Crazy Horse is from the best-selling book that told of the demise of the plains Indians, Dee Brown's 1971 *Bury My Heart at Wounded Knee:* **'Since the time of his youth, Crazy Horse had known that the world men lived in was only a shadow of the real world. To get into the real world, he had to dream, and when he was in the real world everything seemed to float or dance. In this real world his horse danced as if it were wild or crazy, and this was why he called himself Crazy Horse. He had learned that if he dreamed himself into the real world before going into a fight, he could endure anything. On this day, June 17, 1876,** [in the lead up to the battle of Little Bighorn] **Crazy Horse dreamed himself into the real world, and he showed the Sioux how to do many things**

they had never done before while fighting the white man's soldiers' (p.230 of 392). Note the assertion that Crazy Horse named himself and that the name he gave himself acknowledged his unevasive access to the true world.

In the insecure, resigned, evasive world, acts of self-acknowledgment were especially distrusted because in that world they were more than likely to be symptoms of delusions of grandeur than expressions of soundness. Self-acknowledgment and self-commendation in the insecure world *was* a case of arrogance, however in the secure, unresigned, denial-free world it was simply a case of stating the truth. Acknowledging myself as a prophet is not self-glorification, it is simply truthful, necessary self-description. Blake was pointing out that his self-commendation was authority and not arrogance when he said, **'I may praise it, since I dare not pretend to be any other than the Secretary; the Authors are in Eternity.'** Similarly Christ taught **'as one who had authority, and not as their teachers of the law'** (Matt. 7:29). Self-acknowledgment in the secure, denial-free world is simply a case of telling the truth. It is honest, necessary self-description, not arrogant, deluded self-promotion. A prophet acknowledging himself as a prophet is not arrogance but authority.

At this point I might relate an incident that illustrates the gulf between resigned, insecure, denial-complying, mechanistic thinkers, and true (as opposed to 'false') unresigned, secure, denial-free, holistic thinkers or prophets. In January 1983 I met with Professor Anthony Barnett, head of the biology faculty, at the Australian National University in Canberra. At that time Professor Barnett was presenting a weekly radio program called the *Biological Images of Man*. In the program he warned listeners that scientific theories, such as the selfish-gene-emphasising theory of Sociobiology, were not necessarily correct. Impressed by this warning about the limitations of mechanistic science, and other comments that effectively acknowledged the need for science to be more holistic (a rare admission in those days), I arranged a meeting with him. Unfortunately my efforts to have him consider my explanations of the human condition failed. As soon as I took the discussion into the realm of the human condition by pointing out that resistance to holism occurs for the good reason that holism confronts humans with their lack of compliance with holism—that humans are divisively rather than integratively behaved—Professor Barnett became agitated. When I persevered our dialogue became heated. Finally Professor Barnett ended the meeting with this outburst: **'Listen, you are being very arrogant in thinking you can answer questions on this scale; in all written history there are only two or three people who have been able to think on this scale about the human condition, so I'm not about to believe you're another of them'** (in a recorded interview, Jan. 1983).

The first point I would like to make is that this statement was extremely valuable to me at the time because in the heat of the moment Professor Barnett broke the rules or etiquette of denial. He broke the code of silence and acknowledged firstly that the issue of the human condition exists, and secondly that people are living in denial of it. These were truths that I knew but to hear them acknowledged by a professor from the world of denial was very relieving. Living in a world of total silence about what has been going on has not been easy.

Secondly in regard to Professor Barnett's insinuation that I must be deluded and suffering from hubris, I said to him that the intention of my visit was not to arrogantly want him to blindly accept anything I was saying as truth, only to have him consider the merits of the rational explanations I was putting forward. While scepticism about such extraordinary claims, such as that with the help of science I have solved the human condition, are entirely justified and necessary, it does not justify summary rejection and intolerance. The ultimate reason for democracy and the principle of freedom of expression was to allow the human condition to be addressed and ultimately answered. Without tolerance in society, human prejudice, scepticism and fear would stop anyone from ever addressing the human condition or having their explanation of it considered, and humanity could never hope to free itself from the human condition. Despite the odds against someone being able to confront and solve the human condition, there has always been and had to be so-called 'hope and faith', with the unsaid words following this expression being that one day, somewhere, some place, some one will be able to confront and solve the human condition.

The hypocrisy of Professor Barnett's position, and other scientists, who in the last 25 years of the 20th century have sought to make science more holistic, is that unless he and they are prepared to consider explanation of the human condition, which as Barnett demonstrates they are often not, then they have no sincere intention of introducing holism. They are in fact mechanists masquerading as holists.

A prophet's role is to say the truth

The work of a prophet depends on getting the truth up and standing by it, especially when the truth about the human condition is finally found. Despite the sceptical and cynical responses a resigned mind projects onto an unresigned mind, it is vital that an unresigned mind never succumb to the coercive cynicism, but keeps thinking and talking truthfully. Christ never weakened to cynicism. He said, **'I have spoken openly to the world. I always taught in synagogues or at the temple, where all the Jews come together. I said nothing in secret'** (John 18:20). He explained the predicament the unresigned mind is faced with when he said, **'Do you bring in a lamp to put it under a bowl or a bed? Instead, don't you put it on its stand?'** (Mark 4:21).

The confidence and certainty of denial-free, holistic-thinking prophets made them appear to be arrogant to people in the resigned, egocentric, mechanistic world because, to use Carl Jung's term, people tended to 'project' their way of thinking onto others. They tended to think everyone sees the world the way they see it. In this case the resigned mind assumes everyone else is resigned. Projecting their own blind, evasive uncertainty on the world, resigned, mechanistic minds could, if they didn't take the trouble to look into the situation, doubt and attack the certainty of the unresigned thinker or prophet.

The unresigned mind's honesty can certainly be immensely confronting and thus threatening to the resigned mind, but as well as this the authoritativeness of the unresigned mind can, if care isn't taken, be misunderstood by resigned minds as arrogance and symptomatic of the delusion exhibited by dangerous false prophets. If the situation is treated with sensitivity it is not difficult to differentiate between true and false prophets, because there is an immense and easily ascertained difference between the exceptionally sound and the exceptionally deluded. Again, Christ was drawing attention to the difference when he said, **'Satan can't drive out Satan'**. A person simply cannot begin to look into the human condition if he is a deluded, egocentric charlatan, a madman, manipulating people's lives for his own self-aggrandisement . . . .

. . . . If responsible action is not taken, or if the threat of the unresigned mind's denial-free thinking causes deliberate misportrayal of the sound thinking as being the work of a deluded false prophet, then undeserved and indeed . . . . . . misrepresentation of the denial-free thinker can occur. The responsibility of people who are not able to appreciate the world of true prophets is to trust in the democratic principle of freedom of expression that allows new ideas to be openly and fairly assessed on their merit.

The truth is, the greatest care of all needs to be taken in the realm where the subject of the human condition is at issue, because while it is the realm where deluded false prophets have operated, it is also the realm where true prophets operate, the realm where true prophets seek to find understanding of the human condition, the realm from which the liberating understanding of the human condition has to come. The realm where the subject of the human condition resides is the area where most prejudice can occur, yet from where the greatest benefits to humanity also come.

It is a very muddied pool to have to wade into but doing so—addressing the subject of the human condition—is the most important of tasks. If blatant intolerance and prejudice towards different ways of thinking is allowed then humanity can never hope to free itself from the human condition. The fundamental purpose of democracy is to allow freedom of expression, and thus the development of new ideas, and thus ultimately the acceptance of the 2-million-year dreamt of and hoped for arrival of the totally revolutionising, desperately needed, all-important breakthrough understanding of human nature; of ourselves.

The role of contemporary prophets is to resolve the human condition, and by so doing make the need for religions obsolete

The authoritativeness of unresigned thinkers can cause people concern because resigned people can project their way of thinking and mistake it as the arrogance of a deluded charlatan. Similarly, someone recognising themselves as a prophet can lead people to fear that the person is the worst form of deluded false prophet, someone so insecure, and as a result suffering from such delusions of grandeur, that they are fallaciously putting themselves forward as a religious figure of worship.

In pre-scientific times religions were founded around unresigned, denial-free-thinking individuals or prophets. There was a time when unresigned prophets legitimately put themselves forward as sources of soundness for people to associate themselves—to be 'born-again' through—when they became overly corrupted. However, with the development of science, the role of prophets is *the very opposite* to that of creating a religion. Their task now is to bring understanding and amelioration to the human condition and by so doing make the need for deferment of self to a faith obsolete.

There are many contemporary or modern-day prophets, both resigned and unresigned. I have already mentioned those that I have become aware of, for instance, Sir Laurens van der Post, Pierre Teilhard de Chardin, Arthur Koestler, R.D. Laing and Sir James Darling. No contemporary prophet has been concerned with creating a religion around themselves. Having grown up in a scientific age, in a world dedicated to explaining existence, they have only been concerned with bringing understanding to the human condition, and a person simply cannot be concerned with demystifying the human condition, and with it religion, while, simultaneously seeking to create a religion.

The presence of explanation, especially accountable, accredited first principle biological explanation, easily differentiates someone trying to explain the human situation, and in the process demystify religion, from someone trying to establish a religion.

In the same category of concern for people as the claim of being a prophet, is any suggestion by a person of them being 'the messiah'. People fear that it indicates the worst kind of deluded false prophet, someone suffering from extreme delusions of grandeur, someone so insecure that they ultimately represent themselves as a deity. When I first self-published my understanding of the human condition it was in the form of a somewhat naive article—naive in terms of what I now know about the deaf effect and people's inability to respond to denial-free thinking. In it I referred to myself as **'a contemporary "messiah"'** (*National Times,* 24 Feb.–1 Mar. 1980) and it caused some people concern—even though I enclosed the term 'messiah' in inverted commas, and qualified it with the word 'contemporary'. As will be explained below, the inverted commas around 'messiah' was an effort to dissociate myself from the emotive, religious connotations of the word, and the term 'contemporary' was added to emphasise the objective of explaining and demystifying religious concepts, not of creating a religion.

The fact is anybody who dares to grapple with the human condition, anybody who is a truly holistic thinker or prophet, may also be labelled 'messianic' or a 'messiah'. All resigned humans know that it is not safe for humans to confront the subject of the human condition, and that if a person is attempting to have others confront the subject of the human condition then that person is taking on a messianic role, a role of attempting to liberate humanity from its fundamental insecurity and fear of the subject of the human condition.

To illustrate what has just been said, in his London *Times* obituary, Sir Laurens van der Post was described as a **'prophet'** and his work as **'messianic'**. It was mentioned earlier that Arthur Koestler who has been described as a **'prophet'**, was noted as having a **'messiah complex'**. R.D. Laing was also described as a **'prophet'** and labelled a **'messiah'** (*R.D. Laing A Biography,* Adrian Laing, 1994, p.161 of 248).

Since the titles of my first two books, *Free: The End Of The Human Condition* and *Beyond The Human Condition,* clearly state that my work is to do with freeing humanity from the human condition, taking humanity beyond the human condition, my work is clearly messianic.

In terms of people simplistically labelling human condition-confronting, holistic thinkers as 'messiahs', I felt it important that I acknowledge this description. Significantly however, to clearly dissociate myself from any suggestion that I am suffering from delusions of grandeur and putting myself forward as a deity or object of worship, I strongly qualified my use of 'messiah' with the word 'contemporary'. As mentioned above, there are many contemporary or modern-day prophets. The Australian physicist Paul Davies has been described as a **'latter day prophet'** (ABC-TV *Compass, God Only Knows,* 23 Mar. 1997), and Australian biologist Charles Birch has been described as a **'scientist-prophet'** (*Sydney Morning Herald,* 30 May 2000). Latter day prophets or scientist-prophets or contemporary prophets are not concerned with creating religions or offering themselves as objects of worship. They have only been concerned with bringing understanding to the human condition, and as stated, a person simply cannot be concerned with demystifying the human condition, and with it religion, *and,* at the same time, be seeking to establish a religion.

To further dissociate myself from the emotive, religious connotations of the word, and to emphasise the objective of explaining and demystifying religious concepts, I put inverted commas around the term.

Significantly, I also emphasised in the text that **'we can now understand that the possibility of inferiority and superiority does not exist'**, an attitude that is inconsistent with megalomania, with seeing myself as superior to others and desiring of adulation. The very essence of the understanding of the human condition that my books present is that all humans are God-like in the sense of being completely meaningful, worthwhile and wonderful, simply differently corrupted and alienated by the heroic 2-million-year long battle to find understanding of the human condition. The essential insecurity of the human condition was that humans were not sure of their worthiness, they could not understand what was meaningful about being divisively behaved. Understanding of the human condition clarifies that insecurity, it explains why humans have been the way they have been, divisively rather than cooperatively behaved. It explains that all humans *are* a part of God or integrative meaning's plan.

Further still, the sub-heading of my 1980 *National Times* article emphasises the material is about **'the reconciliation of Theology, Philosophy and Behaviour'**. Far from promoting a new religion or faith, my article is about bringing reconciling understanding and demystification to the religious domain, which obsoletes the need for deferment of self to a faith. The first paragraph of the article begins, **'We** [humanity] **have now isolated the principles which unifies the material and spiritual domains, and it accords with sound biological explanations for the origin of our ethics.'**

While the 'contemporary "messiah"' statement uses the simplistic terminology applied to human condition-confronting, holistic thinkers, the real messiah, the real liberator of humanity from the human condition, as I have always emphasised, is science, supported by humanity as a whole. In support of this understanding I have emphasised that Christianity maintains that the third part of the Trinity, the Holy Spirit, the conscious intellect in humans of which science is the ultimate expression of, is the liberator of humanity. Incidentally, the reason Christianity maintains that the Holy Spirit is the messiah is because Christ referred to another who would come after him (the so-called **'second coming'**), a teacher and comforter, as the Holy Spirit. He said **'But the Counsellor, the Holy Spirit, whom the Father will send in my name, will teach you all things** [in particular it will make it possible to explain the riddle of life, the dilemma of the human condition]' (John 14:26).

Although presented in a denial-compliant way, mechanistic science did unearth the first principle understandings of the details and mechanisms of the workings of our world, the understandings that make clarifying explanation of the human condition possible. It is true that the person who synthesises the explanation of the human condition from these hard-won but evasively presented insights that mechanistic science has found, had to be an unresigned prophet, but that does not mean I am promoting myself as a figure of religious worship, or as someone deserving or desiring adoration.

As already explained, there have always been one or two unresigned prophets in the world and any one of them could have produced this synthesis had they lived at the time when science had completed its job of finding the insights into the workings of our world. I simply happen to be the unresigned prophet around when the job of synthesising the truth about the human condition became possible. The jockey Jim Pike happened to be in the right place at the right time to ride Phar Lap, the great Australian racehorse, to his famous 1930 Melbourne Cup victory. Far from promoting myself as any kind of religious figure of worship or even as someone deserving or desiring adoration, I have always gone out of my way to dissociate myself from such a misrepresentation. For example in my book *Beyond* I talk about the **'important but minuscule concluding role** [that the synthesising prophet plays] **in our search for knowledge'** (p.163 of 203).

As time goes by and people begin to digest the understanding of the human condition that is now available they will appreciate more and more this truth that *science is the liberator of humanity*, the messiah.

Part of the reconciliation of science and religion is the inevitable demystification of religious abstractions. As has already been carefully explained, unresigned prophets are not mystical, supernatural, divine beings. They are no more special than the rest of humanity. Nor are prophets gods or deities, although in pre-scientific times some unresigned prophets became revered as such. My work and the work of other contemporary prophets is dedicated to ending the need for worship, which is the very opposite objective of creating a religion for worship. As stated in *Beyond*, **'all the prophets have looked forward to a time when understanding would replace dogma'** (p.186 of 203), a time when, as it says in Genesis, we **'will be like God knowing'** (p.186). **'Religions aren't being threatened, they are being fulfilled'** (p.187) and **'This is the end**

of faith and belief and the beginning of knowing' (p.166).

For its part science has always expected to make religion redundant. To quote again the physicist Paul Davies from his acceptance speech of The Templeton Prize of 1995: **'Yet among the general population there is a widespread belief that science and theology are forever at loggerheads, that every scientific discovery pushes God further and further out of the picture. It is clear that many religious people still cling to an image of a God-of-the-gaps, a cosmic magician invoked to explain all those mysteries about nature that currently have the scientists stumped. It is a dangerous position, for as science advances, so the God-of-the-gaps retreats, perhaps to be pushed off the edge of space and time altogether, and into redundancy.'**

George Bernard Shaw understood this destiny when he said, **'All problems are finally scientific problems'** (*The Doctor's Dilemma* Preface, 1906). Biologist Edward Wilson was more specific when he said **'biology is the key to human nature'** (*On Human Nature,* ch.1, 1978). As Professor Charles Birch has said, **'I think science has done a very important thing for theology—it has shown us what are the false views and the views that were superstitious that necessarily arose in a pre-scientific era'** (*Australian Biography,* SBS TV, 27 Sept. 1998).

We can demystify the human situation now, explain mysticisms, expressions of spirituality and religious metaphysics—make the supernatural natural, including prophets.

To summarise; the three misconceptions regarding the messiah are firstly, that the truth-synthesising prophet is the messiah when the real liberator is science. Secondly, the use of the simplistic label of messiah for the truth-synthesising prophet does not infer that he is superior than other humans since the understandings he is presenting are about explaining the fundamental equality of all humans. Thirdly, the use of the simplistic label of messiah for the truth-synthesising prophet does not mean that he is putting himself forward as a religious figure of worship, in fact the reconciling understanding he is introducing obsoletes the need for humans to defer to a religion or faith.

As has been emphasised, if care is taken, it is not difficult to differentiate between true and false prophets. If care is not taken, or if the threat of the unresigned mind's denial-free thinking attracts deliberate misportrayal as being the work of a deluded false prophet, then undeserved and indeed . . . . . misrepresentation of the denial-free thinker can occur. The responsibility of people who are not able to appreciate the world of true prophets is to trust in the democratic principle of freedom of expression that allows new ideas to be openly and fairly assessed on their merit. Without tolerance in society, prejudice, scepticism and fear would mean that no one would ever be allowed to address the human condition or have their explanation of it considered, and humanity could never hope to free itself from the human condition.

Finally, I want to re-emphasise that prophets, people sound enough to look into the human condition, are the least egocentric of people. To be sound and secure in self is to not suffer from insecurity and delusion. It is certainly true that I am immensely excited that the human condition has been solved, because it means all the suffering in the world can at last be brought to an end, and it is also true that I am desperately keen for that understanding of the human condition to be communicated to others, because I want that suffering to stop as soon as possible. However, having been able to look into the human condition, and being thus unresigned, I have not had to adopt the resigned strategy of seeking reinforcement through power, fame, fortune and glory. I am not ego or self-worth centred or preoccupied. I am not ego-centric. My orientation is selfless not selfish. I want to stand up and tell the world as assertively as I possibly can that the human condition has been solved, but my efforts to do so are not motivated by self-glorification, rather by the desire to stop the terrible suffering in the world.

Denial-free thinkers are not 'brilliant' or 'clever' or 'geniuses'

I might comment on the simple power of denial-free thinking to explain so many mysteries of human life. Firstly, many of the concepts I have put forward are concepts that resigned humans have known but blocked out. The significance of nurturing for example is a truth that all resigned minds are aware of but have had to deny. Such concepts are not discoveries but revelations, truths that resigned humans have known and repressed and all I have done is reveal them.

Coming from a denial-free, honest base my mind is able to synthesise insights, in particular the explanation of the human condition, that resigned minds cannot reach from their dishonest base.

The concepts I am bringing forward are actually very simple and obvious, as long as you are thinking honestly—hence the 16-year-old Lisa Tassone's ability to read and understand my book in one night, as was documented in the *Resignation* essay. It is impossible to build the truth from lies, and the resigned position is fundamentally a false, lying paradigm. To again quote D.W. Winnicott's summary of the situation, **'True intuition can reach to a whole truth in a flash (just as faulty intuition can reach to error), whereas in a** [mechanistic] **science the whole truth is never reached'** (*Thinking About Children,* 1996, p.5 of 343).

My point is that the veritable avalanche of breakthrough insights that I have managed to bring forward is not a brilliant achievement. It is simply what is possible if you are able to think truthfully. In fact I don't have a 'clever', fast-information-processing, 'brilliant' mind. I'm far from a super clever, high IQ 'genius'. When my IQ was measured on a number of occasions at school it was never recorded as being anything higher than average and when I left school their advice was for me to take up a manual occupation. My house master's report of me in my final year at Geelong Grammar School in 1963 said, **'my judgment still is that Jeremy would find a tertiary course at university level very difficult'**. There is a note added to the bottom of this report by the then headmaster, Mr Garnett, saying, **'I am sure Mr. Mappin is right about the Science course'**. To gain entry to university I had to do a correspondence course from home on our family's sheep property in central NSW, and although I had special tutoring I only just managed to pass those exams. I did however achieve first class honours in biology after writing an imaginative essay in which I asked the question 'why don't some ants become lazy and live off the colony?'

It is not clever thinking that I do, it is simple, denial-free, unresigned, soul-directed thinking. To make this point about not needing to be clever, I only need to repeat what Christ said, **'you have hidden these things from the wise and learned, and revealed them to little children'** (Matt. 11:25).

As has been emphasised, the significant difference between people is their different degrees of alienation, not their different IQ levels. As was pointed out in the *Introduction,* the average IQ of humans is quite adequate for understanding. What a high IQ or 'cleverness' was needed for was to deny and evade the truth. That was the real art. Universities have high IQ entrance requirements because they have been the custodians of denial, keepers of 'the great lie'. A student had to be able to investigate the truth and talk about the truth without confronting it or admitting it, which is a very difficult and IQ-demanding undertaking. If humanity had entrusted inquiry to exceptional innocence it would have been continually and dangerously exposed to condemning idealistic partial truths, such as that humans should be cooperative. (They are 'partial truths' because when the full, human condition-explaining truth is found, these 'partial truths' are made non-condemning. For example once it is explained why humans have been divisive then cooperativeness is no longer condemning.) Prophets have been dangerous in the past because they exposed resigned humans to so many condemning truths and were thus necessarily oppressed. Tragically humanity *had to be* mechanistic rather than holistic in its approach to inquiry into the truth about ourselves.

The only thing extraordinary about my life is the nurturing I received from my mother. Any accolades go to her, but in turn her soundness is a product of her background. The point is, with sufficient understanding, both accolades and condemnations become meaningless. With understanding of the human condition, humans are taken beyond the concepts of 'good and evil', or of 'better or less than' or of 'superior or inferior'.

Australia's role in the world

In his celebrated 1931 poem *Australia,* A.D. Hope wrote about Australia's sheltered isolation and resulting relative innocence and lack of sophistication or falseness or alienation. He described how Australia's freshness and raw honesty can produce prophets, people capable of seeing through and defying the superficial 'chatter' of the world of intellectual evasion, denial and delusion and thus capable of reaching the full truth about our divisive condition. These are the words of *Australia:* **'A nation of trees, drab green and desolate grey/In the field uniform of modern wars/Darkens her hills, those endless, outstretched paws/Of sphinx demolished or stone lion worn away// They call her a young country, but they lie/She is the last of lands, the emptiest/A woman beyond her change of life, a breast/Still tender but within the womb is dry//Without songs, architecture, history/The emotions and superstitions of younger lands/Her rivers of water drown among**

inland sands/The river of her immense stupidity//Floods her monotonous tribes from Cairns to Perth/In them at last the ultimate men arrive/Whose boast is not: "we live" but "we survive"/A type who will inhabit the dying earth//And her five cities, like five teeming sores/Each drains her, a vast parasite robber-state/Where second-hand Europeans pullulate/Timidly on the edge of alien shores//Yet there are some like me turn gladly home/From the lush jungle of modern thought, to find/The Arabian desert of the human mind/Hoping, if still from the deserts the prophets come//Such savage and scarlet as no green hills dare/Springs in that waste, some spirit which escapes/The learned doubt, the chatter of cultured apes/Which is called civilization over there.'

That soulful part of humans, that they have lived in denial of, is the only part from which truthful thinking can come. It is the neglected centre of humans—the deserted part of their being, the realm from which they have become alienated—that alone can bring out the reconciling, liberating answers for the human race. That is why metaphorically it is from the **'desert of the human mind'** that **'the prophets come'**. As Sir Laurens van der Post has written, **'He** [Christ] **spoke of the "stone which the builders rejected" becoming the cornerstone of the building to come. The cornerstone of this new building of a war-less, non-racial world, too, I believe, must be...those aspects of life which we have despised and rejected for so long'** (*The Dark Eye in Africa,* 1955, p.155 of 159). In his award-winning 1979 book, *A Woman of the Future,* the Australian author David Ireland expressed the same awareness of where the answers would come from as A.D. Hope, and he used the same metaphor of the desert, recognising that Australians hide along the coast, distanced from the truth that exists in the centre of their being/country: **'The future is somehow/...somewhere in the despised and neglected desert/the belly of the country/not the coastal rind/The secret is in the emptiness/The message is the thing we have feared/the thing we have avoided/that we have looked at and skirted/The secret will transform us/and give us the heart to transform emptiness/If we go there/If we go there and listen/We will hear the voice of the eternal/The eternal says that we are at the beginning of time'** (p.349).

The prophet Isaiah also used the metaphor of the desert for that neglected part of ourselves from which the healing answers would come, when he said, **'A voice of one calling in the desert: "Prepare the way of the Lord; make straight in the wilderness a highway for our God. Every valley shall be raised up, every mountain and hill made low; the rough ground shall become level, the rugged places a plain. And the glory of the**

Lord will be revealed, and all mankind together will see it"' (Isa. 40:3–5).

In an exceptionally prophetic 1995 book, *Edge of the Sacred,* David Tacey, who at the time his book was published was a senior lecturer in English and Australian literature at La Trobe University in Melbourne, anticipated Australia's pivotal role in lifting the siege humanity has been under of the dilemma of the human condition. In the book's final chapter titled 'The Transformation of Spirit', Tacey said: **'Australia is uniquely placed not only to demonstrate this world-wide experience but also to act as a guiding example to the rest of the world. Although traditionally at the edge of the world, Australia may well become the centre of attention as our transformational changes are realised in the future. Because the descent of spirit has been accelerated here by so many regional factors, and because nature here is so deep, archaic, and primordial, what will arise from this archetypal fusion may well be awesome and spectacular. In this regard, I have recently been encouraged by Max Charlesworth's essay "Terra Australis and The Holy Spirit". In a surprisingly direct—and unguarded?—moment, Charlesworth says: "I have a feeling in my bones that there is a possibility of a creative religious explosion occurring early in the next millennium with the ancient land of Australia at the centre of it, and that the Holy Spirit may come home at last to *Terra Australis*". I am pleased that this has already been said, because if Charlesworth had not said it, I would have been forced to find within myself exactly the same prophetic utterance'** (p.204 of 224). (Note, it should be clarified that while the arrival of understanding of the human condition brings about an incredible spiritual awakening in humans—as if from the dead—it does not bring a 'religious explosion' where people transcend their troubled condition and defer to a deity. As previously mentioned, quite the opposite occurs. The arrival of self-knowledge allows humans to at last confront and resolve their troubled condition, become masters of it; as opposed to being tortured victims of it, forced to abandon and transcend their reality and invest their faith and trust in someone or something else as their only means of coping with it.) Tacey goes on to say, **'I would like to complete this book with a last glance at A.D. Hope's "Australia"'** and quotes the final stanzas of the poem that I have previously included. He states that **'The new spirit is** [going to be] **"savage and scarlet as no green hills dare". After** [white Australian] **contact with archaic nature and the red earth, spirit rises again in a form that is qualitatively different from our ancestral English or European spirit. Spirit will be "savage" in the sense of being untamed, primordial, not Wordsworthian, romantic, or consoling.**

"Scarlet" suggests not only the red earth and mountain ranges, but also blood, instinct, passion' (p.206). Tacey here recognises the non-conformity and lack of civility of the defiant, unresigned, denial-free, instinct or soul-directed mind. As the *Bible* says of the extremely defiant personality of true, unresigned prophets, **'zeal for your** [denial-free] **house consumes me'** (Psalm 69:9 & John 2:17).

The Australian character is forged around non-conformity with the sophisticated, false world of denial. Convicts were the basis of European settlement in Australia and many of them were essentially non-conformists; often fey Irish who didn't want to conform to the artificial, intellectual, resigned world of denial and as a result ended up becoming petty criminals. (See the account of the Irish character given on page 353.) Australia's unofficial national anthem, A.B. 'Banjo' Paterson's *Waltzing Matilda,* tells the story of a non-conformist petty criminal swagman who chose to jump into a waterhole or 'billabong' and die rather than fall into line with the bullshit, sophisticated, establishment squatter mounted on his bullshit, sophisticated thoroughbred and backed up by the bullshit, sophisticated, establishment troopers **'one, two, three'**. These are the words of Paterson's 1895 song *Waltzing Matilda:* **'Once a jolly swagman camp'd by a billabong/ Under the shade of a coolibah tree/And he sang as he watch'd and waited till his billy boiled/Who'll come a waltzing Matilda with me//Waltzing Matilda, Waltzing Matilda/Who'll come a waltzing Matilda with me/And he sang as he watch'd and waited till his billy boiled/Who'll come a waltzing Matilda with me//Down came a jumbuck to drink at that billabong/Up jumped the swagman and grabbed him with glee/And he sang as he shoved that jumbuck in his tucker-bag/You'll come a waltzing Matilda with me//** [chorus repeated] **Up rode the squatter mounted on his thoroughbred/ Down came the troopers, one, two, three/Whose that jolly jumbuck you've got in your tucker-bag?/You'll come a waltzing Matilda with me//**[chorus repeated] **Up jumped the swagman, sprang into the billabong/You'll never take me alive said he/And his ghost may be heard as you pass by that billabong/Who'll come a waltzing Matilda with me//**[chorus repeated].'

Australian journalist, Stuart Rintoul, recognised the influence of the Irish in the Australian make-up when he wrote, **'The distinctive Australian identity was not born in the bush, nor at Anzac Cove: these were merely situations for its expression. No; it was born in Irishness protesting against the extremes of Englishness'** (*Weekend Australian,* 7–8 Nov. 1998).

The Australian character is essentially defiant of the world of denial. Australians would rather be ignorant and, as we say, **'fair dinkum'** or **'true-blue'** than be artificial, sophisticated and false. Australia's favourite folk hero is the bushranger Ned Kelly, a man of Irish descent who, in 1880, took on the establishment dressed in a coat of armour he made from plough shears. Ned Kelly symbolises how much uncompromising defiance is necessary to stand up to the world of denial. **'Brave as Ned Kelly'** is a common euphemism in Australia. Kelly's armour, an Australian icon, was his bullshit deflector, his denial resistor. I keep a photo of Ned Kelly's armour beside my desk, and I've even designed an Australian flag (below) featuring the headpiece of his armour. How wonderful it would be if Australians were brave (secure) enough to have this kind of meaningful flag, and, the profundity of *Waltzing Matilda* as our official national anthem.

1993 JEREMY GRIFFITH

Incidentally Tacey not only anticipated the truth about the human condition emerging in Australia.. . . . .
. . . .
. . . .
. . . . He wrote, **'Unfortunately, the televisual and print media thrives on the negative or inferior expressions of the spontaneous religious impulse. Extreme or bizarre elements...are sensationalised and are automatically used to damn everything that seems a bit odd, unusual, or out of the ordinary...The Church will most likely close its doors to the new revelations of the spirit, because its primary task is to defend and support** [the literalist] **orthodoxy'** (*Edge of the Sacred,* 1995, p.125 of 224). Again it has to be emphasised that while the arrival of true self-knowledge is synthesised through the guidance of uncorrupted and thus unresigned and thus unevasive 'spiritual' (ie soul-infused and inspired) introspection, it is fundamentally different to a 'religious impulse'.

Hope, Ireland, Tacey and Charlesworth all intimated that enlightenment of the human condition was going to emerge from the isolated, sheltered, unsophisticated innocent backwater of Australia, and from the backwater within that backwater of Australia's inland or 'bush' as we call it, rather than from the sophisticated ivory towers of intellectualdom in places like Cambridge, Oxford and Harvard. Extraordinary as it may seem, Australia's two most loved poets and essayists, Henry Lawson and the already mentioned Banjo Paterson, both wrote poems expressing sentiments and prophecy identical to those of Hope, Tacey and Charlesworth.

Firstly to quote part of Henry Lawson's 1892 aptly titled poem *When the Bush Begins to Speak:* **'They know us not in England yet, their pens are overbold/We're seen in fancy pictures that are fifty years too old/They think we are a careless race—a childish race, and weak/They'll know us yet in England, when the bush begins to speak** [when innocence makes its contribution]/**..."The leaders that will be", the men of southern destiny/Are not all found in cities that are builded by the sea/They learn to love Australia by many a western creek** [while Australia as a whole is relatively sheltered and thus innocent, it is from the Australian inland countryside or 'bush', rather than from the cities, that exceptional innocence will appear]/**They'll know them yet in England, when the bush begins to speak/...All ready for the struggle, and waiting for the change/The army of our future lies encamped beyond the** [coastal] **range/Australia, for her patriots, will not have far to seek/They'll know her yet in England when the bush begins to speak/...We'll find the peace and comfort that our fathers could not find/Or some shall strike the good old blow that leaves a mark behind/We'll find the Truth and Liberty** [the truth about the human condition that brings liberating understanding to humanity] **our fathers came to seek/Or let them know in England when the bush begins to speak.'**

Banjo Paterson's 1889 poem *Song of the Future* includes strikingly similar chords: **'Tis strange that in a land so strong/So strong and bold in mighty youth/We have no poet's voice of truth/To sing for us a wondrous song** [no exceptionally denial-free thinker or prophet has emerged in the exceptionally innocent country of Australia yet, but such a profound thinker is due]//**...We have no tales of other days/No bygone history to tell/Our tales are told where campfires blaze/At midnight, when the solemn hush/Of that vast wonderland, the Bush/Hath laid on every heart its spell** [of sheltered, soul-drenched innocence]//**Although we have no songs of strife/Of bloodshed reddening the land/We yet may find achievements grand/Within the bushman's quiet life//...For years the fertile west-**

ern plains/Were hid behind your sullen walls [The **sullen walls** referred to eastern Australia's coastal mountain range, aptly called "The Great Dividing Range", that barred the way to the interior during the early days of white settlement. While the ranges had been crossed long before Paterson wrote this poem he was using the crossing as a metaphor for **the future** expedition to take humanity from the alienated bondage of the human condition to the fertile, sun/truth-drenched freedom of a human condition-resolved free world.] **/.../Between the mountains and the sea/Like Israelites with staff in hand/The people waited restlessly:/ They looked towards the mountains old/And saw the sunsets come and go/With gorgeous golden afterglow/That made the west a fairyland/And marvelled what that west might be/Of which such wondrous tales were told//...At length the hardy pioneers/By rock and crag found out the way/ And woke with voices of today/A silence kept for years and years** [brought an end to the silence of the resigned world of denial]**//...The way is won! The way is won!/And straightway from the barren coast/There came a westward-marching host/That aye and ever onward prest/With eager faces to the west/Along the pathway of the sun//...Could braver histories unfold/Than this bush story, yet untold—/The story of their westward march/ /...Our willing workmen, strong and skilled/Within our cities idle stand/ And cry aloud for leave to toil//The stunted children come and go/In squalid lanes and alleys black** [the end-play state of terminal alienation that humanity arrives at just prior to breaking through to self-understanding]**...//And it may be that we who live/In this new land apart, beyond/The hard old world grown fierce and fond/And bound by precedent and bond** [bound up in sophisticated, intellectual denial]**/May read the riddle** [of the human condition] **right and give/New hope to those who dimly see** [those who are embedded in denial/alienation]**/That all things may be yet for good/And teach the world at length to be/One vast united brotherhood** [human condition-ameliorated, reconciled world]**//So may it be, and he who sings/In accents hopeful, clear, and strong/The glories which that future brings/Shall sing, indeed, a wondrous song.'**

Paterson's words, **'The way is won! The way is won!/And straightway from the barren coast/There came a westward-marching host'**, anticipates the time when, with the healing, reconciling understanding of the human condition finally found, humanity begins its great exodus from the false world where it has been incarcerated since consciousness emerged in humans some 2 million years ago. In Plato's metaphor, humanity begins to leave the cave-like state of denial.

It is clear that both Paterson and Lawson were prophetic writers, more so than any other Australian writers that I have encountered. Interestingly Paterson refers to the influence for the coming **'wondrous song'** as being the **'vast wonderland, the Bush'**. Lawson similarly talked about **'the leaders that will be, the men of southern destiny'** learning their love beside **'many a western creek'**, and he also said **'the army of our future lies encamped beyond the range'**. It is no coincidence that Paterson and Lawson spent their early formative, childhood years only 50 kilometres from where I spent mine, over the coastal ranges in the Australian inland, in the central west of the state of New South Wales (NSW). Paterson grew up on a sheep station near a small town called Yeoval, I grew up on a sheep station not far away near another small country town called Mumbil, while Lawson grew up nearby in sheep grazing countryside near the town of Mudgee. Humanity spent its childhood in the savanna country of the Rift Valley of Africa. That is where our soul's home is, and the central west of NSW with its rolling hills of golden grass and clear, humidity-free blue skies, is that part of Australia that most closely corresponds to the environment of our soul's home in the African savanna. To not even see the sun for months on end, like those living near the Earth's poles, is extremely disorientating to our species' instinctive self. It is not surprising that in **'Greenland, depression affects as much as 80% of the population'** (*Time* mag. 16 July 2001). Sir Arthur Streeton's 1889 landscape titled *Golden Summer* held at the Australian National Gallery in Canberra is considered to be an icon of Australian art. It is a painting from our species' psyche, from the subconscious, as much as a painting of Australia because Australia is reminiscent of Africa; it is Africa without all the teeming wildlife so typical of that continent's natural state. Sir Laurens van der Post was struck by the physical similarity of Australia to Africa, observing: **'When I first went to Australia...my senses told me at once that here, beyond rational explanation, was a land physically akin to Africa'** (*The Dark Eye in Africa*, 1955, p.35 of 159). The dominant gum tree in *Golden Summer* almost has the horizontal strata of the acacia trees of Africa's savanna; the eagles in the sky look more like African vultures in their size and flight. Even the Australian magpie in the foreground of the painting has the markings of an African crow rather than the markings of an Australian magpie. As if to compensate for the absent wildlife in the Australian landscape Streeton has included some sheep in the foreground. Standing amongst the sheep is a boy, seemingly there to evoke the presence of a shepherd boy, or possibly even a little Bushman hunter from the landscape of ancient Africa.

On the right hand side in the distance there is a building and a man but they are so faintly painted that you have the feeling that Streeton was subconsciously trying to remove modern man from the landscape, make the landscape pristine, uncorrupted. Just how deeply moved Streeton was by the Australian landscape is apparent in letters he wrote to fellow Australian artists. In a letter to Frederick McCubbin in late 1891 Streeton said, **'My path lies toward the west which is a flood of deep gold. I felt near the gates of Paradise—The gates of the west'** (www.artistsfootsteps.com). In a letter to Tom Roberts postmarked 16 November 1893, Streeton wrote that **'I intend to go straight inland (away from all polite** [dishonest] **society) and stay there 2 or 3 years and create some things entirely new, and try and translate some of the great hidden poetry that I know is here, but have not seen or felt it. It all seems to me like an immense bright sky'** (*Letters from Smike,* eds Ann Galbally & Anne Gray, 1989, p.61). In another letter to Tom Roberts in the first half of 1891, Streeton wrote of **'the great gold plains, and all the beautiful inland Australia and I love the thought of walking into all this and trying to expand and express it in my way. I fancy large canvases all glowing and moving in the happy light. And others bright decorate and chalky and expressive of the hot trying winds and the slow immense Summer'** (ibid. p.30).

In his 1889 poem *Clancy of The Overflow,* Banjo Paterson wrote evocatively of the extraordinarily nourishing, soulful beauty of the Australian bush: **'I had written him a letter which I had, for want of better/ Knowledge, sent to where I met him down the Lachlan, years ago/He was shearing when I knew him, so I sent the letter to him/Just "on spec", addressed as follows: "Clancy, of The Overflow"//And an answer came directed in a writing unexpected/(And I think the same was written with a thumbnail dipped in tar)/'Twas his shearing mate who wrote it, and *verbatim* I will quote it/"Clancy's gone to Queensland droving, and we don't know where he are"//In my wild erratic fancy visions come to me of Clancy/ Gone a-droving "down the Cooper" where the western drovers go/As the stock are slowly stringing, Clancy rides behind them singing/For the drover's life has pleasures that the townsfolk never know//And the bush hath friends to meet him, and their kindly voices greet him/In the murmur of the breezes and the river on its bars/And he sees the vision splendid of the sunlit plains extended/And at night the wondrous glory of the everlasting stars//I am sitting in my dingy little office, where a stingy/Ray of sunlight struggles feebly down between the houses tall/And the foetid air and gritty of the dusty, dirty city/Through the open window floating, spreads its foulness over all//And in place of lowing cattle, I can hear the fiendish rattle/Of**

the tramways and the buses making hurry down the street/And the language uninviting of the gutter children fighting/Comes fitfully and faintly through the ceaseless tramp of feet//And the hurrying people daunt me, and their pallid faces haunt me/As they shoulder one another in their rush and nervous haste/With their eager eyes and greedy, and their stunted forms and weedy/For townsfolk have no time to grow, they have no time to waste//And I somehow rather fancy that I'd like to change with Clancy/Like to take a turn at droving where the seasons come and go/While he faced the round eternal of the cashbook and the journal/But I doubt he'd suit the office, Clancy, of "The Overflow".'

Graziers love their flocks of animals without realising that they are subconsciously relating to Africa with all its herds of animals. It is no wonder **'Clancy rides behind'** his stock in an African-like environment **'singing'**. With regard to the alienating effect of cities that Paterson describes in this poem, I wrote in *Beyond* that **'The truth is cities were not functional centres as we evasively claimed, they were hide-outs for alienation and places that perpetuated/bred alienation'** (p.180 of 203). In Rousseau's 1762 classic, *On Education,* he wrote **'Cities are the abyss of the human species. At the end of a few generations the races perish or degenerate. They must be renewed, and it is always the country which provides for this renewal'** (p.59 of 501). **'The bush'**, as Lawson said, is from where the required innocence will come to stand up and **'begin to speak'** in defence of what is true. As I have said, Australia as a whole is a bastion of innocence compared to other, more alienated, sophisticated and intellectual, human condition-exposed parts of the world, like **'England'** where there now exists, as A.D. Hope described, **'The learned doubt, the chatter of cultured apes/Which is called civilization over there.'**

Interestingly, Sir Laurens van der Post also grew up around the same 30-degree latitude as I did, albeit in Africa, while the unresigned prophets Zarathustra, Buddha, Abraham, Moses, Christ and Mohammed grew up around the 30-degree latitude in the northern hemisphere. Again this is not merely coincidence. The climate of these regions is very similar to the climate of the Rift Valley of Africa where humanity spent its childhood. While the Rift Valley lies on a higher latitude, being extremely elevated it has a similar climate to the lower altitude climate of country at the 30-degree latitude. As we emerge from our alienated states we are going to discover just how sensitive our soul is and these occurrences won't seem so extraordinary. If people don't believe that Africa is our soul's home they only need to visit natural Africa and experience for themselves the mind-bending experience of feeling that you have been there before, that this is where 'you belong' and that everything is 'as it should be'. In her 1967 book, appropriately titled *A Glimpse of Eden,* Evelyn Ames, a poet and novelist, recorded the experiences of a month-long safari undertaken in East Africa. She wrote: **'We thought we knew what to expect. Several friends had been there and told us about it; some, even, had made the same trip we were...going to make, but we discovered that nothing, really, prepares you for life on the East African Highlands. It is life (I want to say), making our usual existences seem oddly unreal and other landscapes dead; that country in the sky is another world...It is a world, and a life, from which one comes back changed. Long afterwards, gazelles still galloped through my dreams or stood gazing at me out of their soft and watchful eyes, and as I returned each daybreak, unbelieving, to my familiar room, I realized increasingly that this world would never again be the same for having visited that one. Nor does it leave you when you go away. Knowing its landscapes and sounds (even more in silence), how it feels and smells—just knowing it is there—sets it forever, in its own special light, somewhere in the mind's sky'** (pp.1–2). **'Each day in Africa my heart had almost burst with Walt Whitman's outcry: "As to me, I know of nothing else but miracles"'** (p.204). In *Henry IV,* Shakespeare wrote of **'A foutra for the world and worldlings base! I speak of Africa and golden joys'** (Part 2, Act V, Scene iii, c.1597). A sign at the entrance to the Serengeti National Park states **'This is the world as it was in the beginning'**. Sir Laurens van der Post wrote that **'There was indeed a cruelly denied and neglected first child of life, a Bushman in each of us'** *(The Heart of The Hunter,* 1961, p.126 of 233), and, **'We need primitive nature, the First Man in ourselves, it seems, as the lungs need air and the body food and water...I thought finally that of all the nostalgias that haunt the human heart the greatest of them all, for me, is an everlasting longing to bring what is youngest home to what is oldest, in us all'** *(The Lost World of the Kalahari,* 1958, p 151 of 253). Natural Africa is our species' spiritual home, it is the most sacred place on Earth. At least once in a lifetime every human should make a pilgrimage there.

In his 1943 poem, *The Stockman,* David Campbell, a sheep grazier, rugby player and another of Australia's best poets, evokes this sense of belonging, this sense of everything being as it should be, that immersion in Australia's African-like landscape generates. He talks of **'that timeless moment'** when **'the sun was in the summer grass'** bringing **'fresh ripples to my brain'**. It is another memory of our species' magic time in Africa bubbling up from our soul. These are the words of *The*

Stockman: **'The sun was in the summer grass/The coolibahs were twisted steel/The stockman paused, beneath their shade/And sat upon his heel/ And with the reins looped through his arm/He rolled tobacco in his palm/ /His horse stood still. His cattle-dog/Tongued in the shadow of the tree/ And for a moment on the plain/Time waited for the three/And then the stockman licked his fag/And Time took up his solar swag//I saw the stockman mount and ride/Across the mirage on the plain/And still that timeless moment brought/Fresh ripples to my brain/It seemed in that distorting air/I saw his grandson sitting there.'**

It should be pointed out that, while this soulful countryside can be so reinforcing, nurturing and nourishing, it can also be confronting and hurtful. As I said above, cities were hide-outs for alienation, they were places where alienated humanity could escape confrontation with the all-exposing and confronting natural world of our soul. I am sitting and writing at this moment in a house in the very heart of the rolling hills of golden grass in the central west of NSW that I have been describing. It is mid-summer and rain fell a few days ago and the beauty of the landscape is so intense that I feel that I can't go out into it, that I must stay inside, that I can't look at it for very long without being overwhelmed. When I was a boy I ran around in this world with such happiness but after a lifetime of defending that happiness, and all the truth it has access to, I'm now sufficiently embattled to feel the pain that the exposing purity of this countryside can produce in humans. Nowadays, when I do venture outside I have to be carrying out a task, like digging out weeds, to distract myself from all the beauty, because I can't face it head on.

Some disturbing quotes from big-game hunters were included in the *Resignation* essay to illustrate the extent to which innocent purity could be condemning and hurtful. The quote from the sport hunter who shot the chimpanzee was particularly revealing. While acknowledging how uplifting, inspiring and healing pure, innocent nature can be, it is also important to acknowledge just *how* confronting, condemning, criticising and hurtful it can be.

As emphasised, it is not only the climate and vegetation of the savanna country of the Rift Valley that we instinctively remember, it is also the wildlife. The vast array of different species and the great herds of animals are imprinted in our soul's memory. Mohammed observed **'that every prophet was a shepherd in his youth'** and it is recorded that, **'Until he was twelve, Muhammad tended his uncle's flocks among the Bedouins of the desert'** *(Eastern Definitions,* Edward Rice, 1978, p.260 of

433). Again it is no coincidence that unresigned prophets were shepherds in their youth. As has been mentioned, the flocks of sheep are like the herds of animals our soul is so familiar with. Of course a shepherd's life is also very close to nature and an occupation that is as far removed and sheltered from the battle and angst of the human condition as one can get. Nature is our instinctive self or soul's original companion and thus growing up with nature is going to be of comfort to, and thus preserving of, our innocent soul. Humans are bereft if they don't have any animals in their life. A shepherd's life is the polar opposite of life in a city.

Zarathustra grew up on the steppes of northern Iran where shepherding is still the main occupation. To quote a description of Zarathustra's origins: **'sometime around or before 600 BC—perhaps as early as 1200 BC—there came forth from the windy steppes of northeastern Iran a prophet who utterly transformed the Persian faith. The prophet was Zarathustra'** (Time-Life History of the World, *A Soaring Spirit 600-400 BC*, 1988, p.37 of 176).

Christ's upbringing is described in the *Bible* thus, **'And the child grew and became strong in spirit; and he lived in the desert until he appeared publicly'** (Luke 1:80). The only occupation in 'the desert' is herding sheep and goats. Someone 'strong in spirit' is a conscious mind infused with a strong conscience, that is, infused with a strong instinctive orientation to cooperativeness.

The exceptional prophet, Moses, grew up in the desert with shepherds and was himself one: **'Moses was tending the flock of Jethro his father-in-law, the priest of Midian, and he led the flock to the far side of the desert'** (Exod. 3:1).

The prophet David, whose pure courage enabled him to kill Goliath and who later became King of Israel, was a shepherd in his youth: **'David went back and forth from Saul to tend his father's sheep at Bethlehem'** (Sam. 17:15).

The prophet Isaiah described the upbringing required to produce a prophet when he said: **'The virgin will be with child and will give birth to a son, and will call him Immanuel. He will eat curds and honey when he knows enough to reject the wrong and choose the right'** (Isa. 7:14-16). The word 'Immanuel' means 'God with us'. It is the description of someone not alienated from their cooperatively orientated instinctive self or soul. 'Curds and honey' is the traditional food of shepherds. They live on curdled milk (yoghurt) and are always on the lookout for the sweet honey from wild bees nests. Someone who 'knows enough to reject the wrong and choose the right' is someone who has grown up sufficiently free of upset to avoid having to choose a resigned life of evasion. To be able to hold onto, and reveal the repressed truths—to be able to **'utter things hidden since the creation of the** [corrupted, resigned] **world'** (Matt. 13:35)—required an exceptionally strong conscience, someone with an unresigned, unevasive, non-soul-repressed mind. The significance of the 'Virgin Mother' as being the metaphor of an innocent mother, a woman capable of nurturing an innocent offspring, has already been explained.

Sir Laurens van der Post grew up in the **'back veldt'** (from Laurens van der Post's Introduction to *Turbott Wolfe* by William Plomer, p.17 of 215) in Africa on a farm that ran **'thousands of sheep'** (*About Blady*, 1991, p.117 of 255).

Banjo Paterson wrote about his upbringing in the Australian bush. In *A.B. 'Banjo' Paterson complete works 1885–1900*, in the chapter titled 'First Impressions', Paterson referred to his **'long days spent out shepherding sheep'**, and said that he was told to go out **'every day and learn to be a shepherd'**. Interestingly Paterson even made the comment that **'Nobody who has anything to do with sheep ever forgets it'** (collected by R.Campbell & P.Harvie, 1983, pp.4,5 of 723).

Henry Lawson was an infant of about two years old when his parents left the goldfields of Grenfell and his father **'took up a selection of land for farming and built a cottage which was to be the family home'** (*The World of Henry Lawson*, ed. W.Stone, 1974, p.11 of 504). The selection of land was in the sheep grazing district of Mudgee.

The family property I grew up on near Mumbil was a 3,500 acre sheep station and when I wasn't at boarding school I was helping muster the sheep on horseback and generally living a life immersed in nature.

While Banjo Paterson and Henry Lawson were both prophetic, Banjo Paterson was exceptionally so. Henry Lawson wrote the exceptionally honest poem included in the *Introduction, The Voice from Over Yonder,* and *When the Bush Begins to Speak,* but Banjo Paterson, in addition to *Song of the Future, Waltzing Matilda* and *Clancy of the Overflow,* penned his most famous literary work with *The Man From Snowy River.* Since it was published in 1895 it has become Australia's favourite and most emblematic poem. An enduring literary work can only be enduring because it contains truths that resonate deeply. In fact, *The Man From Snowy River* is an *extraordinarily* prophetic work. It describes how a **'stripling'** boy (the embodiment of innocence) goes, beyond where the alienated adults dare go, down the **'terrible descent'** of the mountain side where **'any slip was death'**, to confront the issue of the human condition and retrieve the truth about ourselves, symbolised by a thoroughbred horse that has escaped into the impenetrable mountains: '**There was movement at the station, for the word has passed around/That the colt from old Regret had got away/And had joined the wild bush horses—he was worth a thousand pound/So all the cracks had gathered to the fray/All the tried and noted riders from the stations near and far/Had mustered at the homestead over-night/For the bushmen love hard-riding where the fleet wild horses are/And the stockhorse snuffs the battle with delight//There was Harrison, who made his pile when Pardon won the Cup/The old man with his hair as white as snow/But few could ride beside him when his blood was fairly up/He would go wherever horse and man could go/And Clancy of "The Overflow" came down to lend a hand/No better rider ever held the reins/For never horse could throw him while the saddle-girths would stand/He learnt to ride while droving on the plains//And one was there, a stripling on a small and graceful beast/He was something like a racehorse undersized/With a touch of Timor pony, three parts thoroughbred at least/The sort that are by mountain horsemen prized/He was hard and tough and wiry — just the kind that won't say die/There was courage in his quick, impatient tread/And he bore the badge of gameness in his bright and fiery eye/And the proud and lofty carriage of his head//But still so slight and weedy one would doubt his power to stay/And the old man said: "That horse will never do/For a long and tiring gallop—lad, you'd better stop away/The hills are far too rough for such as you"/So he waited sad and wistful, only Clancy stood his friend/"I think we ought to let him come," he said/"I warrant he'll be with us when he's wanted at the end/For both his horse and he are mountain-bred//"He hails from Snowy River, up by Kosciusko's side/Where the hills are twice as steep and twice as rough/Where a horse's hoofs strike firelight from the flintstones every stride/The man that holds his own is good enough/And the Snowy River riders on the mountains make their home/Where the Snowy flows those giant hills between/I have seen full many horsemen since I first commenced to roam/But never yet such riders have I seen"//So he went; they found the horses near the big Mimosa clump/They raced away towards the mountain's brow/And the old man gave his orders: "Boys, go at them from the jump/No use to go for fancy-riding now/And, Clancy, you must wheel them—try and wheel them to the right/Ride boldly, lad, and never fear the spills/For never yet was rider that could keep the mob in sight/If once they gain the shelter of those hills"//So Clancy rode to wheel them—he was racing on the wing/Where the best and boldest riders take their place/And he raced his stock-horse past them, and**

he made the ranges ring/With the stockwhip as he met them face to face/ And they wavered for a moment while he swung the dreaded lash/But they saw their well-loved mountain full in view/And they charged beneath the stockwhip with a sharp and sudden dash/And off into the mountain-scrub they flew//Then fast the horsemen followed where the gorges deep and black/Resounded to the thunder of their tread/And the stockwhips woke the echoes and they fiercely answered back/From cliffs and crags that beetled overhead/And upward, upward ever, the wild horses held their way/ Where mountain-ash and kurrajong grew wide/And the old man muttered fiercely: "We may bid the mob good-day/No man can hold them down the other side"//When they reached the mountain's summit even Clancy took a pull/It well might make the boldest hold their breath/The wild hop-scrub grew thickly and the hidden ground was full/Of wombat-holes, and any slip was death/But the man from Snowy River let his pony have his head/ And swung his stockwhip round and gave a cheer/And raced him down the mountain like a torrent down its bed/While the others stood and watched in very fear//He sent the flintstones flying, but the pony kept his feet/He cleared the fallen timber in his stride/And the man from Snowy River never shifted in his seat/It was grand to see that mountain horseman ride/Through stringy-barks and saplings on the rough and broken ground/Down the hill-side at a racing-pace he went/And he never drew the bridle till he landed safe and sound/At the bottom of that terrible descent//He was right among the horses as they climbed the further hill/And the watchers, on the mountain standing mute/Saw him ply the stockwhip fiercely—he was right among them still/As he raced across the clearing in pursuit/Then they lost him for a moment where the mountain gullies met/In the ranges—but a final glimpse reveals/On a dim and distant hillside the wild horses racing yet/With the man from Snowy River at their heels//And he ran them single-handed till their sides were white with foam/He followed like a bloodhound on their track/Till they halted, cowed and beaten—then he turned their heads for home/And alone and unassisted brought them back/And his hardy mountain pony—he could scarcely raise a trot/He was blood from hip to shoulder from the spur/But his pluck was still undaunted and his courage fiery hot/For never yet was mountain horse a cur//And down by Araluen where the stony ridges raise/Their torn and rugged battlements on high/Where the air is clear as crystal and the white stars fairly blaze/At midnight in the cold and frosty sky/And where, around "The Overflow," the reed-beds sweep and sway/To the breezes and the rolling plains are wide/The man from Snowy River is a household word to-day/And the stockmen tell the story of his ride.'

As this poem expresses so prophetically, only innocence can overcome alienation, tame and return the escaped truth. Incidentally the 'terrible descent' into the issue of the human condition was perfectly described by the poet Gerard Manley Hopkins in his late 1800s poem, *No Worst There is None,* when he wrote, **'O the mind, mind has mountains; cliffs of fall/Frightful, sheer, no-man-fathomed.'**

The same mythology of innocence overthrowing denial occurs in Hans Christian Andersen's 1837 fable, *The Emperor's New Clothes,* where it takes a child/innocent to break the spell of deception and disclose the truth. Similarly in the story of *David and Goliath,* the innocent, 'David', goes out from the besieged ranks of the Israelites (humanity) to defeat the monster 'Goliath', who symbolises the issue of the human condition that the all-dominating state of denial has been blocking access to. Elsewhere in the *Bible,* in Isaiah 11, this truth is more clearly spelt out where Isaiah describes how **'a child will lead them'** to the state where upset and innocence are reconciled, to where, as Isaiah says, the **'wolf will live with the lamb'**. Only innocence, which the child symbolises, is able, as Isaiah says, to **'delight in the fear of the Lord'**—or, as Deuteronomy says, face the cooperative ideals or God **'face to face'**—and by doing so, as Isaiah states, make **'the earth...full of the knowledge of the Lord'** (eliminate the need for denial in the world).

In the great European legend of King Arthur, the wounded (alienated) king whose realm was devastated (humans unavoidably made their world an expression of their own madness) could only have his wound healed, and his realm restored, by the arrival in his kingdom of a simple, naive boy. In the legend the boy's name is Parsifal, which means 'guileless fool'. To the alienated only a naive, guileless fool would dare approach and grapple with the confronting truths about our divisive condition. The American Jungian analyst and prophet, Robert A. Johnson, gave an account of this legend in his 1974 book, *He, Understanding Masculine Psychology.* Johnson says firstly that **'Alienation is the current term for it** [the state of humans today]**. We are an alienated people, an existentially lonely people; we have the Fisher King wound'** (p.12 of 97). He then describes how **'The court fool had prophesied that the Fisher King would be healed when a wholly innocent fool arrives in the court. In an isolated country a boy lives with his widowed mother...His mother had taken him to this faraway country and raised him in primitive circumstances. He wears homespun clothes, has no schooling, asks no questions. He is a simple, naive youth'** (p.90). Johnson goes on to describe how in the myth it is this boy, Parsifal, who, when he becomes an adult, is able to heal the Fisher King's wound of alienation, so that **'the land and all its people can live in peace and joy'** (p.94).

In the *Resignation* essay I quoted Hopkins' 1885 sonnet *No Worst, There is None,* and I included part of the sonnet earlier in this section. The sonnet talked about the state of depression (which, as the title says, there is nothing 'worse' than) that most people experience if they try to look into the human condition. Following on from **'O the mind, mind has mountains; cliffs of fall/Frightful, sheer, no-man fathomed'**, Hopkins added **'Hold them cheap/May** [any] **who ne'er** [have never] **hung there.'** As these words intimate, only innocent people, those free of upset, could investigate the human condition without becoming depressed—**'hung'** being the perfect description for the depressed state. For innocents it is not costly, rather, as Hopkins says, it is **'cheap'** for them.

Sir Laurens van der Post clearly saw the truth of innocence having to lead humanity home when he wrote, **'Whatever happens, I shall be there in the end, for I, child that I am, am mother of your future self'** (*Jung and the Story of Our Time,* 1976, p.167 of 275).

It is guidance from the long-repressed unevasive, innocent, soulful clarity that humans had before the upset, alienated human condition developed that is needed to synthesise the liberating truth that frees humanity from its alienated state. Again to quote from Sir Laurens van der Post, **'One of the most moving aspects of life is how long the deepest memories stay with us. It is as if individual memory is enclosed in a greater which even in the night of our forgetfulness stands like an angel with folded wings ready, at the moment of acknowledged need, to guide us back to the lost spoor of our meanings'** (*The Lost World of the Kalahari,* 1958, p.62 of 253).

In Paterson's *The Man From Snowy River* the character Clancy of the Overflow persuaded the station owner Harrison to let the boy join their expedition to retrieve the escaped thoroughbred by saying, **'I warrant he'll be with us when he's wanted at the end'**. While innocence was unbearably confronting while the search for understanding was being undertaken, it was needed at the end to synthesise the unevasive explanation of the human condition from science's hard-won, but evasively presented insights.

The alienated simply cannot open the door to the human condition—or easily assist in holding it open—because that door *is* our block-out or denial or evasion or alienation. In the human condition-confronting new world the order is reversed. Soundness or innocence leads whereas for the last 2 million years it has been oppressed because of its unjust condemnation of humans. Innocence comes to the fore now to lead us back home to soundness and away from the alienated state. The following quotes illustrate the point: **'The meek...will inherit the earth'** (Matt. 5:5) and **'many who are first will be last, and many who are last will be first'** (Matt. 19:30,20:16; Mark 10:31; Luke 13:30). Artful, sophisticated, evasive, intellectual cleverness was needed to establish, defend and maintain the safe, non-confronting, escapist, alienated state but artless, simple, honest, soulful, instinctual soundness is needed to retrieve humanity from the alienated state.

Just as the sophisticated, cleverly evasive, esoteric, cryptic, intellectual world benefited from a high level or quotient of intelligence, so the unevasive, honest, unsophisticated, direct, open and plain new world depends on soundness. It is the more innocent who are able to lead humanity home to the full, compassionate, reconciling, dignifying, upset-subsiding, ameliorating truth about ourselves.

In Australian mythology, there is yet another truthsayer who has written about the finding of the Holy Grail of understanding of the human condition, and the exciting liberation of humanity that follows. Mark Seymour, the lead singer of the Australian rock band Hunters & Collectors wrote the following words to the band's 1993 popular song *Holy Grail* (some of this song was mentioned earlier in this essay, and it was also included in the *Introduction*): **'Woke up this morning from the strangest dream/I was in the biggest army the world had ever seen/We were marching as one on the road to the Holy Grail//Started out seeking fortune and glory/It's a short song but it's a hell of a story/When you spend your lifetime trying to get your hands/on the Holy Grail//Well have you heard about the Great Crusade?/We ran into millions but nobody got paid/Yeah we razed four corners of the globe for the Holy Grail//All the locals scattered, they were hiding in the snow/We were so far from home, so how were we to know?/There'd be nothing left to plunder/When we stumbled on the Holy Grail//We were so full of beans but we were dying like flies/And those big black birds, they were circling in the sky/And you know what they say, yeah nobody deserves to die//Oh but I've been searching for an easy way/To escape the cold light of day** [I've lived a life of resigned evasion]**/I've been high and I've been low** [suffered the consequences of recurring depression]**/But I've got nowhere else to go/There's nowhere else to go!//I followed orders, God knows where I've been/but I woke up alone, all my wounds were clean** [our psychosis was cleared up]/**I'm still here, I'm still a fool for the Holy Grail/I'm a fool for the Holy Grail.'**

Before concluding this section some explanation should be given as to why Australia is so devoid of animals compared to Africa. It is thought that Australia's marsupial mega fauna—we even had a marsupial lion—died out because, having not emerged with humans, as the mega fauna of Africa did, they weren't sufficiently adapted to humans to survive when humans finally arrived in Australia. In particular they were too easily hunted. The same reason is given for the extinction of the mega fauna of South America. In Australia's case I suspect there is another contributing factor that to date hasn't been acknowledged. Australia has been taken over by gum trees or eucalypts which are extremely fire-encouraging (because of their very waxy, oily leaves and bark) and extremely fire-adapted (because of their epicormic buds which are kept protected by the outer bark but grow quickly after fire). The fires that now erupt every 10 to 20 years in the all-dominating gum forests of Australia incinerate virtually all other wildlife, animal and vegetable. Interestingly, in the history of Australia's flora **'the gums are...all but absent until a few tens of thousands of years ago'** (from a review of Ashley Hay's 2002 book *Gum, Bulletin* mag. 19 Nov. 2002). It was the arrival of humans to Australia a few tens of thousands of years ago, with their practice of burning off the scrub to both trap game and later attract game to the short regrowth, that apparently enabled the gum trees to become so pervasive. Fires started from lightning strikes are evidently too infrequent to allow the fire-weed, gum tree monoculture to develop the way it has in Australia. If fires from lightning strikes had been numerous enough to allow gums to proliferate then surely gums would have proliferated much earlier. You can tell gums are an upstart variety of tree because they still haven't refined the distribution of their branches to efficiently catch the light; they are extremely disorganised and messy trees. Eucalypts are so successful now in Australia that it is said that every variety of plant community will be dominated by a variety of eucalypt, with the one exception perhaps being the very dry inland which still seems to be dominated by acacias. I have read that, be it heathland, scrub, open woodland or forest, **'eucalypts always come out on top'**. Australians have come to love their eucalypts but in some ways eucalypts are like dangerous crocodiles planted tail-down everywhere.

Expressions of anticipation of the arrival of understanding of the human condition

The insights expressed in Mark Seymour's *Holy Grail* as to what it will be like when understanding of the human condition arrives are exactly the same as those expressed in Bob Dylan's song *The Times They Are A-Changin',* the words of which were included at the end of the *Resignation* essay. Other prophetic anticipations of the arrival of the human condition-ameliorated new world were included in the *Introduction,* in particular Jim Morrison's song *Break On Through,* U2's *I Still Haven't Found What I'm Looking For,* John Lennon's *Imagine* and the Rolling Stones' *I Can't Get No Satisfaction.*

The immense paradigm shift that occurs with the arrival of the understanding of the human condition has been anticipated throughout history. The awareness is embedded in all mythologies as illustrated in the following examples.

Notice that although these prophetic works come from vastly different times they share the same imagery. This coincidence arises because all humans are acutely aware, albeit subconsciously, of what occurs when understanding of the human condition arrives.

Cat Stevens' 1971 *Peace Train:* **'Now I've been happy lately/thinking about the good things to come/and I believe it could be/something good has begun/oh, I've been smiling lately/dreaming about the world as one/and I believe it could be/some day it's going to come/cause out on the edge of darkness/there rides a peace train/oh peace train—take this country/come take me home again...everyone jump upon the peace train/come on now peace train/get your bags together/go bring your good friends too/cause it's gettin' nearer/it soon will be with you/come and join the living...now I've been cryin' lately/thinking about the world as it is/why must we go on hating/why can't we live in bliss/cause out on the edge of darkness/there rides a peace train/oh peace train take this country/come take me home again...'**

Bob Dylan's 1964 *When The Ship Comes In:* **'Oh the time will come up when the winds let up and the breeze will cease to be breathing/Like the stillness in the wind before the hurricane begins/The hour that the ship comes in** [when the liberating but also confronting truth about humans arrives]**/And the sea will split and the ships will hit/And the sands on the shoreline will be shaking/And the tide will sound and the waves will pound/**

And the morning will be a-breaking//The fishes will laugh as they swim out of the path/And the seagulls they'll be a-smiling/And the rocks on the sand will proudly stand [nature is going to be immensely relieved by the arrival of the peace-bringing reconciling understanding for humans]**/The hour that the ship comes in//And the words that are used for to get the ship confused/Will not be understood as the spoken** [truth]**/Or the chains** [holding the truth back] **of the sea will have busted in the night and be buried on the bottom of the ocean** [the denial will try to reimpose itself but it won't be allowed to succeed]//**...Oh the foes will rise with the sleep still in their eyes/And they'll jerk from their beds and think they're dreaming/But they'll pinch themselves and squeal and they'll know that it's for real/The hour that the ship comes in/And they'll raise their hands saying "we'll meet all your demands"/But we'll shout from the bow "your days are numbered"/And like the Pharaoh's tribe they'll be drowned in the tide/ And like Goliath they'll be conquered.'**

Cat Stevens' 1971 *Changes IV:* **'Don't you feel a change a coming/from another side of time/breaking down the walls of silence/lifting shadows from your mind/Placing back the missing mirrors/that before you couldn't find/filling mysteries of emptiness/that yesterday left behind//And we all know it's better/Yesterday has past/now let's all start the living/for the one that's going to last...//Don't you feel the day is coming/that will stay and remain/when your children see the answers/that you saw the same/ when the clouds have all gone/there will be no more rain/and the beauty of all things/is uncovered again...//Don't you feel the day is coming/and it won't be too soon/when the people of the world/can all live in one room/ when we shake off the ancient chains of our tomb...'**

The *Bible,* Micah 7:4,6,9,16,17,18,19 (parts of this quote also appear in Matthew 10:35): **'The day of your watchmen has come...Now is the time of their confusion...For a son dishonours his father, a daughter rises up against her mother...a man's enemies are the members of his own household...I will bear the Lord's wrath** [bear condemnation from cooperative idealism], **until he pleads my case and establishes my right** [until the knowledge that explains who I am arrives]. **He will bring me out into the light; I will see his justice...Nations will see and be ashamed, deprived of all their power. They will lay their hands on their mouths and their ears will become deaf...They will come trembling out of their dens...Who is a God like you, who pardons sin and forgives the transgression...You do not stay angry forever but delight to show mercy. You will again have compassion on us; and will tread our sins underfoot and hurl all our iniquities into the depths of the sea.'**

The *Bible* quotes many other anticipatory references to the arrival of understanding of the human condition. A powerful description is given in the Book of Joel, a quote from which has already been included in this essay. Isaiah gave a clear description of the arrival of the reconciling, peace-bringing understanding of the human condition when he talked of the time when, **'the earth will be full of the knowledge of the Lord...** [and, as a result] **the wolf will live with the lamb'** (11:6,9). Daniel anticipated the time that has now arrived when alienation and its pseudo forms of idealism—referred to as **'the abomination that causes desolation'** (9:27, 11:31 & 12:11)—would threaten to destroy humanity. Daniel described the arrival of the exposing truth that would bring an end to all the artificiality of human life as, **'a time of distress such as has not happened from the beginning of nations'** (12:1). He also anticipated the difficulty resigned minds would have in hearing the denial-free truth, saying, **'None of the wicked will understand, but those who are wise will understand'** (12:10). The prophet Hosea described the arrival of the cleansing honesty of **'the day of reckoning'** (5:9) as going to be so relieving it **'will come to us like the winter rains, like the spring rains that water the earth'** (6:3). Like Daniel, Hosea anticipated the deaf effect the truth will cause when he said, **'Who is wise?...He will understand...** [while the rest will] **stumble'** (14:9).

The song *Aquarius* from the 1960s rock musical *Hair,* lyrics by James Rado and Gerome Ragni: **'When the moon is in the Seventh House/and Jupiter aligns with Mars/Then peace will guide the planets/And love will steer the stars//This is the dawning of the age of Aquarius/The age of Aquarius/Aquarius! Aquarius!//Harmony and understanding/Sympathy and trust abounding/No more falsehoods or derisions/Golden living dreams of visions/Mystic crystal revelation/And the mind's true liberation/Aquarius! Aquarius!//** [chorus repeated] **As our hearts go beating through the night/We dance unto the dawn of day/To be the bearers of the water/Our light will lead the way//We are the spirit of the age of Aquarius/The age of Aquarius/Aquarius! Aquarius!//Harmony and understanding/Sympathy and trust abounding/Angelic illumination/Rising fiery constellation/Travelling our starry courses/Guided by the cosmic forces/Oh, care for us; Aquarius.'**

A poem I wrote in 1965 at the age of 20: **'This is a story you see, just a story—but for you/Um—I remember a long time ago in the distant future a timeless day/a sunlit cloudless day when all things were fine/when we all slow-danced our way to breakfast in the sun//You see the day awoke with music/Can you imagine one thousand horses slow galloping towards**

you across a vast plain/and we loved that day so much/We all danced like Isadora Duncan through the morning light//We skipped and twirled and spun about/Fairies were there like dragonflies over a pool/Little girls with wings they hovered and flew about/their small voices you could hear/You see it was that kind of morning//When the afternoon arrived it was big and bold and beautiful/In worn out jeans and bouncing breasts we began/to fight—our way—into another day/into something new—to jive our way into the night/from sunshine into a thunderstorm//We all took our place, rank upon rank we came/as an army with Hendrix out in front/and the music busted the horizon into shreds/By God we broke the world apart/The pieces were of different colours and there were so many people/We danced in coloured dust, we left in sweat no room at all/We had a ball in gowns of grey and red/There were things that happened that nobody knew/Bigger and better, I had written on my sweater/Where there was sky there was music, huge clouds of it/and there were storms of gold with coloured lights/It was so good we cried tears into our eyes/In a tug of war of love we had no strength left at all/Dear God we cried but he only sighed and/whispered strength through leaves of laughter//On and on we came in bold ranks of silvered gold/to lead a world that didn't know to somewhere it didn't care/It couldn't last, it had to end and yet it had an endless end/We were so happy in balloons of coloured bubbles that wouldn't bust/and we couldn't, couldn't quench our lust/There we were all together for ever and ever/and tomorrow had better beware because/when we've wept and slept we will be there to shake its bloody neck.'

One cold, pure February night

With regard to the culminating role Australia has had to play in synthesising the understanding of the human condition, it may seem extraordinarily coincidental, or plagiarism, that these Australians—Hope, Paterson, Lawson, Tacey and Charlesworth and, to an extent Seymour—could have experienced such an identical awareness of this breakthrough occurring here. However all they were doing was tapping into the deeper awareness about the real nature of the human journey that is buried within us all. Below the superficial chatter and deliberately created distracting mess and confusion of our evasive world we all know that it is the issue of the human condition that is the real problem on Earth, and that only an exceptional innocent can synthesise the reconciling, liberating understanding of this condition. Further we know within our deeper selves who is, and what countries are, relatively innocent and therefore where in the world the answers about our divisive nature are going to emerge.

For those who might not believe what I have just said I will quote an extract from the writings of the French unevasive thinker or prophet, Albert Camus, that clearly reveals this deeper knowledge. In this remarkable piece Camus acknowledges that innocence has to lead humanity home to the state of freedom from the human condition. He talks of the **'whiteness and its sap'** needed to **'stand up to'** all the fraudulent evasion and denial that has accumulated on Earth, and he recognises that this innocence is going to come from fresh, sheltered realms that still survive on the periphery of the great, evasive, artificial, intellectual establishments.

Camus, who won a Nobel Prize for literature, wrote these words in 1940 in an essay titled *The Almond Trees:*

'All we then need to know is what we want. And what indeed we want is never again to bow down before the sword, never more to declare force to be in the right when it is not serving the mind.

This, it is true, is an endless task. But we are here to pursue it. I do not have enough faith in reason to subscribe to a belief in progress, or to any philosophy of History. But I do at least believe that men have never ceased to grow in the knowledge of their destiny. We have not overcome our condition, and yet we know it better. We know that we live in contradiction, but that we must refuse this contradiction and do what is needed to reduce it. Our task as men is to find those few first principles that will calm the infinite anguish of free souls. We must stitch up what has been torn apart, render justice imaginable in the world which is so obviously unjust, make happiness meaningful for nations poisoned by the misery of this century. Naturally, it is a superhuman task. But tasks are called superhuman when men take a long time to complete them, that is all.

Let us then know our aims, standing steadfast on the mind, even if force dons the mask of ideas or of comfort to lure us from our task. The first thing is not to despair. Let us not listen too much to those who proclaim that the world is ending. Civilizations do not die so easily, and even if this world were to collapse, it will not have been the first. It is indeed true that we live in tragic times. But too many people confuse tragedy with despair. "Tragedy", Lawrence said, "ought to be a great kick at misery." This is a healthy and immediately applicable idea. There are many things today deserving of that kick.

When I lived in Algiers, I would wait patiently all winter because I knew that in the course of one night, one cold, pure February night, the almond trees of the Vallée des Consuls would be covered with white flowers. I was then filled with delight as I saw this fragile snow stand up to all the rain and resist the wind from the sea. Yet every year it lasted, just long enough to prepare the fruit.

This is not a symbol. We shall not win our happiness with symbols. We shall need something more weighty. All I mean is that sometimes, when life weighs too heavily in this Europe still overflowing with its misery, I turn towards those shining lands where so much strength is still untouched. I know them too well not to realize that they are the chosen lands where courage and contemplation can live in harmony. The contemplation of their example then teaches me that if we would save the mind we must pass over its power to groan and exalt its strength and wonder. This world is poisoned by its misery, and seems to wallow in it. It has utterly surrendered to that evil which Nietzsche called the spirit of heaviness. Let us not contribute to it. It is vain to weep over the mind, it is enough to labour for it.

But where are the conquering virtues of the mind? This same Nietzsche listed them as the mortal enemies of the spirit of heaviness. For him they are the strength of character, taste, the "world", classical happiness, severe pride, the cold frugality of the wise. These virtues, more than ever, are necessary today, and each can choose the one that suits him best. Before the vastness of the undertaking, let no one in any case forget strength of character. I do not mean the one accompanied on electoral platforms by frowns and threats. But the one that, through the virtue of its whiteness and its sap, stands up to all the winds from the sea. It is that which, in the winter for the world, will prepare the fruit' (*Summer,* 1954, pp.33–35 of 87).

To examine what Camus has said, he began by stating that the fundamental priority and responsibility of humanity is to solve the human condition, liberate the human mind from its underlying upset and by so doing replace the need for **'force'** to control our upset, troubled natures with the ability to explain, understand and pacify the upset. He acknowledged the human condition, **'we live in contradiction'**, and that we have to live in denial of this condition, **'we must refuse this contradiction'**. He also acknowledged the reality of the current extremely upset, depressed human state, **'nations poisoned by the misery of this century...** [a world] **utterly surrendered to that evil which Nietzsche called the spirit of heaviness'**. He then went on to emphasise the need for the clarifying, first principle reconciling biological explanation to **'overcome our condition'**, saying we need **'to find those**

first few principles' that will **'stitch up what has been torn apart'**.

Significantly, Camus acknowledges that these answers are not going to come from the ivory towers of intellectualdom, but from outlying realms where there is still sufficient innocence, **'strength still untouched'**, **'whiteness and its sap'**, to overcome all the evasion, denial and dishonesty, to **'stand up to all the winds from the sea'**, and find the reconciling understanding of the human condition, **'prepare the fruit'**. Importantly, in terms of what I have been saying about where the answers about ourselves would emerge, Camus says, **'I turn towards those shining lands where so much strength is still untouched. I know them too well not to realize that they are the chosen lands where courage and contemplation can live in harmony.'**

Conclusion

What has been revealed in this essay is that almost the whole of the human race is suffering from an immense psychosis. Humanity has been living in a state of very deep denial or block-out, of which it is almost completely oblivious.

The human race is suffering from a state of mental block-out so great almost no one who has lived during recorded history has been able to be free of it and thus able to look into it and expose it. Our concept of God, of something divine overseeing us, and of a mystical world beyond ourselves is the closest humanity has been able to come to talking about this denied other, non-alienated, integrative state and place.

As R.D. Laing said, we **'desperately'** need **'to explore the inner space and time of consciousness...We are so out of touch with this realm that many people can now argue seriously that it does not exist.'**

Now with the human condition at last explained, that realm has been demystified. It is as if a rocket probe into our inner space has finally returned with the first photos, non-mystical views, of this previously impregnable, uncharted realm. With the path found through the morass of our blocked out state of denial the greatest of all explorations into the real 'dark continent' of ourselves can begin in earnest.

In the televised 1990 Royal Geographical Society of London Lecture of the Year, titled 'Exploration', Sir Laurens van der Post emphasised this other great exploration that had to take place: **'It's been said that the explorers in mankind must be singularly unemployed because there's nothing left in this world to explore...**[however] **in the sense to which exploration is both an exploration into the physical *and* into the spirit of man there is a lot ahead in your keeping...We must go sharply into reverse. We must get our old natural selves to join with our other conscious, wilful, rational, scientific selves. This is the other act of exploration that will fulfil the exploration you've had in the past. Your job is not over.'**

In his 1991 book, *About Blady,* Sir Laurens van der Post said, **'There is, somewhere beyond it all, an undiscovered country to be pioneered and explored, and only a few lonely and mature spirits take it seriously and are trying to walk it'** (p.87 of 255). That trail the explorers blazed through the wilderness of our inner selves will now become humanity's highway to freedom. The way forward now is back through all the layers of denial and resulting psychosis.

The Foundation For Humanity's Adulthood

A profile written by FHA Vice-President Tim Macartney-Snape AM

(for more details about the FHA visit www.humancondition.info)

The Foundation for Humanity's Adulthood (FHA) was established in 1983 to promote the biological explanations of the human condition put forward by Australian biologist Jeremy Griffith in his books *Free: The End Of The Human Condition,* published in 1988, and *Beyond The Human Condition,* published in 1991, and now *A Species In Denial.* The aim of the FHA, as stated in its memorandum of association, is to **'bring forward understanding of the human condition and through doing that ameliorate that condition'**. The Foundation is registered as a charity in NSW, Australia, and is a company limited by guarantee. The Foundation has six directors, Jeremy Griffith, Simon Griffith, Stacy Rodger, Tim Watson, Annie Williams and myself. It also has Members and Supporters.

Jeremy has mentioned that he, his brother Simon, myself and others involved in the FHA attended Geelong Grammar School in Victoria, Australia, where we were the beneficiaries of the unique approach to education of the late Sir James Darling. In the essay about Plato's cave allegory, Jeremy referred to Sir James' emphasis in education on preserving and cultivating the sensitive, instinctual, soulful side of students—as opposed to the usual emphasis on intellectual achievement; on academic prowess and success in competition. In a highly competitive and pragmatic world this was a very brave approach to take. In fact it bordered on heresy, but I believe the benefits are fully on display in this book, for only innocence can confront and then reach all the way to the bottom of the issue of the human condition. The inclusion of Sir James as one of the 200 'great Australians', the only headmaster on the list, in Australia's Bicentennial year in 1988, was testament to his stature as an educator. It is also significant, especially in the context of this book, that in Sir James' full-page obituary in *The Australian* newspaper in 1995 he was described as, **'a prophet in the true Biblical sense'** (3 Nov. 1995).

Sir James Darling was himself a product of very special education in England. His teacher and mentor was William Temple, an inspired educator who went on to become Archbishop of Canterbury and to be considered by many as the most enlightened of all leaders of the Anglican Church. Indeed these insights into the human condition that Jeremy has synthesised are the fruition of a carefully cultivated enterprise that goes right back to Plato, a sequence of inspired educators who have sought to cultivate and preserve soul against the always threatening demands of a pragmatic, compromising world in denial. All of us involved in the FHA regard it a very great privilege to participate in the final stage of this enterprise. Now that the truth about the human condition has finally been dug up from its historically repressed state and is scientifically explained our job is firstly to defend the all-precious explanation against the historic denial, and secondly to educate the world at large about the explanation. The following is a summary of the work of the Foundation.

I should first explain the Foundation's 'key held aloft' logo and name. The discovery of the cause of humans' capacity for good and evil is the *key* that ameliorates that troubled condition, and our task in the FHA is to hold that key aloft. Since such understanding matures humanity from insecure adolescence to secure adulthood we are laying *The Foundation for Humanity's Adulthood.*

Initially the FHA offered formal membership subscriptions and published regular lengthy newsletters, the first in December 1988, the last, Newsletter 32, in March 1997. By late 1997 the FHA had so developed that it was decided to replace subscription and newsletters with a website that in time will be complemented with university-level computer courses in the study of the human condition, together with public lectures.

Originally the FHA recognised what it called 'Founding Members of the FHA'. They were the people who had been active members

. . . .

. . . .

In 2002 the FHA decided to redefine the 'Founding Members' category as those who are dedicated to bringing understanding of the human condition to the world by supporting the FHA on a daily basis, and to broaden it to include everyone who is doing this, regardless of when they joined, and further to rename them as a 'Member of the FHA'. This limits the term 'member'. For those people that the FHA accept as being appreciative and supportive of its work, but who are not directly involved, the FHA now terms them a 'Supporter of the FHA'.

Those supporting the FHA on a daily basis *are* the Foundation for Humanity's Adulthood, its true members, the people who have taken it upon themselves to pioneer the new, human condition-ameliorated world into being despite the immense resistance from an existing world deeply habituated to living in a resigned, egocentric state of denial of the issue of the human condition.

This group of what are now termed 'Members' and 'Supporters' numbered more than 100 in the early 1990s and . . . .

. . . .

. . . .

. . . .

. . . .

Apology: As explained in Notes to the Reader, the dotted lines indicate text that has temporarily been withdrawn because it deals with issues before the courts. Once the legal restrictions end the fully restored pages will be available on the FHA's website, or from the FHA.

. . . .

. . . .

. . . .

. . . . I note that it is to the steadfast commitment, selflessness and love of the FHA Members that Jeremy has dedicated this book.

The defenders of a new idea play a critical role, as noted by science historian Thomas Kuhn: **'In science** [according to Kuhn] **ideas do not change simply because new facts win out over outmoded ones...Since the facts can't speak for themselves, it is their human advocates who win or lose the day'** (Shirley C. Strum, *Almost Human,* 1987, p.164 of 294). Similarly, John Stuart Mill, in his 1859 essay *On Liberty,* emphasised that, **'the**

dictum that truth always triumphs over persecution is one of those pleasant falsehoods which men repeat after one another till they pass into commonplaces, but which all experience refutes. History teems with instances of truth put down by persecution. If not suppressed for ever, it may be thrown back for centuries' (*American state papers; On liberty; Representative government; Utilitarianism,* 1952, p.280 of 476).

Sir James Darling described the courage needed to defend a new paradigm when he wrote the following words about the very early Christian Church: **'Under the strain of danger and persecution, the society** [the early Christian Church] **was tested like gold tried in the fire. Inside the fellowship we can imagine that the bonds of fellowship were very strong and the pride in membership high. Such a fellowship, though only the brave would join it, would naturally attract the best, but, even so, only if its members were convinced themselves and anxious to convince others that they had found...a pearl of great price'** (*The Education of a Civilized Man,* 1962, p.117 of 223). I should emphasise that what we are involved in is not a new church or religion. What we are involved in is not concerned with 'faith' and 'belief', rather it is concerned with first-principle-based, biological understanding. In many ways knowledge is both the opposite of faith and what ultimately replaces the need for faith. It is nevertheless true that this is a new paradigm that is being introduced, just as the early Christian Church was a new paradigm for humans.

The majority of FHA Members are, in 2003, around 30 years of age. Most became interested in the understanding of the human condition in Jeremy's books through attending study groups that were established by fellow university students who had already become interested in the understanding. Periodically Jeremy was invited to attend these meetings. The main study groups that formed were at universities in Sydney, Brisbane, and the city of Armidale in the New England district of northern NSW.

The FHA is a highly organised and motivated corporation. It has its own offices in Sydney from where its many projects are managed. The FHA Members live nearby in neighbouring suburbs, most sharing various rented accommodation, while Jeremy and his partner Annie spend approximately 10 months of the year at a retreat five hours drive from Sydney where Jeremy is able to concentrate on his writing—a whole new paradigm is being opened up for humans and there is a great deal to explain about it. Many Members have professional qualifications which they are putting to good use within the various FHA departments. We even have our own rugby team called the Ned Kellys, named after the legendary Australian bushranger who courageously defied the establishment. Since 2001 the Foundation's exhaustive records have been progressively converted to electronic archives so that anyone in the future can know its full history.

Most Members have long-standing partners within the FHA but because of their commitment to the immense task they have undertaken of ending humanity's denial most are not as yet formally married, or have children. With the priority so clearly being to defend this information against the entrenched denial of the issue of the human condition, getting married and having children has very much been a secondary concern. In the case of having children, with billions of children in the world and no answers to the overwhelming problems facing humanity, the Members, in their denial-aware state, are able to see that getting these answers out into the world logically comes first. Also, the more a person understands the importance of nurturing the more they appreciate how consuming and important a task parenting is. Essentially, the greater the need for the new, all-important but extremely confronting insights, and the greater the resistance to them, the more selfless and committed we have to be prepared to be.

Our undertaking is infinitely more important and difficult than winning an international yacht race; it demands even more of the dedication, sacrifice and commitment that was demonstrated in Australia's 1983 victory of the America's Cup: **'The Australians had never won the world's most prestigious sailing event, the America's Cup. Australian captain John Betrand was more than willing to change all of that. His motivation was to something that he loved dearly, sailing to victory with a team totally committed to excellence and to each other. Instead of focusing on years of past losing, he gave his a crew a simple vision. "We are going to sail this boat as boats will be sailed in the year 2000, as no crew has ever sailed a boat before." Knowing that teamwork required a willingness to get off of your own position for the sake of a bigger vision, he had his team commit themselves to living together for two years and to having a visualisation session each day, of winning the Cup together. They would see, hear, taste, and feel every aspect of the race. Two years of this made the vision a reality—they won the 1983 America's Cup. In 1987, the American team, under the direction of Dennis Conner, had a similar vision and victory'** (Thomas F. Crum, *The Magic of Conflict,* 1998, p.164 of 251).

With regard to resistance to the new ideas that we are presenting,

. . . .

Apology: As explained in Notes to the Reader, the dotted lines indicate text that has temporarily been withdrawn because it deals with issues before the courts. Once the legal restrictions end the fully restored pages will be available on the FHA's website, or from the FHA.

. . . .

The FHA has the ultimate product for humans. The reconciling understanding of the human condition is the information that humans have been searching and anticipating for 2 million years! Having the ultimate product for humans means the FHA is the ultimate business prospect and offers the ultimate in meaningful careers, however, while all new products typically have to overcome difficulties before their potential can be realised, no new product has as many initial obstacles to overcome as the understanding of the human condition. In particular, before we can market our product we firstly have to overcome the 'deaf effect', people's inability to take in or 'hear' analysis of the human condition. Secondly we need to survive the backlash from the shock that the world of denial feels from having its denial exposed. Above all what these extreme start-up difficulties demonstrate is that the FHA has to initially be *self-sufficient.* The foundation capital for the FHA has come from the sale of Jeremy's half share of a furniture business he established, and from the sale of a successful youth hostel business established by Jeremy's brother Simon. It is the commitment and support of the Foundation Membership, in particular Members donating their time and developing business enterprises that are dedicated to supporting the FHA, that has enabled the FHA to become a strong organisation capable of launching the most worthwhile and important of all products.

The following is a comprehensive profile of the FHA Membership assembled by the FHA administration team. The ages and details given are for 2003.

Annabel Armstrong, BA Dip Ed, 29, grew up in Brisbane and was educated at St Aidan's Anglican Girls School where she was vice-house captain. Annabel was introduced to the understanding of the human condition in 1993 by school friend and FHA Member Emma Cullen-Ward. After graduating from the Queensland University of Technology Annabel moved to Sydney in 1998 to more directly support the FHA and is working part-time as an art teacher while helping develop the FHA's University of Denial-Free Studies.

Susan Armstrong, BRurSc (Hons), 30, grew up on a sheep station near Cumnock in central west NSW and was educated at Abbotsleigh Girls School, Sydney where she was Vice Head Boarder. Given a copy of *Beyond* by Steven van Hemert, a close friend of Jeremy, in 1991, Susan followed up her interest in the book's ideas with FHA Member Annabelle West while attending the University of New England. After graduating in 1996 Susan moved to Sydney to more directly support the FHA and is working full-time helping to prepare the FHA's court cases.

Sam Belfield, BRurSc (Hons), 30, grew up on a sheep station near Armidale in central north NSW and was educated at The Kings School, Sydney. He was introduced to the understandings of the human condition in 1993 by a number of other students who were also attending the University of New England, in particular FHA Members Susan Armstrong and Annabelle West. Sam began working full-time for the FHA in 1995 and moved to Sydney in 1996. As the FHA CEO Sam is involved with all aspects of the FHA, including heading the FHA's legal team with John Biggs. The legal team is currently preparing the FHA's court cases.

John Biggs, LLB BCom, 32, grew up on Queensland's Gold Coast and in Vanuatu in the South Pacific before moving to Brisbane where he was educated at the Anglican Church Grammar School achieving a leaving score in the top 2.5 percent of the state. In 1990, while attending the University of Queensland, where he was elected secretary of the student union, John became interested in the FHA through long-time friend and FHA Member James West. After graduating in 1994 John worked full-time for the FHA in 1995 before moving to Sydney to more closely support the FHA. John is a lawyer in a city law firm and, with Sam Belfield, heads the FHA's legal team helping prepare the FHA's court cases.

Richard Biggs, BCom, 31, grew up on Queensland's Gold Coast and in Vanuatu in the South Pacific before moving to Brisbane where he was a prefect at the Anglican Church Grammar School. In 1991, while studying at the University of Queensland he was introduced to the understandings of the human condition by his good friend James West and his brother John. After graduating Richard moved to Sydney in 1996 to more directly support the FHA. He is completing a course to become a certified practicing accountant and helping develop the University of Denial-Free Studies while working full-time helping to develop business enterprises dedicated to supporting the FHA.

Kate Campbell, BEd, 28, grew up on a sheep station near Baradine in north west NSW. After leaving school at Loreto College, Normanhurst, Sydney, Kate studied at the Queensland University of Technology in Brisbane where, in 1994, she was introduced to the understandings of the human condition by FHA Member Rick Biggs. In 1997 Kate moved to Sydney to more directly support the FHA and works full-time helping prepare the FHA's court cases.

Lachlan Colquhoun, BE (Hons) DipEngPrac, 32, grew up in Brisbane and was educated at Brisbane Grammar School. He was introduced to the understandings of the human condition in 1990 through his best friend FHA Member John Biggs while attending Queensland University, where he was a student union vice-president. Lachlan moved to Sydney in 1997 to more directly support the FHA. He completed his engineering degree at the University of Technology in Sydney and worked for a large engineering company before working full-time helping to develop business enterprises dedicated to supporting the FHA.

Eric Crooke, BE (Mech) BCom, 32, grew up in Brisbane where he attended the Anglican Church Grammar School, achieving a leaving score in the top 1 percent of the state. He was introduced to the understandings of the human condition by his friend and FHA Member John Biggs in 1991 while attending the University of Queensland. After moving to Sydney in early 1998 to more directly support the FHA, Eric rose to be the NSW Manager for his division in a leading Australian packaging company before leaving to play a leading role developing business enterprises dedicated to supporting the FHA.

Emma Cullen-Ward, BCom, 29 years, grew up mostly in Brisbane and was educated at St Aidan's Anglican Girls School where she was a vice-house captain. Emma was introduced to the understanding of the human condition in 1991 while at the University of Queensland by her cousin, FHA Member James West. Emma moved to Sydney in 1996 to more directly support the FHA and is now a director at a leading public relations firm in Sydney while overseeing FHA PR.

Fiona Cullen-Ward, BA, 27, grew up in Brisbane where she attended St Aidan's Anglican Girls School. Fiona was introduced to the understanding of the human condition in 1991 by her cousin, FHA Member James West. After graduating from the University of Queensland, Fiona worked in hospitality and childcare before moving to Sydney in 2001 to more directly support the FHA. She currently works full-time as a consultant in a leading public relations firm in Sydney, while editing Jeremy's writing.

Anthony Cummins, BArch, 33, grew up at Noosa Heads on Queensland's Sunshine Coast and attended the Anglican Church Grammar School in Brisbane achieving a leaving score in the top 3 percent of the state. Anthony was introduced to the understanding of the human condition in 1992 by his school friends and FHA Members John Biggs and Eric Crooke whilst at The University of Queensland. He moved to Sydney in 1996 to more directly support the FHA where he works part-time in accounting while helping manage the FHA's accounts.

David Downie, BCom LLB CA, 31, grew up in Brisbane and was educated at Brisbane Boys College where he was a school prefect and a vice-house captain and achieved a leaving score in the top 1.5 percent of the state. David was introduced to the understandings of the human condition in 1993 by FHA Member James Mollison whilst attending the University of Queensland. Since graduation in 1995 he has worked for chartered accounting firms and in commerce. David moved to Sydney in 1997 to more directly support the FHA where he works part-time as a chartered accountant while playing a leading role in the financial management of the FHA.

Eliza Easterman, BAsianStudies(s)/LLB, 26, grew up in Tamworth and was educated at Calrossy Girls School in Tamworth where she was a prefect. Eliza was first introduced to the understanding of the human condition in 1994 through her friendship with Tim and Ali Watson and their four daughters, and became fully appreciative of the understanding in 1999 through her friendship with FHA Member James Press. Eliza completed her law degree at the Australian National University in 2001 and works part-time as a paralegal while assisting in the preparation of the FHA's court cases.

Sally Edgar, BA, 30, grew up in Dalby in southern Queensland and was educated at New England Girls School in Armidale before attending the University of Queensland in Brisbane. She was introduced to the understanding of the human condition in 1998 by FHA member Nick Shaw while working as a journalist with a rural newspaper. In May 2000 Sally moved to Sydney to support the FHA and works as a communications manager with an agribusiness company while running FHA PR.

Bronwyn FitzGerald, BAg Ec (Hons), 29, grew up on a farm near Goondiwindi in southern Queensland and was educated at St Peters in Brisbane and Goodiwindi High School where she was vice-captain and achieved a leaving score in the top 2 percent of the state. Bronwyn was introduced to the understandings of the human condition at the University of New England in 1994 by FHA Member Susan Armstrong. She moved to Sydney in 1996 to more directly support the FHA and works as a paralegal while developing the University of Denial-Free Studies and assisting with FHA PR.

Jeremy Griffith, BSc, 57, grew up on a sheep station near Mumbil in central west NSW and was educated at Tudor House School, Moss Vale and Geelong Grammar School in Victoria before attending the University of New England, where he played representative rugby, making the trials for the national team, the Wallabies, in 1966. Jeremy deferred his studies in 1967 to try to save the Thylacine or Tasmanian Tiger from extinction. The search, which concluded the Thylacine was extinct, lasted for six years and attracted international scientific and popular media coverage. After completing his degree at Sydney University in 1971 Jeremy began a furniture manufacturing business which employed some 45 people and was a major tourist attraction when Jeremy sold his half share of it in 1991. Jeremy began actively writing about the human condition in 1975, established the Foundation for Humanity's Adulthood in 1983, and published his first book, *Free: The End Of The Human Condition,* in 1988. While Jeremy is an FHA director his main role is in developing the understanding of the human condition.

Simon Griffith, 50, grew up on a sheep station near Mumbil in central west NSW and was educated at Tudor House School, Moss Vale and Geelong Grammar School in Victoria. Simon began a degree in architecture at the University of Newcastle but left after injuring his leg in a car accident and deciding to pursue his interest in travel and the hospitality industry. These interests culminated in Simon designing, building and running a successful youth hostel at Avalon on Sydney's northern beaches, a business that he sold in 1995 to help fund the FHA. Simon is Jeremy's youngest brother and has always been interested in, and supportive of his study of the human condition. An FHA director, Simon has many roles in the FHA, however his main role is in developing the University of Denial-Free Studies and creating the FHA's film studio.

Damon Isherwood, BArch, 36, grew up in Sydney and was educated at Sydney Grammar School where he was a prefect. Damon attended the University of NSW and became interested in the understanding of the human condition after reading a review of *Beyond* in the *Bulletin* magazine in 1991. He works part-time helping develop business enterprises dedicated to supporting the FHA while playing a key role in preparing the FHA's court cases.

Felicity Jackson, BSc, 30, grew up in Bairnsdale in eastern Victoria and was educated at Geelong Grammar School where she was a house prefect. While attending the University of New England Felicity was introduced to the understanding of the human condition in 1993 by other fellow students, including FHA Members Susan Armstrong and Annabelle West. She moved to Sydney in 1996 to more directly support the FHA's work and works part-time as a business administrator while managing FHA administration.

Charlotte James, BAg Ec (Hons), 28, grew up in Melbourne and attended Lauriston Girls School in Melbourne where she was a house vice captain. Charlotte was introduced to the understanding of the human condition in 1994 by FHA Member Susan Armstrong while attending the University of New England. Charlotte moved to Sydney in 1996 to more directly support the FHA and works full-time helping manage FHA finances.

Heulwen (Lee) Jones, BA, 29, grew up mostly on a sheep station near Barraba in north west NSW and was educated at Calrossy Girls School, Tamworth where she was boarders captain. Lee was introduced to the understandings of the human condition in 1990 by FHA Member and good friend Prue Watson and moved to Sydney in 1997 to more directly support the FHA. Lee works full-time with Marcus Rowell running the FHA's IT department.

Sally Kaufmann, 57, born in Western Australia. Sally's family moved to Victoria when she was two, and she subsequently attended nine different public schools, including Maribyrnong High School. Sally completed a year at Melbourne University in conjunction with training as a journalist with the Murdoch News Group, where she worked for 10 years before joining *Choice* magazine for six years, the last three as editor. Sally has since worked as a freelance publicist and editor, editing Jeremy's first book *Free* in 1987, after which she became supportive of the understandings of the human condition and has played an important role editing FHA publications.

Monica Kodet, DipTeach, 31, grew up in Sydney where she was educated at the Presbyterian Ladies College, Sydney before attending Sydney's Institute of Early Childhood. She was introduced to the understanding of the human condition through her friend and FHA Supporter Howard Saunders whilst overseas in 1995 and works part-time in hospitality while helping manage the FHA accounts.

Anthony Landahl, BAg Ec, 29, grew up in Sydney where he was educated at Barker College before attending Sydney University. He was introduced to the understanding of the human condition in 1992 by a number of FHA Members prior to undertaking a four-wheel-drive expedition around Australia with school friends. Anthony works full-time developing the University of Denial-Free Studies and business enterprises dedicated to supporting the FHA.

Tim Macartney-Snape AM OAM, BSc, 47. A twice honoured Order of Australia recipient, Tim was born in Tanzania and moved to Australia in 1967 to a small farm in north eastern Victoria. Educated at Geelong Grammar School, Tim attended the Australian National University in Canberra where his passion for mountaineering developed, becoming the only summiteer on the ANU's 1978 Dunagiri expedition in the Indian Himalaya. Tim went on to climb dozens of peaks in the Himalaya, including making the first Australian ascent of Everest in 1984, which was achieved without the aid of oxygen and via a new route. He returned to Everest in 1990 by climbing the mountain from sea level on the Bay of Bengal. Tim was introduced to the understanding of the human condition in 1987 by Jeremy and is a director of the FHA. Tim has his own outdoor equipment manufacturing business, Sea to Summit, is a consultant to World Expeditions, assists with FHA PR, and is helping prepare the FHA's court cases.

Manus McFadyen, BA, 46, grew up mainly in Sydney where he was educated at St Pius X College before attending Sydney University. Manus has worked in Human Resource Management within the Australian Public Service since 1982 where he is a HR Manager. He played ruby union with the Gordon Rugby Union Club for 13 years before coaching there for another 7 years. Manus became a friend of Jeremy in 1989 and actively interested in the understanding of the human condition in 1991. He helps the FHA as a personnel adviser and in administration.

Marie McNamara, BA, 27, grew up in Auckland, New Zealand where she was educated at Baradene College. Marie was introduced to the understanding of the human condition in 1997 by FHA Member Simon Mackintosh, and moved to Sydney with Simon in 1999 to more directly support the FHA, where she now works full-time helping to develop business enterprises dedicated to supporting the FHA.

Simon Mackintosh, 33, grew up in Auckland, New Zealand and was educated at Wanganui Collegiate where he was a prefect. Simon was selected from New Zealand's 1987 school leavers to be one of 32 junior ambassadors for New Zealand at the 1988 World Expo held in Brisbane. Simon was introduced to the understanding of the human condition in 1994 by a family friend. Simon, who is a musician—his CD, *On The Road To Find Out,* contains songs about the work of the FHA—moved to Sydney to more directly support the FHA in 1999 and is working full-time helping to develop business enterprises dedicated to supporting the FHA.

Damian Makim, 30, grew up in Croppa Creek in northern NSW and Toowoomba in Queensland and attended Nudgee College in Brisbane. Damian was introduced to the understanding of the human condition in 1994 by FHA Member Bronwyn FitzGerald. Damian moved to Sydney in 1996 to more directly support the FHA. He established his own fencing company before working full-time helping to develop business enterprises dedicated to supporting the FHA.

Sean Makim, 22, grew up on a farm near Croppa Creek in northern NSW and educated at Toowoomba Grammar. He was introduced to the understanding of the human condition by his cousin, FHA Member Damian Makim, and moved to Sydney in 2000 to support the FHA. Sean is working full-time helping to develop business enterprises dedicated to supporting the FHA.

Sally Miller, BEc SocSc, 29, grew up in Sydney where she was educated at Ravenswood School for Girls. In 1994, while in her second year at Sydney University, Sally became interested in the understanding of the human condition through her older brother and FHA Supporter Richard. Sally works part-time as a personal assistant to a company executive and assists with FHA administration and business enterprises dedicated to supporting the FHA.

James Mollison, BCom LLB, 31, grew up mostly in Brisbane and attended Brisbane Boys Grammar achieving a leaving score in the top 3 percent of the state. He was introduced to the understanding of the human condition in 1993 by FHA Member Eric Crooke while studying at the University of Queensland. James moved to Sydney in 1996 to more directly support the FHA. After working for leading investment banks, he commenced work at the FHA in 2000, concentrating on finance, investment and the development of the University for Denial-Free Studies. James still consults in the finance industry while playing a leading role developing business enterprises dedicated to supporting the FHA.

Rachel O'Brien, BA Dip Ed, 28, grew up in Cumnock in central west NSW near the family farm and was educated at Hurlstone Agricultural High School in Sydney where she was a school house captain. She attended the University of New England where she was introduced to the understanding of the human condition in 1994 by FHA Members Prue Watson, Lee Jones and her childhood friend, and FHA Member, Susan Armstrong. Rachel moved to Sydney in 1998 to more directly support the FHA. She works as a school teacher while helping with FHA administration.

James Press, BSc, 31, grew up in Sydney and attended Knox Grammar School. In 1991, during his first year at Sydney's Macquarie University, James was introduced to the understanding of the human condition by school friends who had met FHA director Tim Macartney-Snape whilst travelling around Australia. James works part-time as a business manager while playing a key role in preparing the FHA's court cases.

Amanda Purdy, BEd, 29, grew up in Baradine in northern NSW near the family farm and was educated at Coonabarabran High School where she was a prefect. She was introduced to the understanding of the human condition in 1994 by FHA Member Richard Biggs while at the Queensland University of Technology. After teaching for several years Amanda moved to Sydney in 2001 to more directly support the FHA and now works full-time helping to develop business enterprises dedicated to supporting the FHA.

Stacy Rodger, BEd GradDip Special Education, 37, grew up in Sydney and was educated at Killara High School, St George Institute of Education and the University of Canberra. Stacy was introduced to the understanding of the human condition in 1991 and has been a member since then. She is an FHA director, is part of the PR team and she works as both a special education teacher and as a regular primary teacher. She lives in the southern highlands of NSW with her partner Tim Macartney-Snape with whom she also helps conduct trekking expeditions in the Himalaya.

Marcus Rowell, BSc (Hons), 31, grew up mainly in Hobart, Tasmania, where he was educated at Hobart College before attending the University of Tasmania. As a Queen's Scout with a love of the outdoors, Marcus purchased *Beyond* from the Uni bookshop in 1992 after seeing Tim Macartney-Snape had written the foreword. Marcus completed his honours degree in biology with the results being published in the *Australian Journal of Botany*. He worked as a biologist and made regular trips to the FHA in Sydney before moving there in 1997 to support the FHA full-time. Marcus heads the FHA's Information Technology section and is helping develop business enterprises dedicated to supporting the FHA.

Genevieve Salter, 28, grew up in Tamworth and attended Calrossy Girls School where she was a prefect. Genevieve was introduced to the understanding of the human condition in 1991 through FHA Member Lee Jones and through her friendship with Tim and Ali Watson and their four daughters. In 1996, Genevieve moved to Sydney to more directly support the FHA and works full-time helping to prepare the FHA's court cases.

William Salter, BAg, 24, grew up in Tamworth and was educated at Knox Grammar School in Sydney. He studied Agriculture at the University of Western Sydney, Hawkesbury campus where he was President of the Students Union. Will became interested in the FHA through his sister and FHA Member, Genevieve, and through his lifelong friendship with Tim and Ali Watson and their four daughters. Will became actively involved in the FHA in 2001. He works full-time helping to develop business enterprises dedicated to supporting the FHA and is helping to develop the University of Denial-Free Studies.

Nicholas Shaw, BCom LLB CA, 30, grew up in Goondiwindi in southern Queensland and was educated at the Anglican Church Grammar School in Brisbane, where he was school vice-captain, before attending Queensland University, and then qualifying as a chartered accountant. He was introduced to the understanding of the human condition in 1992 by school friend and FHA Member Richard Biggs. Nick moved to Sydney in 1999 to more directly support the FHA and plays a leading role in managing FHA finances while also helping to develop business enterprises dedicated to supporting the FHA.

Peter Storey, BA (Hons), 30, grew up in Brisbane where he was educated at Nudgee College. Peter was introduced to the understanding of the human condition in 1994 by FHA Member Annabel

Armstrong while at Queensland University of Technology. After graduation Peter pursued his interest in art and music and in 2000 moved to Sydney to more directly support the FHA. He works full-time helping to develop business enterprises dedicated to supporting the FHA.

Ali Watson, RN, 51, grew up on a farm near Cowra in southern NSW and was educated at the Methodist Ladies College, Burwood, in Sydney. Wife of Tim Watson, Ali became interested in the understanding of the human condition through Tim's involvement in the FHA. While Ali is a nursing sister she now works with Tim in their own business while helping to develop business enterprises dedicated to supporting the FHA.

Anna Watson, 27, grew up in Tamworth in northern NSW where she was educated at Calrossy Girls School. Anna was introduced to the understanding of the human condition by her father, FHA director Tim, and her older sister and FHA Member, Prue. After completing an Office Administration and IT course Anna moved to Sydney in 1998 to more directly support the FHA. She works part-time as a paralegal while assisting with the preparation of the FHA's court cases.

Prue Watson, 29, grew up in Tamworth in northern NSW where she was educated at Calrossy Girls School and in her senior years at the performing arts school, McDonald College, Sydney. Prue was introduced to the understanding of the human condition by her father, FHA director Tim. She began a degree at the University of New England before leaving to further her studies of the understandings of the human condition and support the work of the FHA. In 1996, Prue moved to Sydney where she works as a business administrator while helping prepare the FHA's court cases.

Tim Watson, 56, grew up in Sydney where he was educated at Knox Grammar School. Tim is a long-time friend of Jeremy and first became interested in and supportive of the understanding of the human condition in the late 1970s. Tim is a computer analyst and successful businessman and his dedication to all aspects of the work of the FHA, including the development of business enterprises dedicated to supporting the FHA, has made him an ideal director of the FHA.

Annabelle West, BA DipEd, 29, grew up in Mackay, Queensland, then Nigeria and Thailand, before attending the New England Girls school in Armidale where she was a school prefect. Annabelle was introduced to the understanding of the human condition by her brother and FHA Member James in 1991 in her final year at school.

After graduating from the University of New England, Annabelle moved to Sydney in 1996 to more directly support the FHA. She works as a school teacher while managing the FHA's archives.

James West, BE (Mech) (Hons), 31, grew up in Mackay, Queensland, then Nigeria and Thailand, before attending the Anglican Church Grammar School in Brisbane where he was a house captain and achieved a leaving score in the top 1 percent of the state. James' father has been a friend of Jeremy since attending the University of New England together and James was introduced to the understanding of the human condition in 1989 by Jeremy's brother Simon at a shared family Christmas party. James began his engineering degree at the University of Queensland where he interested his friends John and Richard Biggs in the understanding of the human condition. In 1992 James moved to Sydney to be able to support the FHA more directly and he completed his degree at Sydney University. James works in engineering design while helping develop business enterprises dedicated to supporting the FHA.

Stirling West, Asc Dip AppSc, 26, born in Nigeria and at the age of 6 returned to Australia and was educated at Brisbane Boys College. He was introduced to the understanding of the human condition by his brother James in 1989. In 1997 Stirling finished his tertiary education at University of Queensland's Gatton Campus in Queensland and moved to Sydney to more directly support the FHA. Stirling works part-time as a tennis coach while helping develop the University of Denial-Free Studies.

Prue Westbrook, 22, grew up mainly in Orange in rural NSW where she was educated at James Sheahan Catholic High School. Prue is completing a Bachelor of Business degree at the University of Western Sydney where in 2000 she was a first year student representative for the student union. She became interested in the understandings of the human condition in early 2002 through discussions with Will Salter who also attended UWS. Prue works full-time in the hospitality industry in Sydney for a leading international hotel chain and is dedicated to helping support the work of the FHA.

Annie Williams, 41, grew up in Panania in Sydney and was educated at Sir Joseph Banks High School where she was a prefect. Jeremy met Annie in 1980 while she was attending Alexander Mackie Art College and they have lived and worked together on this project every day ever since. Annie has been Jeremy's partner and personal assistant now for 23 years and is an FHA director.

Index

A Far Off Place 193
A Glimpse of Eden 480
A Story Like the Wind 193
A View from the Ridge 301, 359
AA (Alcoholics Anonymous) 303
abomination 255, 372, 404
Abt, Theodor 128, 448
acceptance 303
ADD (attention deficit disorder) 281
Adolescence the Essential Guide to, The 229
adolescence 217, 357
Adolescentman 358; Born-Again, Sophisticated 360
adolescents: resist denial 233
adulthood 217, 357
adults: older: suffer most from deaf effect 57
adventure 355
Africa: dsyfunction of 367; landscape of 479; left 119; natives of 367
African fable: basket 336
African Safari 204
African Saga 338
afterlife 423–425
age: increases alienation 152
Age of Extremes 312
Albright, Madeleine 325
Alexander, Peter 169
alienated state: loneliness of 287–290
alienation: age factor 379; differences in 135, 376–378; different lifestyles 380; dreaded A word 382; genetically adapted to 380; is personality 382; now instinctive 110; origin of 121; people will be free of 425; spectrum of 99, 123–130, 384; revealed 400; twice alienated 118
Almond Trees, The 230, 494
altruism 114
America's Cup 503
Ames, Evelyn 480
Andersen, Hans Christian 22, 141, 148, 486
anger: origin of 121
Angles 351
Anglo-Saxons 352, 365
Anglosphere 373
angry response 146
animal condition 111
An Olive Schreiner Reader 223, 327
Another Brick in the Wall 216
answers: avalanche of 268
Antichrist 116, 404
Apocalypse 398–404; Book of Revelation 399
apostles 156
Aquarian Conspiracy, The 292
Aquarius 492
Aquinas, Thomas 181
Arabs 362
arcadia 262
Ardrey, Robert 117, 249
Ariadne's Thread 140, 285
Aristotle 84
Armageddon: Battle of 152, 158, 398–405, 400
Armstrong, Annabel 505
Armstrong, Susan 505
artists: great 265; pain of being 265
Aryan 351, 362
Ascent of Man, The 40, 132, 386
Athenian: society 84, 128, 448
Atlantic Monthly, The 115
attractiveness 332
Australia 470; devoid of animals 489; gum trees 489; isolation of 354; like Africa 478; role of 470–482
Australian Aborigines 426
Australian character 473
Australopithecines 357
autism 258–260; understanding 269–276
Autism Research Unit Flinders University 259
autistic state: shows how mind dissociates 269–276
Autobiography in Five Short Chapters 304
Aztecs 364
Baber, Asa 324
Bacon, Francis 36, 344
Badrian: Alison and Noel 109
Baldwin, James 220
Banjo. *See* Paterson, A.B.
Bardot, Brigitte 329
Barnett, Anthony 387, 389, 459
Bass, Ellen 148, 298
Beale, Bob 346
Beard, Peter 205
Beatles 215
Beautiful Girls 330
Beautiful Losers 429
beauty 266; in humans 109; of women 238
Beckett, Samuel 295, 385
Belfield, Sam 505
Bell, W.D.M. 204
Benford, Gregory 88
Bennett, James 373
Berdyaev, Nikolai 32, 44, 139, 389, 447
Berman, Morris 287
Beyond Good and Evil 286, 331
Bible: great fire 95, 381, 408; lightning 401; little children 102; man in image of God 110; NIV translation 444; nought to 447; one will be taken 401; prophets without honour 442–443

Big Bad Wolves: Masculinity 336
big crunch 421
Biggs, John 505
Biggs, Richard 446, 505
bin Laden, Osama 373
biologists: responsibility of 141
Biology and The Riddle of Life 41, 140, 417
bipedalism 123
Birch, Charles 19, 40, 139, 183, 370, 389, 411, 413, 416, 447, 464, 467; as prophet 139, 464; books 91, 417; mechanism dead end 139, 285; on subjectivity 140; self-organisation 140; Templeton Prize speech 421
Blair, Tony 373
Blaise 288
Blake, William 28, 31, 36, 37, 53, 138, 311, 389, 447, 457; pictures on cover 36, 344; terrifying honesty 45
blondes 331
body hair: loss of 109, 329
Bohm, David 91
Bone Games 263
Bono 77, 215
bonobos 109
Booker, Christopher 166
Born Of A Woman 106
boys: sailing boats 215; growing up 347
Boys of Syracuse, The 290
Bradley, General Omar 139
Brahma the Creator 427
brain: highways 304; how works 299–300; volume stabilised 119
Brando, Marlon 339
brave new world 56, 184, 308, 398
Break on Through 67, 75, 310
Breakfast in America 217
British Isles 351
Bronowski, Jacob 40, 109, 133, 180, 386
Brown, Dee 458
Brown, Max 63
Buddha 180, 250, 415
Buddhism 122, 364
Buddhist scripture 291, 449
bullshitter, double 241
bullshitting 241
bullying 204
burden of guilt 154
Bury My Heart at Wounded Knee 458
Bush, George W. 373
Bushmen of the Kalahari 254, 359, 361, 368, 426; dance 263
Cain and Abel 360, 380
Campbell, David 480
Campbell, Jeremy 83
Campbell, Joseph 139, 389, 447
Campbell, Kate 506
Camus, Albert 139, 230, 389, 447, 494
Canada 374
canary 267
cancer 291
cancer patient 43, 65, 68
Canon Raven 431
Capra, Fritjof 285
caritas 89, 107, 114, 418
Casals, Pablo 312
Caucasian Steppes 354
Caucasians 362
cave: departure from 176–182; liberation from 144–152; resistance to leaving 145
cave allegory: what happened in practice 158–170
cave prisoners: try to kill him 146, 151
celibacy 333
Celts 351, 352, 362
Chandler, Raymond 331
change: adjusting to 300; difficulty if older 56; difficulty of 147–149, 151; initial shock 301; main shock phase 301; procrastination 302
Changes IV 491
chanting and singing: rhythmic 426
Chaos Theory 117
Chapman, Tracy 450
Charles Darwin 113, 153
Charlesworth, Max 472
Chatwin, Bruce 104, 427
child: breaks spell 141, 148; will lead them 141; within 245
child prodigies 212
childhood 217, 357; stress 196
Childman 357
Children 206
children: ask real questions 26, 195; born innocent 385; innocence of 205; vulnerability of 271
child's-heart 103, 261
chimpanzee: pygmy. *See* bonobos; shot 204
Chinese 364
chivalry 333
Christ 162, 250, 372, 455; authority of 456; be like God knowing 413; betrayed 171; challenge denial 147; childhood of 482; demystified 428–430; male disciples 392; exceptionally sound 392; false witness against 170; family sided against him: NIV quotes 443–444; how get such learning 282; little children 59, 282, 428; man no-one knows 428; miracles demystified 433–436; new wine 153; no greater love 114; not weak 429, 451; not yet fifty 432; overcome the world 142, 252; quote from 59, 63, 128, 130, 246, 284, 360, 469; re martyrdom 436; resurrection demystified 433–436; servant of all 179; soul strength 431; soundness condemning 433; soundness healed 433; speaking figuratively 413; spoke openly 461; truth set you free 433; without honour 442
Christianity 156, 364
Christie, Agatha 347
Cimbrians 327, 350; women 327
cities 481
civilisations: peak and decadent 363
Civilisedman 359
Civilization in Transition 122
Clancy of The Overflow 478
Clark, Mary E. 140, 285
Cline, Patsy 289
clothes 332, 359
cloud on the horizon 183
Cocker, Joe 330
codependent 197

Cohen, Leonard 429
Cohen, Stanley 46
Coles, Robert 207
collective unconscious 49, 100
Collins Persse, M.D. de B. 171
Colour of the Clouds, The 196
Colquhoun, Lachlan 506
Coming to Our Senses 287
communism 409
competitive world: belief in 232–239
concrete: fifty feet of solid 288. *See also* Laing, R.D.
Conrad, Joseph 42
consciousness: 2 million years ago 119; how emerged 117–119
Consider Me Gone 220
Consilience 54, 115, 133
Continental Drift: theory of 90
contrived excuses 113–116; for competitiveness 235
Convergence of Science and Religion, The 412
convicts 473
cooperative ideality: what it took to block out 190
cooperative world: belief in 232–239
Copernicus 149
Corian, The 171
Cosmic Blueprint, The 417
Counsellor, The 465
Coupland, Douglas 281
Courage to Heal 148, 298, 302, 304
Cowley, Jason 381
Crazy 289
Crazy Horse 458
Crooke, Eric 211, 506
Crum, Thomas F. 503
Cry, the Beloved Country 33
Cullen-Ward, Emma 506
Cullen-Ward, Fiona 507
Cummins, Anthony 507
Cunningham, Richie 336
Curthoys, Jean 247
Dahl, Roald 199
damage: psychological 202
Dame Edna Everage 271
Daniel 255, 372, 492
Dante 248
Darian, Joe 210
dark night of the soul 236
Darling, James 57, 139, 164, 171, 389, 412, 447, 453, 502; as prophet 499; one of 200 great Australians 499; peak vision at 51 432; preserve sensitivity 499; re Christ 442; re England 352; science and religion 418; sensitive and tough 431
Darwin, Charles 113, 138, 150, 389, 447; hostile reviews 153
Darwin, Francis 153
David: childhood of 482
David and Goliath 141, 410
Davidson, Iain 145
Davies, Paul 92, 285, 389, 411, 417, 421, 447, 464; as prophet 139, 464; books 91, 417; god of the gaps 410; mechanism sterile 140; on God 88, 406; recent books 93; Templeton Prize speech 467
Davis, Laura 148, 298
deaf effect 42–51; age and alienation factors 62; age increases 56–57; alienation increases 154; blame presentation 52–54; degrees of 439; erode 69; increase with IQ 58–62; overcoming 64–66; psychologically desperate: don't suffer 438–439
DeBlois, Tony 258
Deconstructionism 182, 404, 409
Delaheve, Diane 342
democratic principle 58, 155, 174, 468
denial: exactly how achieved 233–236; fortress walls of 176; humans' historic 33–34; masters of 60; psychological principles in overcoming 294, 298
Denial-free books 447–448
denial-free thinkers 389. *See also* prophets
denials: examples of 145
Depression: National Institute in Australia 230
depression: of the human condition 206–208; suicidal 233
Derrida, Jacques 182
Descartes, René 181
despair 62, 264; shatter denial 62
Destiny of Man, The 32, 44
development of order 87, 113, 426
Didn't You Used To Be R.D. Laing?, 250. *See also* Laing, R.D.
die: as one prepares to 228
Diogenes 397
Disraeli, Benjamin 312
Divine Comedy, The 248
dog: new tricks 56, 153
dogma: pure forms of 409
Donaghy, Bronwyn 229
Doors of Perception, The 311
Downie, David 507
dream 100; flying 290
drugs: hallucinatory 264
Dworkin, Andrea 321
Dylan, Bob 164, 490
dysfunctional: families 199
earliest work: most honest 83
Easterman, Eliza 507
Edgar, Don 246
Edgar, Sally 507
Edge of the Sacred 472
Education of a Civilized Man, The 57, 164, 172, 353, 431, 502
ego: definition 121
egocentricity: origin of 121
Egypt 362
Egyptians 364
Einstein, Albert 143, 301
elephant 9, 12, 34, 44; in living room 21–26, 188–194; shot 204
Eliot, George 196
Eliot, T.S. 26, 45, 342, 345, 457
Elliott, Anthony 47
Emperor's New Clothes 22, 141, 148, 486
Empiricism 181
End of the Game, The 205
end-game 230
England 351, 352
English 352
enthusiasm: God within 263
environmentalism 241
Essay on Man 28, 40

Eucalypts 489
Evans, Marian 196
Everest from Sea to Summit 263
evil: not sanctioned 123
Evolution: television series 134
evolution 113, 134
Evolutionary Psychology 114, 117, 235
Executive Woman's Report 62
Existentialism 182
Exodus: Book of 448
Ezekiel 273, 438
faith 40, 443; and belief 23, 73; end of 180, 398, 443, 467; and hope 141, 215, 460; and reason 181, 184; lack of 440; obsolete 138, 463
fall from grace 104
Farewell, My Lovely 331
Farrell, Warren 337, 341
fasting 264
fatigue 263, 360
Faust 48
Fear and Loathing in Las Vegas 344
Feminism: limitation of 409
Feminist Movement 182, 409
Ferguson, Marilyn 292
fertile crescent 364
Fetner, P. Jay 204
Fetzer Foundation 24
FHA 499; has ultimate product 504; logo and name 500; Member 501; Membership profiles 504; Supporter 501
Fields, W.C. 206
Fijians 366
fire: great 38–39; metaphor 35–37, 85–88
first cut is the deepest 327
First Essay on Interest 384
First Time Ever I Saw Your Face, The 289
first will be last 128, 307, 488
Fisher King 230, 486
Fisher King & The Handless Maiden, The 344
FitzGerald, Bronwyn 508
Fonzie 336
forty-year-old: reality of 366
forty-year-old-equivalent 359
Fossey, Dian 139, 223, 389, 391, 447
Foucault, Michel 182
four horsemen 405
Fouts, Roger 272
foxes have holes 252, 387
free enterprise 370
Freud, Sigmund 139, 247, 332, 389, 447; work resisted 150
fuck: powerful swear word 450
fundamentalism: rise of 405
future: species' fabulous 311
Future Shock 69, 151
future shock 56, 184, 398
Gaddafi, Colonel 373
Galileo 149
Gamaliel 156
Garden of Eden 34, 37, 48, 104, 121, 221, 332, 377
Garland, Judy 290
Gauls 351
Geelong Grammar School 171, 172, 370, 432, 469; FHA Members attended 499
gene-based learning 120, 376, 426
generation: ACES 281; X 281; XX 281
Genesis 320, 321, 333, 392; banishment 310; Cain and Abel 380; flaming sword 95; like God knowing 180; man in image of God 377
genetic reciprocity 235
genetic refinement 113, 329; limitation of 376
genetics 106; limitation of 107, 117
Germanic races 362
Gibran, Kahlil 139, 389, 430, 447, 451
gifted people: threaten others 440
Giovanni's Room 220
girls: horses 215
glandular fever 194
goby fish 267
God 430; abstraction gap 407–408; as negative entropy 89; concerns about 406; cooperative ideals 37, 377; demystification of 406–416; exposing 407–414; face to face 125, 250, 388, 430; fearing 37, 72, 90, 381; is love 419; is negative entropy 416; loves you 27, 40, 133
God and the New Physics 417
God the Father 427
God the Holy Ghost or Spirit 427
God the Son 427
Goethe 48
golden age 34, 48, 104
Golden Summer 478
good die young 363
goose: lays golden egg 155, 174
Goths 350
Gould, Stephen Jay 411
Gramsci, Antonio 312
Gray, John 323
great lie, the 218
Greeks 362
Green Movement 182; empty rhetoric 221
Greenland 478
Greer, Germaine 322
gridiron 138
Griffith, Jeremy 502, 508; and Irishman 353; as contemporary prophet 453; biography 528; books 50; repetitive 54; books deaf effect response 50–52; brothers 445; childhood of 477, 481; degree 528; family sided against 445; father 445; influenced by van der Post 53; mother 445; National Times article 463; poem 492; school report 469; standard talk 70; student of Charles Birch 419; support derived from van der Post 173
Griffith, Simon 171, 445, 446, 508
guardians: more alienated see themselves as 155
Gurdjieff, George 260, 451
Gurdjieff: An Approach to His Ideas 451
Guys and Dolls 339
Hair, the musical 185, 492
hairless 109
Hand-Me-Down Blues 229
Handless Maiden 344
Happy Days 336
happy ending 349
Hart, Charles 212
Hart, Larry 289
Harvard Magazine 54
Hawking, Stephen: God is laws of physics

422; impersonal God 416; on God 88, 406
Hayman, Ronald 243
He: Understanding Masculine Psychology 230, 486
Health & Survival in the 21st Century 337
Heart of the Hunter, The 255
heaven 262, 290, 424
Hebrews 362; collected prophets 448
Heinberg, Richard 105
hell 424
Hemingway, Ernest 31
heresy 65
Hesiod 104, 291
Hillman and Ventura 297
Hinduism 364, 427
history: taught at schools 350
Hitler 368
Hobsbawn, Eric 312
holism 88
Hollow Men, The 342
Holy Ghost or Spirit 427
Holy Grail 76, 401, 488
Holy Grail: of the human journey 76
Holy Spirit 465
Homer 48, 136, 178, 447
Homo 358
Homo erectus 358
Homo habilis 358
Homo sapiens 358
Homo sapiens sapiens 358
homosexuality 336
honesty: come like summer rains 218; is therapy 350
Hooper, Ella 214
Hope, A.D. 470
Horne, Ross 337
Hours By The Window 249
House of Cards 211, 280
Howard, John 373
Howard, Robert 149
human condition 25–27; agony of 27–33; anticipations of solution 74, 490–491; contradictions of 27; denial of 33–34; exists within 232; extent of fear 243–250; glancing references to 54; resistance to 163; to solve objective of human journey 292–293; trying to confront 200–204; ultimate threat of 155; understanding of: liberates beauty 420
human race: denial-free history 349–370
humans: don't want to think 192; elderly 291; how became corrupted 376–378
Hume, David 181
humour 359; demystified 449–450
Humphries, Barry 271
Huns 363
Hunter, J.A. 205
Hunters & Collectors 76, 401, 488
hunting 320, 359; big game 204
Hussein, Saddam 373
Huxley, Aldous 120, 284, 311
Huxley, Thomas 113
I Can't Get No Satisfaction 77, 490
I Still Haven't Found What I'm Looking For 77, 215
I Who Have Nothing 330
idealism: how confronting 205
ideals: condemning 35–37
ignorance: threat to group 237
Ikhnaton 93
Imagine 77, 215, 490
Impossible Dream, The 210
In Search of the Miraculous 260
Incas 364
incest: victims of 298, 455
Indian Fijians 366
Indians 364, 367
infancy 217, 357
Infantman 357
Inherit the Wind 252
innocence: doesn't exist 221; face truth 149; hated 221; image of 289
insecurity: degrees of 379
instinctive selves: hurt or damaged 378
instincts: ignorant 120
integration 426; two great tools 427
integrative meaning 106; seen as dangerous truth 134; humans fear 422
intellectuals 61, 183
Intelligent Design 134
Intercourse 321
Intimations of Immortality 103, 121
IQ 61, 469; high needed for denial 307; limit 120; tests 128
Ireland 351
Ireland, David 471
Irish 352, 353, 473
Is Your Child Depressed 229
Isaiah 44, 45, 51, 62, 141, 471, 482
Isherwood, Damon 509
Islam 364
Islamic art 266
Islands in the Stream 31
Italians 363, 365
Jackson, Felicity 509
Jacob 388
Jagger, Mick 78
James, Charlotte 509
Janus: A Summing Up 87
Japanese 251
Japanese proverb 355
Jaspers, Karl 273, 438
Jesuit saying 202
jewellery 359
jigsaw: pieces of 137; pieces side down 414
Job: deep shadow 38, 95, 389
Joel 492
John the Baptist 170
Johnson, Robert A. 139, 230, 344, 389, 447, 486
Jones, Heulwen (Lee) 146, 509
Jones, J.D.F. 166, 251, 458
Jowett, Benjamin 131
Judaism 364
judgment day 184, 398; fear of 403
Jung and the Story of Our Time 244
Jung, Carl 100, 139, 170, 243, 250, 256, 389, 447; as prophet 247; collective unconscious 49; journey into unconscious 243; let himself drop 245; projection 461; quote from 122, 396, 433; Salome 324, 391; swearing 452
Jungian principles 230
Kamiya, Gary 248
Kauffman, Stuart 91, 139, 389, 447
Kaufmann, Sally 509

Kazantzakis, Nikos 222, 342, 344
Kelly, Ned 63, 474
Kennett, Jeff 229
Khan, Genghis 144
Kierkegaard, Søren 31, 32, 45, 46, 138, 389, 447
Killing Heidi 214
King Arthur legend 231, 249, 486
King, Martin Luther 436
kissing disease 194
Kodet, Monica 510
Koestler, Arthur 139, 170, 389, 447; negative entropy 87, 418; on science 90; quote from 285; suicide 249
Koko: the gorilla 290
Kronemeyer, Robert 337
Kuhn, Thomas 153, 165, 501
Laing, Adrian 249
Laing, R.D. 46, 139, 249, 273, 389, 447; alienation goes to roots 287; as prophet 249; each child potential prophet 385; exiled truth 96; families cause madness 279; famine in land 48; fifty feet of solid concrete 288, 397; quote from 158, 192, 211, 216, 247, 277, 282, 438, 496, 260–261, 296, 397; swearing 452; turn the mind around 178
lambs among wolves 170
Landahl, Anthony 446, 510
Landahl, Peter 81
language: Indo-European 363
Last Two Million Years, The 396
Lawrence, D.H. 105, 361, 386
Lawson, Henry 31, 37, 208, 475; childhood of 483
Leakey, Louis 139, 389, 447
Leakey, Richard 447, 449
Leaving Early 229
Left Behind series 403
left wing 370
left wing and right wing 184
legal system: denial-compliant 173
Lennon, John 77, 215, 490
Lessic, Michael 211
Let It Be 215
Let Me Explain 330
Lewin, Roger 91, 449
liars 61
Liar's Tale: A History of Falsehood, The 83
liberated position (LP) 294, 304
life: magic of 425
Life of Jung 243
lightning 183; flashes 399
Lindbergh, Anne 61
Little Prince, The 196, 199
Lizard Lounge, The 345
Llewellyn, Kate 323
loaves and fishes 434
Locke, John 181
Logical Song, The 217
loneliness 288
Longfellow, H.W. 206
Lord Mountbatten 251
Lorenz, Konrad 117
Lost World of the Kalahari, The 255
love: fall in 239, 288, 331; unconditional 201
Love, Let's Fall In 288
love-indoctrination 137, 237, 270, 329; explained 107–110
Luddites 148
Luke 399
Luther, Martin 447
lying 284
Macartney-Snape, Tim 81, 171, 420, 446, 510; first Australian climb Mt Everest 263
MacColl, Ewan 289
Macken, Deidre 92, 391
Mackintosh, Simon 511
MacLaine, Shirley 323
Magic of Conflict, The 503
Makim, Damian 511
Makim, Sean 511
Mallarmé, Stéphane 260
man: image of God 110
Man From Snowy River, The 141, 410, 483
Man of La Mancha, The 210, 321
Manley Hopkins, Gerard 42, 207, 244, 486
Marais, Eugène 121, 139, 389, 447; suicide 249
marriage 333, 359
Marx, Karl 181
matriarchy 109, 320, 324
maturation stages 354; explained 217
maturity 357
McCarthy, Mary 349
McCubbin, Frederick 478
McFadyen, Manus 510
McNamara, Marie 511
mechanistic science 139
media: much superifical talk 312
meditation 262
meek inherit earth 128, 307, 488
Meir, Golda 325
Mellen, Joan 336
Memories & Visions of Paradise 105
Memories, Dreams, Reflections 243
men: are worthless 347; as warrior 348; boot into the dirt 346; crumpled 324; movements 347
men and women: different roles 237–241
Men are from Mars, Women are from Venus 323
Men: From Stone Age to Clone Age 346
Mencius 103, 261
Merlin 249
messiah 143, 463
messiah complex 249
messianic 464
metaphysical descriptions 419
Mexican stand-off 436
Micah 491
Middle East 369
might ruled over right 128
Miles, Siân 60
Mill, John Stuart 157, 501
Miller, Bruce 258
Miller, Fiona 209, 213, 233, 257, 395
Miller, Sally 511
Milligan, Spike 434
mind: twice alienated 118
Mind of God, The 417
mind-control 163
miracles 434
Mohammed 128, 250, 386, 481
Mollison, James 511

Mongols 360
Monkey Trial 252
monotheism 419
Monroe, Marilyn 325
Moral Animal, The 114, 235
Moral Intelligence of Children, The 207
morality: basis of 115
Morrison, Jim 67, 75, 196, 310, 311
Morton, John 1, 139, 389, 447, 453; Beyond commendation 413
Moses 125, 250, 388, 448, 457, 482; childhood of 482
mothers: stroke brow 193
Mount Sinai Medical School 196
mountains of the mind 244
Mousetrap, The 347
muddied pool 462
Mugabe, Robert 373
Murray, Les 384
music: 1960s: optimistic 426; late 20th century: manic 426
My Friends the Baboons 249
Myth of Male Power, The 337, 341
mythology 78
Nagel, Thomas 72
National Review 373
natural selection 113, 134
nature 267; friend of soul 221
near-death experience (NDE) 261
Ned Kelly: Australian Son 63
negative entropy 86, 137, 140; if ends with heat death 421; where spirituality 416–418
Nelson, Willie 289
neotenous features 109, 238, 329
neoteny 109
nerve-based learning 120, 299, 377, 426
Nesse, Randolph 116
Neumann, Erich 139, 231, 243, 389, 447; Jung's most gifted student 231; quote from 296
New Age Movement 23, 72, 75, 77, 153, 182, 286, 292, 404, 412, 453
New Zealand 374
Next of Kin 272
Nietzsche, Friedrich 138, 230, 248, 262, 329, 331, 334, 347, 348, 370, 389, 447; largely unread 282; lying 47; quote from 286, 392; 'superman' 286
Nineteen Eighty-Four 372
Niven, David 46
No Worst There is None 207, 486
Noah's Ark: explained 393–394
noble savage 115
noisy nines 214
non-falsifiable situation 63–64
Noonday Demon, The 229, 279
Norseman 351
nurturing 135, 237, 270, 329, 333; main influence 376; now safe to admit significance of 280; significance of 107–110, 272; took back seat 377
nutcracker: bird 267
O'Brien, Rachel 512
Odysseus 49, 136
Odyssey, The 48, 136
omega point 292
On Education 319, 479
On Human Nature 55, 133
On Liberty 157, 501
On Purpose 417
On the Contrary 349
One Solitary Life 431
Operant Conditioning 117
orienteering 267
Origin of Species, The 113; criticised 150
Origins 449
Origins and History of Consciousness, The 231
Origins of Virtue, The 116
Orwell, George 372
Out of My Later Years 143, 301
Outside magazine 360
Over the Rainbow 290
Overcoming Homosexuality 338
Paglia, Camille 322, 325
palm trees 204
paradigm shift 56, 184, 398; degrees of difficulty 148
Parsifal 487
Passion of the Western Mind, The 149
Paterson, A.B. (Banjo) 139, 141, 389, 447, 473, 475, 478; childhood of 483
patience and perseverance: need for 67–70
Paton, Alan 28, 30, 31, 33, 36, 222
patriarchy 237, 320
patterns of behaviour 148, 302
Peace Movement 182
Peace Train 490
Perennial Philosophy, The 120
Phaedo 101, 102, 103
Phantom of the Opera, The 212
Phar Lap 466
Phenomenon of Man, The 134
philosophy: definition 84
Pink Floyd 216
Planck, Max 153
Plato: absolute justice 172; anticipated psychiatry 296; cave allegory 35–41, 83; cave prisoners 43–45, 67, 95, 96; crisis in life 249; departure from cave 176–183; Encarta 35, 39, 85, 96; Forms 85, 101; four states of mind 98; human potential once liberated from cave 297; humans denial 43–45; ideal education 126; liberation from cave 96–97; middle period 101; nature of soul 101–104; object of knowledge 40; on education 131; on innocence 124, 383; on nurturing 131; on Socrates 146; philosopher rulers 126–129; predicted how human condition solved 142; someone returns to cave 159–172; universal first principle 86; wrote Republic at 50 432
PMT 334
pocket the win 297
Pogo comic strip 45, 246
Politically Correct Movement 371, 404, 409
Ponnuru, Ramesh 373
Pope, Alexander 28, 31, 37, 40, 132, 133
Porter, Cole 289
Postmodernism 83, 182, 371, 409
poverty 371
power-fame-fortune-glory obsessed 236
prayer 262
Press, James 512
Prigogine, Ilya 91, 139, 389, 447

Prince Charles 167, 172
Prison Notebooks 312
Progress Without Loss of Soul 128, 448
prophets 104, 125, 130, 349; authority of 430, 454–457; classes of 430; confront integrative meaning 388; contemporary 389; Jeremy Griffith 138; role of 138, 462–465; defiant 451; demystified 383–389; different to a saint 429; false 72, 162, 404–405; first place go is own family 440, 441; fundamental struggle 253; historically condemning 138; maturation long 432; not clever 468; persecute 126, 130; shepherd in youth 128; small concluding role 466; some suicided 265; sufficiently loved 245; took on world of lies 429; unbearably confronting 386; unresigned and resigned 389–391, 430; unresigned people 384; without honour 439–444; female 391
pseudo-idealism 182, 240, 371, 404–406; forms of guilt free 409; movements 75; rise of 405
psyche 150
psychiatry: black majic 297; Plato anticipated 177; soul healing 297
psychological stages 357
psychosis: definition 150
puberty 194, 229
Public Enemy 293
Purdy, Amanda 512
Pygmy Chimpanzee, The 109
Quantock, Rod 31, 73, 247
Queen Victoria 339
questions: fundamental 227, 350; key re humans denial 187
R.D. Laing A Biography 249
races: cynical 369
racism 349
Rain Man 257
Ramanujan 258
Rand, Ayn 325
Ransdell, Eric 360
rationalism 181
reciprocity 114
reductionism 139
religion: beginnings of 396; origins 115
religions 130; civilised humanity 157; demystifying role of 395–396; too confronting 241; value of 395
Republic, The 35, 83, 101
resignation 52, 187–208; cost of 256–263; effects on men and women 239; end of 309–310; history of analysis of 229–231; life leading to 211–215; moment of 232–242; most important psychological event 229; Noah's Ark metaphor for 394; poetry 208–210; renegotiating 294–305
resigned mind: dysfunction of 282–285; necessary dishonesty of 277–280
resurrection 435, 436
Rhodes, Cecil 367
Ricciardi, Mirella 338
Richard, Keith 78
Richards, Renée 337
Ridley, Matt 116
right wing 370
Ring, Kenneth 262
Rintoul, Stuart 473
River of Second Chances 360
Roberts, Tom 478
rock climbers 263
Rocks Of Ages 411
Rodger, Stacy 512
Roethke, Theodore 208
Rolling Stones 77, 490
Roman society 328
Romans 352, 362
Rome 351
Rousseau, Jean-Jacques 105, 115, 138, 319, 389, 447, 479
Rowell, Marcus 513
Ruth 393
Saint, Francis of Assisi 429
Saint-Exupéry, Antoine de 61, 139, 196, 447 199, 389
Salome 391
Salter, Genevieve 513
Salter, William 513
Samuel 454
Samutchoso 394
Santa Fe Institute 91
Sardar, Ziauddin 283
Sartre, Jean-Paul 182
Satan: can't drive out Satan 162, 429, 454, 461
savant syndrome 257–260
savants: insight into human abilities 268
Saxons 351
Scandinavia 351, 365
school teachers 213
schools: pass on great lie 216
Schopenhauer, Arthur 138, 389, 447; journey of new ideas 147; quote from 261, 283
Schreiner, Olive 63, 138, 202, 264, 319, 326, 328, 339, 345, 389, 391, 447; first 6 years make us 386; pre-resigned state 223–226; railed against men 393; vulnerability of children 279
Schroder, Jeremy 308
Schultheis, Rob 263
science: makes understanding possible 143; mechanistic 283; eg of denials 135; ground work 137; hard-won insights 135; maintain denial 136; stalled 139; mechanistic not holistic 90, 133–135; peak expression of human effort 143; the liberator 131–135, 143, 415; the messiah 143, 415
science and religion: different perspective on one subject 412; divide between 411; reconciliation of 183
Science Friction 92, 391, 411, 417
scientific method 283
scientists: resist new ideas 153
Scopes, John 252
Scotland 351
scree jumping 263
sea: metaphor of 136
Sea Kingdoms, The 351
second coming 465
secularism 408
self-organisation 86, 140
self-restraint 123
self-selection 108
selflessness: meaningful 87; unconditional 110, 116, 329
Semites 362, 364

Separate but Equal 312
Seven Up documentaries 202
sex 321, 335; demystified 449–451; explained 238
sex-object attention 327
sexes: war between 317–333
sexual conquest 327
Sexual Personae 322
Seymour, Mark 76, 215, 488
shadow: to own 349
Shakespeare 60
shattered defence 273
Shaw, George Bernard 151, 175; quote from 467
Shaw, Nicholas 513
sheep among wolves 172
Shelley, Percy Bysshe 261
shepherds 128, 327
ships at sea 218–221; letter from 219
Shiva the Destroyer 427
sickness 291
Sickness Unto Death, The 31
Siegler, Dr 229
silence: destructive 194–197
Silicon Valley 258
Simmons, Jean 339
Simon and Garfunkel 199
Simone Weil, An Anthology 60
Simpsons, The 347
sin: origin of 117
sins of the father 45, 196, 378
Sistine Chapel 27
Skinner, B.F. 117
Slattery, Luke 169
Slavs 362
Smuts, Jan 88
SNAGs 325
Snyder & Mitchell 259
Social Darwinism 113, 235
socialism 181
Sociobiology 114, 117, 235
Sociobiology: The New Synthesis 114
Socrates 84, 146, 149, 170; corrupting the young 146; unexamined life 146
Solomon, Andrew 229, 279
Somewhere 288, 331
Somewhere, Some Time, Some Place 223
Sondheim, Stephen 288, 331
Song of the Future 475
Songlines, The 104
sophisticated 182
soul: how acquired 106–112; how corrupted 117–120; language of 290; mythology 104–105; nature friend of 192; subtle form of selfishness 115; the synthesiser 136–143, 415
Soul of the Ape, The 121, 249
Soul of the White Ant , The 249
Sound of Silence, The 199
soundness: humanity waiting for 141; leading 129; winnow truth 137
South Africa 374
Spencer, Herbert 113
spirit 260, 419; of humans 423
Spong, John Shelby 106, 110
sportsperson: only receive credit when die 440
stages with ages 217
States of Denial 46
Steadman, Ralph 344
Steinem, Gloria 323
Stengers, Isabelle 91
Stephen 147; false witness 147, 160
Stevens, Cat 490
Stevenson, Robert Louis 196
Sting 220
Stockman, The 480
Storey, Peter 513
Story of an African Farm, The 63, 202, 264, 279, 319, 326, 339, 393
Strahan, Ronald 453
Streeton, Arthur 478
Strum, Shirley 109
Summa Theologica 181
sun: makes intelligible 94; metaphor 35–37, 85–88; people who could live in 123–129
sunglasses 204
superheroes 286
Superman 261
supermodels 330
survival of the fittest 113
survival of the fittest world: belief in 232–236
Susman, Randall L. 109
swearing 452; demystified 449–451
T-model Ford 359
Tacey, David 472
Talking Heads TV program 417, 421
Tarnas, Richard 149
Tassone, Lisa 53, 59, 188
teenage-equivalent 354
Teilhard de Chardin, Pierre 22, 134, 139, 292, 330, 389, 447, 463
Teiresias 48, 136, 335
teleology 88, 140
Temple, William 500
Templeton Prize 41, 92, 417, 467
Terra Australis and The Holy Spirit 472
terrorism 369
Testament to the Bushmen 263
Thatcher, Margaret 325, 251
Theogony 104, 291
Theories of Everything, The 258
Thinking About Children 193, 274–276
thirty-year-old equivalent race 369
Thompson, Hunter S. 344
thunder 183
Thus Spoke Zarathustra 329, 334, 348, 370, 393
Thylacine. *See* Tiger: Tasmanian
Tiger: Burning Bright 28; Tasmanian 508
Times Square 327
Times They Are A-Changin', The 164, 490
Toffler, Alvin 69, 151
tolerance: in society 468
toughening process 201, 355
Tow, David Hunter 258
Townes, Charles H. 412
Treasured Writings of Kahlil Gibran, The 431
Trimurti 427
trinity 426; Christian 427; demystification of 426–427; fundamental forces of 427; Trimurti 427
truth: adjusting to 305; assemble outside cave 137; difficulty of confronting 145–153; live at different distances 379–381; replaced with lie 236; set you free 66, 185, 350

truths: self-evident 55
twenty one year old: tradition 355
twenty-year-old-equivalent 355
U2 77, 490
Ulysses 136
Uncommon Genius 257, 272
unconscious counterposition 243
University For Denial-Free Studies 294
unresigned: had to survive persecution 387; loneliness of 387; state ultranatural 397
unevasive thinkers. *See* prophets
unresigned thinkers. *See* prophets
upset: origin of 120
Valentino, Rudolph 336
Vamps & Tramps 325
van der Post, Laurens 62, 139, 193, 251, 256, 389, 447; as prophet 50, 53; Athenian society 448; childhood of 483; compassion 41; formative writing at 51 432; great journey of life 111; live not only our own lives 426; persecution of 166–173; quote 105, 122, 128, 129, 162, 183, 270, 293, 312, 328, 334, 345, 350, 361, 368–369, 413, 437, 457, 471, 496; re Freud 150; re Jung 61, 244; re Jung's journey into unconscious 291; re pain of artists 265; The Other Journey 49; victory parade 423; Voice of the Thunder 49
Van Gogh, Vincent 265, 390; only sold 1 painting 440
View From Nowhere, The 72
Vikings 351, 360
Virgil 202, 248, 262, 447
Virgin Mary 333
Virgin Mother: Christian emphasis on 443; demystified 383–389; symbol of nurturing 443
Vishnu the Preserver 427
Voice of the Thunder, The 49, 183
von Weizsäcker, Carl 370
Waiting For Godot 295, 385
Wald, George 26, 195
Waldrop, M. Mitchell 91
Wales 351
Wall Street Journal, The 55
Wallace & Gromit 118
Wallace, Alfred Russel 113
Waltzing Matilda 473
Watson, Ali 514
Watson, Anna 514
Watson, Prue 514
Watson, Tim 514
Weil, Simone 60, 139, 223, 389, 391, 447
West, Annabelle 514
West, James 81, 515
West, Mae 217
West, Morris 30, 31, 301, 359
West Side Story 288, 331
West, Stirling 515
Westbrook, Prue 515
We've Had a Hundred Years of Psychotherapy 297
What Am I Doing Here 427
Wheatcroft, Geoffrey 370
When the Bush Begins to Speak 475
When The Ship Comes In 490
white hunter 205
White, Patrick 29, 31, 222, 266
Whitehead, Alfred North 84, 185, 449
Whitman, Walt 480
Why 450
will to power 348
Will To Power, The 262
William Blake Selected Poems 457
William the Conqueror 351
Williams, Annie 171, 446, 515, 528
Wilson 467
Wilson, Edward O. 54, 114, 133, 235
Winfrey, Oprah 229
Winnicott, D.W. 139, 192, 272, 283, 389, 447; quotes re autism 274–276; quote from 469; significance of nurturing 277–279
winnowing: shovel 49, 137
wise: thoughts futile 60, 177
Witkin, Georgia 196
Wittgenstein, Ludwig 25, 67
Wizard of Oz 290
wolf: live with lamb 141
Wolfman Jack 195
Woman of the Future, A 471
Woman Question, The 326
women: ageing 342; beauty of 330; bimbo, breeder, etc 343; blindness 391; feel 334; magazines 332; mystery of 289, 331; not mainframed 391; not sympathetic to battle 391; pubescent body shape 332; take over 324; unmainframed 323; victim of a victim 333
Wordsworth, William 29, 31, 103, 121, 125, 138, 202, 266, 389, 447
Wright, Robert 114, 235
Yapko, Michael 229
Yin and Yang 184
You Can Leave Your Hat On 330
Young, Robyn 257
youth: suicide and depression 229
Zarathustra 250, 387, 429, 482
zeal 451
Zen 122
Zorba The Greek 222, 342
Zoroastrian religion 37, 95

About the Author

Jeremy Griffith was born in Albury, NSW, Australia in 1945 and raised on a sheep station in central west NSW. He was educated at Tudor House School in NSW and Geelong Grammar School in Victoria before attending the University of New England, where he played representative rugby, making the trials for the national team, the Wallabies, in 1966. Jeremy deferred his studies in 1967 to attempt to save the Thylacine or Tasmanian Tiger from extinction. Attracting international scientific and popular media coverage, the six-year search, the most thorough yet carried out, concluded the Thylacine was extinct. After completing a science degree, majoring in biology, at Sydney University in 1971, Jeremy began manufacturing furniture to his own simple and natural designs. The business employed some 45 people when Jeremy sold his half share in 1991.

Jeremy began actively writing about the human condition in 1975, established the Foundation for Humanity's Adulthood in 1983, published his first book, *Free: The End Of The Human Condition,* in 1988, and his second book, *Beyond The Human Condition,* in 1991. While he is an FHA director Jeremy's main role is developing understanding of the human condition. With his partner of 23 years, Annie Williams, Jeremy spends most of his time writing at his brother Simon's property in central west NSW.